Lecture Notes in Civil Engineering

Volume 13

Lecture Notes in Civil Engineering (LNCE) publishes the latest developments in Civil Engineering - quickly, informally and in top quality. Though original research reported in proceedings and post-proceedings represents the core of LNCE, edited volumes of exceptionally high quality and interest may also be considered for publication. Volumes published in LNCE embrace all aspects and subfields of, as well as new challenges in, Civil Engineering. Topics in the series include:

- Construction and Structural Mechanics
- Building Materials
- Concrete, Steel and Timber Structures
- Geotechnical Engineering
- Earthquake Engineering
- Coastal Engineering
- Hydraulics, Hydrology and Water Resources Engineering
- Environmental Engineering and Sustainability
- Structural Health and Monitoring
- Surveying and Geographical Information Systems
- Heating, Ventilation and Air Conditioning (HVAC)
- Transportation and Traffic
- Risk Analysis
- Safety and Security

More information about this series at http://www.springer.com/series/15087

Anirudhan I. V. · V. B. Maji
Editors

Geotechnical Applications

IGC 2016 Volume 4

Springer

Editors
Anirudhan I. V.
Geotechnical Solutions
Chennai, Tamil Nadu
India

V. B. Maji
Department of Civil Engineering
Indian Institute of Technology Madras
Chennai, Tamil Nadu
India

ISSN 2366-2557 ISSN 2366-2565 (electronic)
Lecture Notes in Civil Engineering
ISBN 978-981-13-0367-8 ISBN 978-981-13-0368-5 (eBook)
https://doi.org/10.1007/978-981-13-0368-5

Library of Congress Control Number: 2018939462

Printed on acid-free paper

This Springer imprint is published by the registered company Springer Nature Singapore Pte Ltd.
part of Springer Nature
The registered company address is: 152 Beach Road, #21-01/04 Gateway East, Singapore 189721, Singapore

Foreword

The Indian Geotechnical Society (IGS) was started as the Indian National Society of Soil Mechanics and Foundation Engineering in the year 1948, soon after the Second International Conference on Soil Mechanics and Foundation Engineering held at Rotterdam. The Society was affiliated to the International Society in the same year, and since then, it has strived to fulfil and promote the objectives of the International Society. In December 1970, the name 'Indian Geotechnical Society (IGS)' was adopted.

The Society conducted several symposia and workshops in different parts of India since its inception in 1948. It was in the year of 1983, the Indian Geotechnical Society organized its first annual conference IGC 1983 in Indian Institute of Technology Madras.

Several local chapters of the Society were established over the years, and gradually, the annual conferences were held in different cities under the leadership of the respective local chapters. The conferences were well utilized as the venue for showcasing the research works and the case studies on geotechnical engineering and geoenvironmental engineering, and the papers presented during the deliberations were being published as conference proceedings volume.

The annual conferences IGC 1996 and IGC 2006 were organized by IGS Chennai Chapter, and the papers were published as printed volumes. The practice of printing the volumes was later discontinued in view of large expenditure involved and also because of the easy access to the electronic storage media. The papers then became not accessible to the rest of the geotechnical community.

The responsibility of organizing the annual conference of 2016 was taken up by IGS Chennai Chapter, and the conference was held during 15–17 December 2016. It was felt necessary to publish selected papers for the benefit of the entire geotechnical engineering community as well as of the contributing authors. The Chapter approached Springer with a request to take up this responsibility, and with the great help of Springer, about 150 selected papers are being published in four volumes under the series 'Lecture Notes in Civil Engineering'. The selected papers are distributed among these volumes depending upon the topic of discussion.

Volume 1—Geotechnical Characterisation and Geoenvironmental Engineering
Volume 2—Ground Improvement Techniques and Geosynthetics
Volume 3—Soil Dynamics and Earthquake Geotechnical Engineering
Volume 4—Geotechnical Applications

I am thankful to all the general editors and the reviewers of these papers who provided valuable review comments for the betterment of the papers published in these volumes. The support provided by the former President of the Indian Geotechnical Society, the late Prof. Sreerama Rao, and the former Honorary Secretary of IGS, Shri Jai Bhagwan, is gratefully acknowledged here.

Bringing out these volumes under the banner of prestigious publishers, Springer will immensely improve the visibility of these conference proceedings volumes. The Chennai Chapter of the Indian Geotechnical Society places on record its acknowledgement of the efforts put forth by Springer for bringing out these volumes for the benefit of the geotechnical engineering community.

Chennai, India

Prof. A. Boominathan
Chairman, Indian Geotechnical Conference 2016
Department of Civil Engineering IIT Madras

Preface

The annual conferences of the Indian Geotechnical Society bring out research papers and case histories on various topics in geotechnical engineering and geoenvironmental engineering. The conference IGC 2016 held at IIT Madras showcased more than 350 papers.

Springer is bringing out conference proceedings with selected papers from the conference. There are four volumes covering different conference themes. This fourth volume of the proceedings is presenting the papers on geotechnical applications and case histories. The major themes covered under this volume are (i) shallow and deep foundations; (ii) stability of earth and earth retaining structures; (iii) rock engineering, tunnelling and underground constructions; (iv) forensic investigations and case histories; (v) reliability in geotechnical engineering; (vi) special topics: offshore geotechnics, remote sensing and GIS, geotechnical education, codes and standards.

Major geotechnical engineering applications are seen in the foundations of very light structures to very heavy ones. A good number of papers on geotechnical engineering of shallow and deep foundations are included in this volume. Deep excavations are very common in the present-day construction. Case histories and analytical studies on deep excavation problems are presented in some of the papers.

Slope stability studies cover a significant volume of geotechnical engineering application. There are 11 papers on the slope stability problems in this volume. Some papers cover the dynamic response of slopes under seismic conditions. One of the key areas where the modern geotechnical procedures are adopted is the reinforced earth wall constructions. There are two papers on this topic.

Geotechnical engineering applications on offshore constructions require complex studies and analytical procedures. Three papers on the subject are finding place in this volume.

Chennai, India

Mr. Anirudhan I. V.
Dr. V. B. Maji

Contents

Part VI Theme 12: Special Topics: Offshore Geotechnics, Remote Sensing and GIS, Geotechnical Education, Codes and Standards

About the Editors

Anirudhan I. V. is a geotechnical consultant based in Chennai, India. His career as an independent consultant started in 1987 after associating with a leading geotechnical consultancy firm. His major fields of expertise are foundation design and construction and ground improvement techniques. He has been involved in the design of foundations and ground improvement schemes for the power and water projects in India and abroad. His organization, Geotechnical Solutions, Chennai, has executed more than 1500 small and medium geotechnical investigation jobs and provided foundation recommendations for small- to high-rise buildings. He graduated in Geotechnical Engineering from IIT Kanpur, India, in 1981, after obtaining a B.Sc. Engg. (Hons) degree in Civil Engineering from the College of Engineering, Thrissur, Kerala, India. He is actively involved in the activities of the Indian Geotechnical Society and has conducted conferences and seminars. He has authored several conference papers on geotechnical engineering applications and a manual on lateritic soils. He was a Member of TC-302 of ISSMGE on Forensic Engineering and the Convener of TC-04 of the Indian Geotechnical Society on Geotechnical Investigation. He is the Vice Chair of the Deep Foundations Institute, India.

V. B. Maji joined the Department of Civil Engineering at the Indian Institute of Technology (IIT) Madras in 2008 and is currently working as an associate professor there. He received his Ph.D. degree from the Indian Institute of Science, Bangalore. He has a Master of Technology degree from the Institute of Technology (now IIT), Banaras Hindu University (BHU), India, and a Bachelor of Science Engineering degree from BIT Sindri, Dhanbad, India. Prior to joining IIT Madras, he worked as Associate Consultant at Advanced Technologies and Engineering Services (ATES), a consultancy division of AIMIL Ltd., New Delhi, India. His teaching and research interest includes rock mechanics and geotechnical engineering.

Part I
Theme 2: Shallow and Deep Foundations

Measured and Predicted Settlement of Shallow Foundations on Cohesionless Soil

Kedar C. Birid and Ramvir Singh Chahar

Abstract Various methods are available to evaluate the settlement of shallow foundations on cohesionless soil. Fifteen numbers of such commonly adopted methods have been selected for the settlement calculations. These methods have been categorized as empirical (nine numbers), semi-empirical (two numbers), and theoretical (four numbers). The settlements of foundations with different footing sizes have been calculated using these methods on improved sandy soil (with the stone columns or vibro compaction) and on unimproved soil. The calculated settlement values are compared with the plate/footing load test results carried out on corresponding plate sizes and full scale RC footings. Among various studied methods, empirical method proposed by Terzaghi and Peck, semi-empirical method proposed by Schmertmann and a theoretical method proposed by Janbu appeared to predict reasonably accurate and reliable settlement values. The stiffness of soil does not appear to be nonlinear with strain and stress level as observed based on the actual settlement values at different stress levels. Almost all the methods overestimated the footing settlements at lower as well as higher stress levels. Though semi-empirical and theoretical methods are preferred over empirical methods due to lack of theoretical study in later, it is also suggested to appropriately select the geotechnical design parameters rather than trying to select the best method for the settlement analysis.

Keywords Settlement · Shallow foundations · Young's modulus
Cohesionless soil

K. C. Birid (✉) · R. S. Chahar
Toyo Engineering India Pvt. Ltd., Mumbai, India
e-mail: kedar.birid@toyo-eng.com

R. S. Chahar
e-mail: ramvir.chahar@toyo-eng.com

A. I. V. and V. B. Maji (eds.), *Geotechnical Applications*, Lecture Notes in Civil Engineering 13, https://doi.org/10.1007/978-981-13-0368-5_1

1 Introduction

Shallow foundations are designed to satisfy bearing capacity and settlement criteria. Out of both, the permissible settlement is often the controlling design criterion for larger footing widths. Numerous methods are available in the literature to estimate the settlement of shallow foundations. Some of these methods are based on laboratory parameters and some on the field test parameters. Thus these methods vary from empirical, semi-empirical to theoretical in nature. Attempts have already been made in the past to compare the different methods and to evaluate the best available method.

The views of different authors who had analyzed various settlement methods in past are as quoted below;

During last fifty years or so, a number of procedures have been developed to predict elastic settlement; however, there is a lack of reliable standardized procedure (Das 2013).

Remarkable number of methods has been developed to estimate the settlement of shallow foundations on sand, yet consistent success in accurately predicting such settlements remains elusive. These methods range from purely empirical methods developed originally for conservative footing design to complex category three nonlinear finite element method (Poulos 1999).

Many of the new methods rely on in situ SPT or CPT data; hence it is not possible to satisfactorily examine the theoretical relationship between the various methods (Poulos 1999).

Das and Sivakugan (2007) also reported wide range of settlement predictions during Settlement ’94 prediction sessions held in Texas due to deficiencies in the current state of the art.

Thus there is a common supposition about the inconsistency and inaccuracy of the present methods.

Authors, also in this paper, have attempted to assess the performance of various such methods on the basis of its comparison with the measured settlements using plate load test and full scale footing load test.

2 Review of Literature

As per Reza and Abdolhosain (2013), the most popular methods for settlement predictions, discussed commonly in textbooks, are the ones proposed by Terzaghi and Peck (1967), Schmertmann (1970), Schmertmann et al. (1978), Burland et al. (1985), Meyerhof (1956) and Peck and Bazaraa (1969) methods are similar to the one proposed by Terzaghi and Peck (1967).

Various other researchers such as D'Appolonia et al. (1970), Hough and Schmertmann (1970), Briaud (1992), Berardi and Lancellotta (1991) and Mayne and Poulos (1999), Sargand et al. (1997), Shahin et al. (2002), Sivakugan and Johnson (2004), Duzceer (2009), Reza and Abdolhosain (2013) etc. suggested different models and formulae to arrive at the settlement. Methods for evaluation of settlement are also provided in Canadian Foundation Engineering Manual, CFEM (1992) and NAVFAC Design Manual (1986).

Das (2013) divided various methods to calculate the elastic settlement available at the present time into three general categories such as empirical, semi-empirical and theoretical.

3 Testing Program and Settlement Methodology

An assessment of the reliability of 15 different methods as listed in Table 1 has been carried out by comparing calculated and measured settlements for 12 cases where footing are resting on sand with stone columns/vibro compaction and two cases where footings are resting on sand without stone columns/vibro compaction.

Load tests were carried out on steel plates of 1.0, 1.8, 2.4, and 3.0 m square sizes and a 4.5 m square RC footing. The load tests were carried out on different plate/footing sizes as no consensus methods exist for extrapolating the settlement of standard size plate to the settlement of prototype footing and the outcome of load test imposes severe limitations on the interpretation of loading tests on both cohesionless and cemented soils (Consoli et al. 1998). The soil stratification consisted of fine to medium grained and medium dense sand. The soil was improved by virtue of stone columns and vibro compaction to mitigate the liquefaction potential and did not have any other purpose towards settlement specific studies.

The modulus of elasticity of soil with stone columns was calculated based on the area ratio of treated (stone columns) to untreated soil. The correlations from SPT value and CPT cone resistance by Bowles (1996) were adopted for evaluation of Young's modulus of soil. The Young's modulus of both the materials, i.e., soil and stone columns is then combined on weighted average basis to arrive at the composite Young's modulus of treated soil mass. The identical philosophy was adopted to arrive at composite Young's modulus value in case of vibro compacted soil.

All the settlement methods were evaluated in terms of (1) accuracy (the ratio average calculated to measured settlement), and (2) reliability (the percentage of cases in which the calculated settlement equaled or exceeded the measured settlement as suggested by Tan and Duncan (1991).

Table 1 List of different methods considered for settlement calculations (Resin et al. 2009)

S. No.	Method	Expression	Parameters	Category
1	De Beer (1965)	$S = H/C \log[(\sigma_0' + \Delta\sigma)/(\sigma_0')]$	$C = 1.5 * q_c/\sigma_0'$ q_c = cone resistance	Empirical
2	Meyerhof (1956)	$S = C_D Cw \times 1.25q/N$ (For $B \leq 1.22$ m) $S = C_D Cw\ [2q/N] \times [(B/(B + 0.3))^2]$ (For $B \geq 1.22$ m)	N = SPT value	Empirical
3	Schultze and Sherif (Anderson et al. 2007)	$S = \frac{f * q}{N^{0.87}\left(1 + \frac{0.4Df}{B}\right)}$	N = SPT value	Empirical
4	Peck-Hanson-Thornburn (D'Appolonia et al. 1970)	$S = q/(0.44 * Cw \times N_1)$	N_1 = Corrected N value	Empirical
5	Terzaghi and Peck (1967)	$S = C_D \times Cw\ [3q/N] \times [(B/(B + 0.3))^2]$	N = SPT value	Empirical
6	Peck and Bazaraa (1969)	$S = C_D \times Cw\ [2q/N] \times [(B/(B + 0.3))^2]$	N = SPT value	Empirical
7	Burland et al. (1985)	$S = \alpha_1\alpha_2\alpha_3\ [(1.25 * L/B)/(0.25 + L/B)]^2 *$ $B * q'$	–	Empirical
8	Anagnostropoulos et al. (1991)	$S = 2.37 * q^{0.87} * B^{0.7}/N^{1.2}$	N = SPT value	Empirical
9	Hough (1969)	$S = \sum H_c \times 1/C' \times \log\left[(\sigma_0' + \Delta\sigma)/\sigma_0'\right]$	$C'' = (1 + e_0)/Cc$	Empirical
10	Schmertmann (1970)	$S = C_1 \times C_2 \times (q - q') \times \sum(Iz \times \Delta z/E_s)$	E_s = Modulus of elasticity	Semi-empirical
11	NAVFAC (1986)	$S = 4q/Kv_1 * [B/(B+1)]^2$	Kv_1 = Modulus of subgrade reaction	Semi-empirical
12	Janbu (CFEM 1992)	$S = \sum H(1/mj) \times \left[(\sigma_1'/\sigma_r') - (\sigma_0'/\sigma_r')\right]$	–	Theoretical
13	D'Appolonia et al. (1970)	$S = \mu_0\mu_1 \times q \times B/M$	M = Modulus of compressibility	Theoretical
14	Bowles (1996)	$S = (1 - \mu^2) \times q \times B' \times Is \times If/E_s$	μ = Poisson's ratio E_s = Modulus of elasticity	Theoretical
15	Schleicher (1926)	$S = p \times B \times (1 - \mu^2) \times I/E_s$	E_s = Modulus of elasticity	Theoretical

4 Results and Discussion

The calculated settlement and the actual settlement based on the plate/footing load test under different stress levels are plotted graphically as calculated settlement against the measured settlement at stress level of 165 kPa as shown in Fig. 1. It can be observed from Fig. 1 that, all the methods overestimate the footing settlement at all stress levels. The elastic-based methods, based on CPT, SPT gives reasonable predictions.

The settlement values are also plotted against normalised pressure, q/q_u, (see note Fig. 2) for the selected methods of Terzaghi and Peck, Schmertmann, Janbu and load test results as presented in Fig. 2. It can be observed from Fig. 2 that, the settlement values are more sensitive to q/q_u ratio for large footing sizes (B) as the small variation in q/q_u varies the settlement values by large extent. Due to this reason more accuracy is required in evaluation of soil stiffness parameters in case on larger footing sizes.

The nonlinear variation of soil stiffness with strain and stress level has been discussed in various literatures. Due to such soil behavior, the settlement of plate or footing during the load test was expected to reduce with the increased stress levels. In order to verify this, the settlement values at varying stress levels of 50, 100 and 165 kPa were studied.

A plot of reliability versus accuracy as shown in Fig. 3 indicates that the ratio of calculated to measured settlement is much more than 1 and moves towards 1 with the increased stress level.

The increase in Young's modulus (and decrease in settlement) with the increased stress is not observed from these results. This behavior could be attributed to the pre-densification of soil mass due to the stone column installation and vibro compaction of subsoil.

It can be also observed that the ratio of calculated to measured settlement at lower stress of 50 kPa is more than 2.0 for all the methods except for Schmertman, Peck and Bazaraa and Janbu method. As the stress level has increased to 100 kPa, the accuracy has increased and is between 1 and 2. With further increase in the stress level to 165 kPa, most of the methods indicated high accuracy near 1.

It can also be observed from this plot that as the accuracy decreases (i.e., higher the ratio of calculated to measured settlement), the reliability increases in the sense that they underestimate the settlement relatively infrequently. This behavior in line with the observations made by Tan and Duncan (1991) who found that there is generally a trade-off between accuracy and reliability among all methods.

The plot of average value of accuracy and reliability at different stress level is indicated in Fig. 4. It can be observed from this plot that, the variation in accuracy with different methods decreases with the increase in stress level. However the reliability still has a higher range irrespective of the stress level.

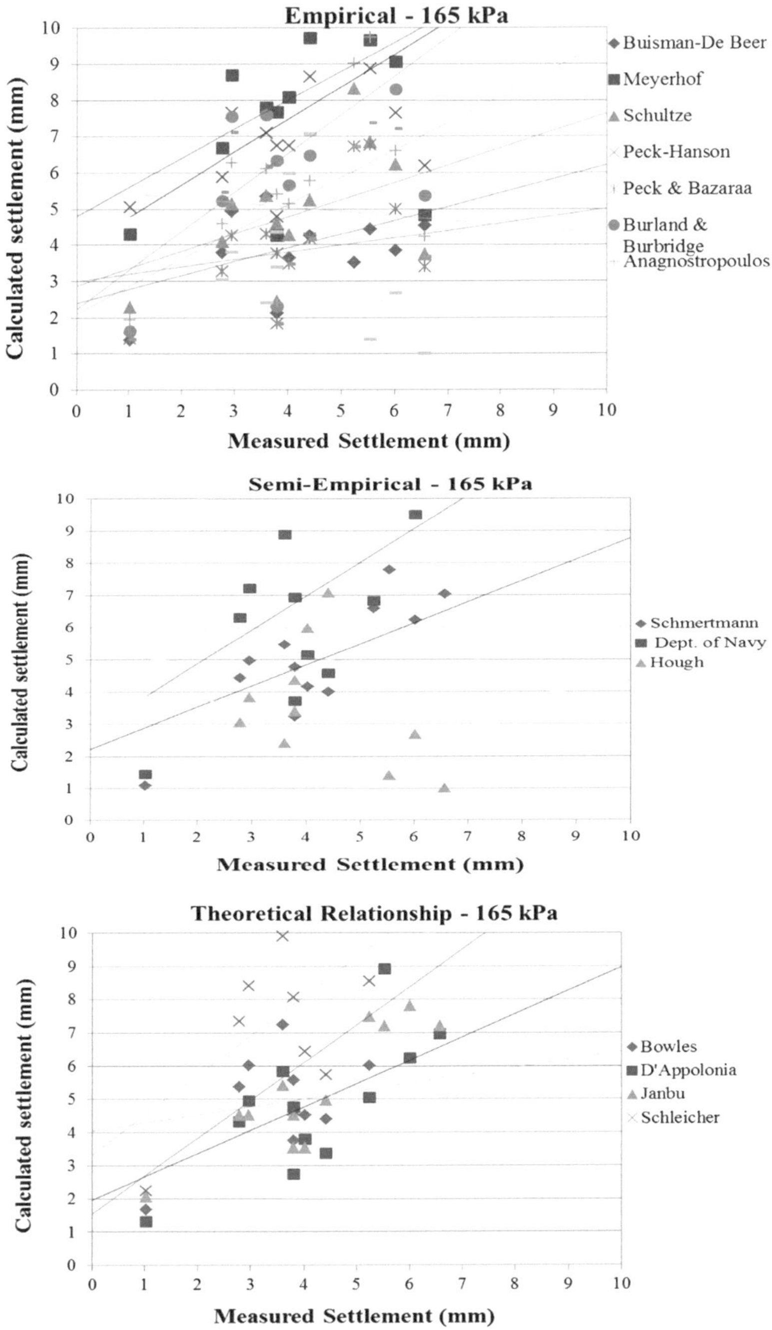

Fig. 1 Calculated settlement versus actual settlement (for load intensity of 165 kPa)

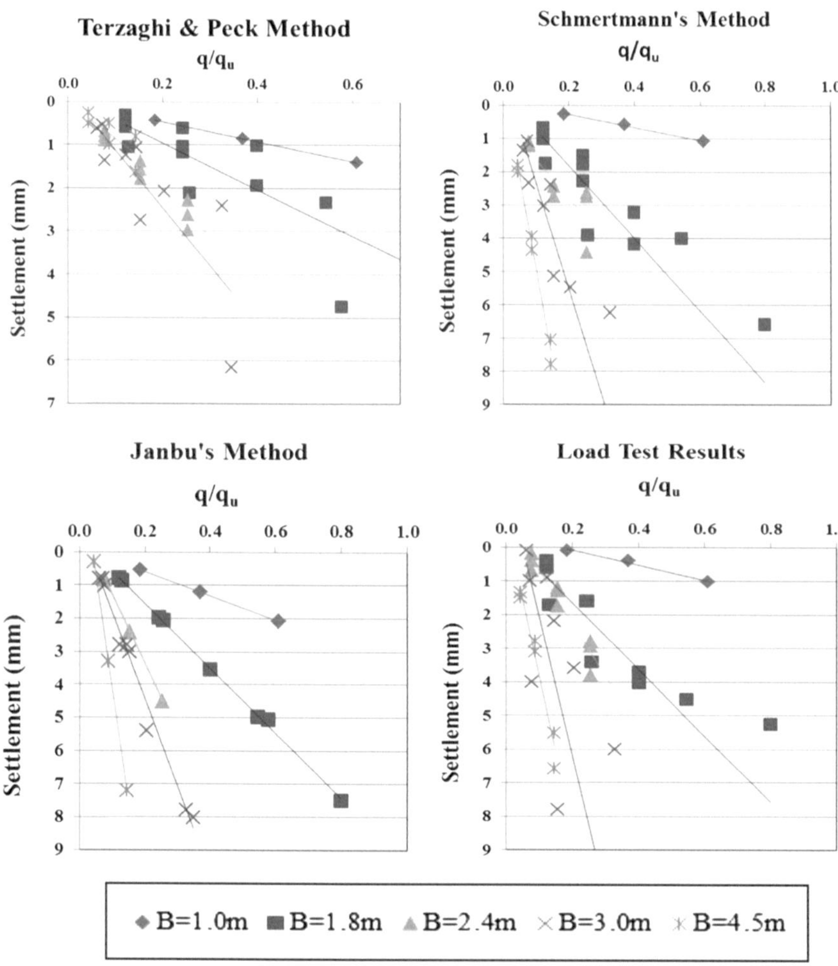

Fig. 2 Settlement versus normalised pressure (q/q_u). *Note* q = pressure applied on footings; q_u = ultimate bearing pressure of soil

As stated by Poulos (1999), accurate prediction of the settlement is dependent as much on the experience of the predictor, and a good amount of luck, as on the method employed. The selection of geotechnical parameters also plays a major part in the success or otherwise of a prediction, and may outweigh or mask any shortcomings of the method used.

Das and Sivakugan (2007) also had an opinion that, one of the main factors that contribute to the uncertainty in settlement predictions is our inability to quantify the soil stiffness correctly.

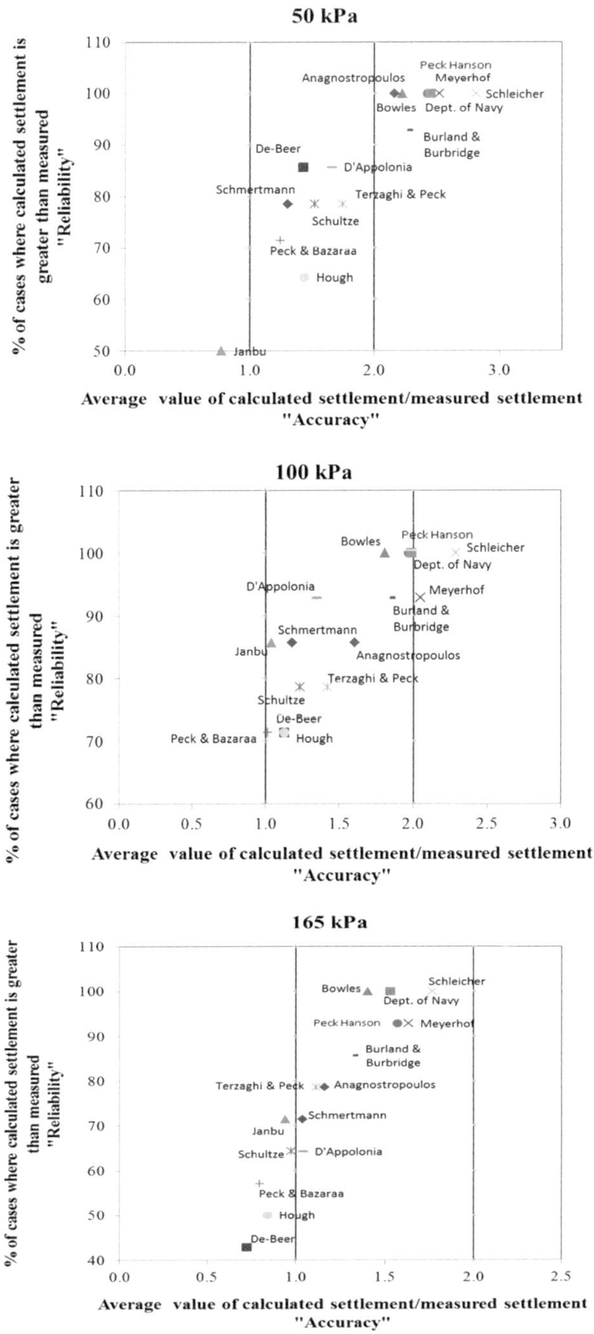

Fig. 3 Average (calculated settlement/actual settlement) versus % of cases where calculated settlement is greater than the actual settlement (i.e. accuracy vs. reliability)

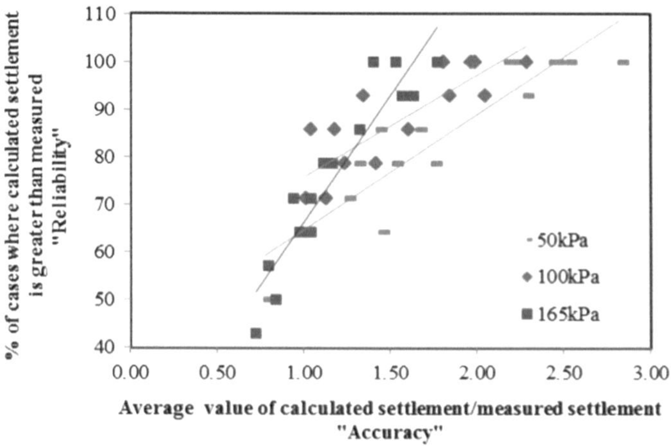

Fig. 4 Reliability versus accuracy for all loads and methods

The discrepancies between the observed and the predicted settlement are primarily due to the inability to estimate the modulus of elasticity of soil using the results of the standard penetration tests and/or cone penetration tests, Das (2013).

The authors, based on this study and past experience agrees with the views stated by Poulos (1999), Das and Sivakugan (2007), Das (2013).

5 Conclusions

The settlement prediction by all the methods overestimates the settlement values irrespective of stress values. The empirical, semi-empirical and theoretical methods proposed by Terzaghi and Peck, Schmertmann and Janbu respectively provided reasonable settlement estimates. For these methods, the settlement values are also sensitive to normalized pressure q/q_0 for larger footing sizes.

The accuracy of calculated settlement increases with the increase in stress level. Terzaghi and peck, Anagnostropoulos et al., Schmertmann, Janbu, Schultze and Sherif and D'Appolonia methods appear to have accuracy but poor reliability. On the other hand Bowles, Schleicher, Department of Navy, Meyerhof and Peck-Hanson-Thornburn methods are more conservative but have less accuracy. It was also observed that the decrease in accuracy increases the reliability and vice versa. The accuracy in settlement predictions also lies in appropriate modeling and selection of Young's modulus of sand than in the type of method employed.

The other methods can also be adopted depending on the designer's experience and appropriate correction factors if any. These methods can be used for preliminary estimates of the settlement values. However most of the empirical equations are soil and foundation type dependent and hence shall be adopted with caution to all types of soil and foundations.

References

Anagnostopoulos, A. G., Kalteziotis, N., Tsiambaos, G. K., & Kavvadas M. (1991). Geotechnical properties of the Corinth Canal marls. *Geotechnical and Geological Engineering, 9*(1), 1–26.

Berardi, R., & Lancellotta, R. (1991). Stiffness of granular soil from field performance. *Geotechnique, 41*(1), 149–157.

Bowles, J. E. (1996). *Foundation analysis and design* (5th ed.). Singapore: The McGraw-Hill.

Brian Anderson, J., Townsend, F. C., & Rahelison L. (2007). Load Testing and Settlement Prediction of Shallow Foundation. *Journal of Geotechnical and Geoenvironmental Engineering, 133*(12), 1494–1502.

Briaud, J.-L. (1992). *The pressuremeter.* Brookfiled, Vt: Balkema.

Burland, J. B., Burbidge, M. C, & Wilson E. J. (1985). Settlement of foundations on sand and gravel. *Proceedings of the Institution of Civil Engineers, 78*(6), 1325–1381.

CFEM, Canadian Foundation Engineering Manual, (1992). (3rd ed.) Bitech Publishers Ltd. Richmond, Canada.

Consoli, N. C., Schnaid, F., & Milititsky, J. (1998). Interpretation of plate load tests on residual soil site. *Journal of Geotechnical and Geoenvironmental Engg, 124*(9), 857–867.

D'Appolonia, D. J., D'Appolonia, E., & Brissette, R. F. (1970). Settlement of spread footings on sand: Closure. *Journal of the Soil Mechanics and Foundations Division, ASCE, 96*(2), 754–762.

Das, B. M. (2013). Elastic settlement of shallow foundations on granular soil: A critical review. *GLE video seminar.* http://gle.wisc.edu/wp-content/uploads/2013/07/Elastic-Settlement-Shallow-Foundations_A-Critical-Review-2.pdf.

Das, B. M., & Sivakugan, N. (2007). Settlements of shallow foundations on granular soil—an overview. *Journal of Geotechnical Engineering, 2007*(1), 19–29. https://doi.org/10.3328/IJGE.2007.01.01.19-29.

De Beer, E. E. (1965). Bearing capacity and settlement of shallow foundations on sand. *Proceedings, Symposium on Bearing Capacity Settlement of Foundations* (pp. 15–33), Durham, N.C.: Duke University.

Duzceer, R. (2009). Observed and predicted settlement of shallow foundation. In 2nd International conference on new developments in soil mechanics and geotechnical engineering (pp 590–597).

Hough, B. K. (1969). *Basic soils engineering.* New York: Ronald Press.

Meyerhof, G. G. (1956). Penetration tests and bearing capacity of cohesionless soils. *Journal of the Soil Mechanics and Foundations Division, ASCE, 82*(1), 1–19.

NAVFAC (1986), Design Manual 7.01, Soil Mechanics, Naval Facilities Engineering Command, Department of the Navy, Washington, D.C.

Peck, R. B., & Bazaraa, A. R. S. S. (1969). Discussion of Settlement of spread-footings on sand. *Journal of Soil Mechanics and Foundation Division, ASCE, 95*(SM3), 305–309.

Poulos, H. G. (1999). Common procedures for foundation settlement analysis-Are they adequate? *Australian geomechanics*, March 1999.

Rasin, D., & Kasktas, A. S. (2009). Observed and predicted settlement of shallow foundation. In *2nd International Conference on New Developments in Soil Mechanics and Geotechnical Engineering*, 28–30 May 2009, Near East University, Nicosia, North Cyprus.

Reza, A. A., & Abdolhosain, H. (2013), A non-linear method to estimate elastic settlement of shallow foundations using small-strain stiffness. In *International Conference on Case Histories in Geotechnical Engineering.* Paper 2, http://scholarsmine.mst.edu/icchge/7icchge/session01/2.

Sargand, S. M., Hazen, G. A., & Masada, T. (1997). *Field and laboratory evaluations of spread footings for highway bridges,* FHWA Report no: OH/98-017.

Schleicher, F. (1926). Zur Theorie Des Baugrundes, *Der Bauingenieur, 48*, 49.

Schmertmann, J. H. (1970). Static cone to compute static settlement over sand. *Journal of Soil Mechanics and Foundation Division, ASCE, 96*(3), 1011–1042.

Schmertmann, J. H., Hartman, J. P., & Brown, P. R. (1978). Improved strain influence factor diagrams. *Journal of Geotechnical and Geoenvironmental Engineering, 104*(8), 1131–1135.
Shahin, M. A., Maier, H. R., & Jaksa, M. B. (2002). Prediction settlement of shallow foundations using neural network. *Journal of Geotechnical and Geoenvironmental Engineering, 128*(9), 785–793. ASCE.
Sivakugan, N., & Johnson, K. (2004). Settlement prediction in granular soils: a probabilistic approach. *Geotechnique, 54*(7), 499–502.
Tan, C. K., & Duncan, J. M. (1991). Settlement of footings on sands: Accuracy and reliability. *Proceedings of Geotechnical Engineering Congress 1991, 1*, 446–455. Colorado.
Terzaghi, K., & Peck, R. B. (1967). *Soil mechanics in engineering practice* (2nd ed.). New York: Wiley.

Prediction of Bearing Capacity and Settlement from SPT Values Using Genetic Algorithm

C. R. Athira and N. Sankar

Abstract Many empirical and semi-empirical methods are existing to predict bearing capacity and settlement, but majority of them are inconsistent and not user friendly. In this work, a genetic algorithm approach is used for predicting bearing capacity and settlement of shallow foundations on C-soil, φ-soil and C-φ soil, separately, based on those input parameters which can be easily find out from simple experiments. The development and verification of the genetic models were done using a large database containing about 832 datasets from 167 soil investigation reports. The equation developed for bearing capacity and settlement thus obtained can be used in prediction of new cases that were not used for the development of the genetic model. The results of the model obtained were compared with various conventional equations available for calculating bearing capacity and settlement and was found to be superior. The correlation of predicted data with actual field measurements was determined and it was found out that the genetic algorithm approach have high degree of accuracy. Parametric study was also done to evaluate the effect of varying each of the input parameters on the corresponding output.

Keywords Bearing capacity · Settlement · SPT · Genetic algorithm

1 Introduction

Two important soil parameters which determine the type of foundation are settlement and bearing capacity. Usually existing methods to measure them, including plate load tests, can only give a rough idea due to a number of limiting conditions.

C. R. Athira (✉) · N. Sankar
Department of Civil Engineering, National Institute of Technology Calicut, Kozhikode 673601, India
e-mail: ammuelarajan@gmail.com

N. Sankar
e-mail: sankar@nitc.ac.in

A. I. V. and V. B. Maji (eds.), *Geotechnical Applications*, Lecture Notes in Civil Engineering 13, https://doi.org/10.1007/978-981-13-0368-5_2

So we need an integral method of prediction. Genetic algorithms are a part of evolutionary computing, which is a rapidly growing area of artificial intelligence (Chan et al. 2009; Fischer 2008; Porta et al. 2013; Rezania and Javadi 2007; Momeni et al. 2014; Shahnazari et al. 2013; Aggour 2002). In this study an approach is made to create an integral method to form a genetic algorithm which is trained to predict bearing capacity and settlement of a soil in a particular condition with given values of some practical input parameters that can be easily find from construction sites, which are SPT value, water content, depth of water table and bulk density of soil, cohesion and angle of internal friction.

2 Methodology

GA is mainly an optimization technique and to use this algorithm for prediction we need to collect a number of actual values of input and the resulting output parameters and that is the first step. For bearing capacity GA model, the input parameters are Standard Penetration Test (SPT) values, depth of water table, bulk density of soil, cohesion, angle of friction and moisture content. For settlement GA model, the input parameters are SPT values, bulk density, cohesion, angle of friction, and bearing capacity. The data were collected from bore log details and other related reports regarding each site in different parts of Kerala, India, from Geotechnical Laboratory of NIT Calicut, Kerala. Separate dataset were created for C-soil, φ-soil and C-φ soil for easiness of comparison of results from GA models with conventional methods. The output parameters in the dataset, both bearing capacity and settlement will be treated as actual values of output parameters with which the results of GA models compared.

Next step is dividing dataset of each soil into subsets such as training and validation set. In this paper 80% adopted as training and 20% as testing or validation data and is shown in Table 1. Division of data was done using a statistically consistent approach so that both sets possess similar statistical properties including mean, standard deviation, maximum and minimum.

Then GA models were developed using dataset with the help of Scilab 5.5.2 software until the Objective function = $[\Sigma(\text{predicted value-experimental value})^2]/N$, is minimized, where N is the number of data available in training set. In order to arrive at the best solution, the program was run several times by changing parameters like number of generations, crossover and mutation probabilities and the expression of bearing capacity and settlement corresponding to the best solution was selected for each model. Table 2 shows the features of developed models.

Table 1 Division of database

Item	Training data	Testing data	Total
C-soil	273	68	341
φ-soil	110	28	138
C-φ soil	282	71	353
Total data			832

Table 2 Features of developed genetic algorithm models

Model	Crossover Pro.	Mutation Pro.	Population size	No. of generations	Variable array size	Operator array size
BC C-soil	0.8	0.05	300	5000	9	10
BC φ-soil	0.8	0.05	300	5000	9	10
BC C-φ soil	0.8	0.05	200	5000	11	12
Settlement C-soil	0.8	0.1	300	5000	7	8
Settlement φ-soil	0.8	0.09	200	5000	7	8
Settlement C-φ soil	0.8	0.09	300	5000	9	10

3 Results and Discussions

The relationships obtained for each of the models to predict bearing capacity and settlement are shown below.

For C-soil

$$\mathrm{BC} = 106.394N^{0.0937} + \frac{82.532W^{0.0036} + 427.785M^{-2.6276}}{7.0589\Upsilon^{-2.3011} + 418.323C^{-1.7184}} \tag{1}$$

$$\begin{aligned} s = {} & 224.4172N^{-3.6921} + 197.1212\Upsilon^{-2.5558} \\ & + 195.0985C^{-5.3672} + 18.0130\mathrm{BC}^{-7.9628} \end{aligned} \tag{2}$$

For φ-soil

$$\begin{aligned} \mathrm{BC} = {} & 0.0649N^{0.1067} \times 14.1645W^{0.1162} \\ & \times \left(472.4094M^{-0.5723} - 3.6548\Upsilon^{1.4720} + 8.9611\varphi^{0.5214}\right) \end{aligned} \tag{3}$$

$$s = 82.2013N^{-2.6483} + 436.4909\Upsilon^{-3.0253} + 5.7952\varphi^{-6.5541} + 277.663\mathrm{BC}^{-2.3415} \tag{4}$$

For C-φ soil

$$\begin{aligned} \mathrm{BC} = {} & \left(26.1341N^{0.3488} \times 11.6380W^{1.0454} \times 19.4894M^{-6.1385} \right. \\ & \left. \times 323.0065\Upsilon^{0.5279} \times 2.0989C^{1.1524}\right) + 141.3081\varphi^{0.6697} \end{aligned} \tag{5}$$

$$\begin{aligned} s = {} & 711.3923N^{-3.7469} + 52.6576\Upsilon^{-2.2392} \\ & + 59.0848C^{-7.9594}\left(452.5694\varphi^{-1.2921} + 626.8442\mathrm{BC}^{-2.6938}\right) \end{aligned} \tag{6}$$

where

W	Depth of water table (m)
M	Water content (%)
N	SPT value
γ	Bulk density of soil (kN/m^3)
C	Cohesion of soil (kN/m^2)
Φ	Angle of internal friction (Degree)
BC	Bearing capacity (kN/m^2)
S	Settlement (m)

The correlation of results from six GA models for bearing capacity and settlement for C-soil, φ-soil, and C-φ soil, with the actual values were checked in terms of RMSE, Correlation Coefficient, MAE values and found to be superior (Akpila 2013, 2014).

The performance of the models for training dataset was found out by plotting predicted values against actual values and found that for all models majority of data fall near 1:1 normal line. As an example performance of bearing capacity and settlement models of C-soil are shown in Figs. 1 and 2 respectively.

Then the six GA models were tested with the testing or validation dataset to show that GA model can predict over a wide range of, even, unseen cases with accuracy, and the correlation of obtained results from GA models with actual values were checked in terms of RMSE, Correlation Coefficient, MAE values and found to be superior. As an example performance of bearing capacity and settlement models of C-soil with testing or validation dataset are shown in Figs. 3 and 4 respectively.

Next step is comparison of results obtained from these six GA models with conventional methods, separately for C-soil, φ soil and C-φ soil and the error were calculated for each data, as the difference between actual value and the value given by each of the method (including GA model). The same was plotted in a graphical form. For each model, the conventional methods used were selected based on nature of database available. From the obtained graphs, it can be seen that all GA models gives the lesser error than any other conventional method. As an example,

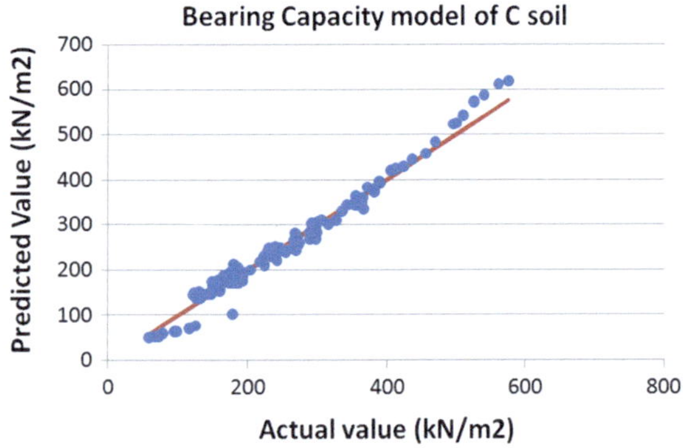

Fig. 1 Bearing capacity model of C-soil with training data

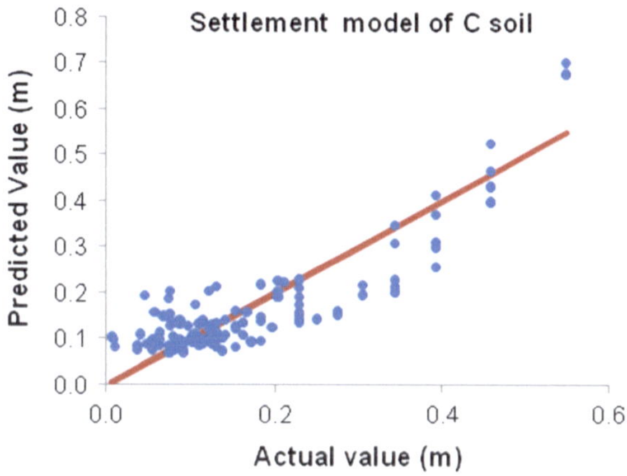

Fig. 2 Settlement model of C-soil with training data

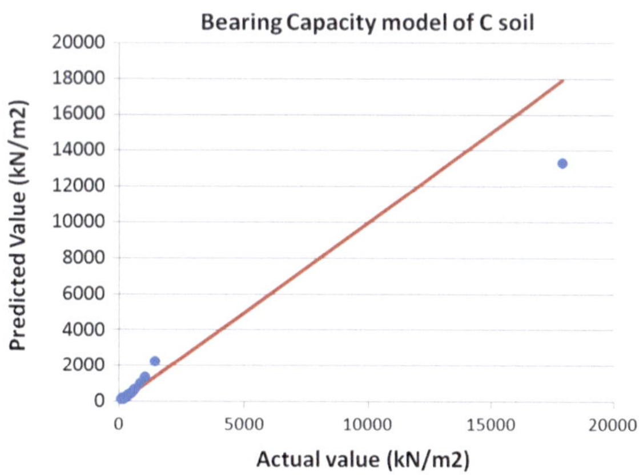

Fig. 3 Bearing capacity model of C-soil with testing data

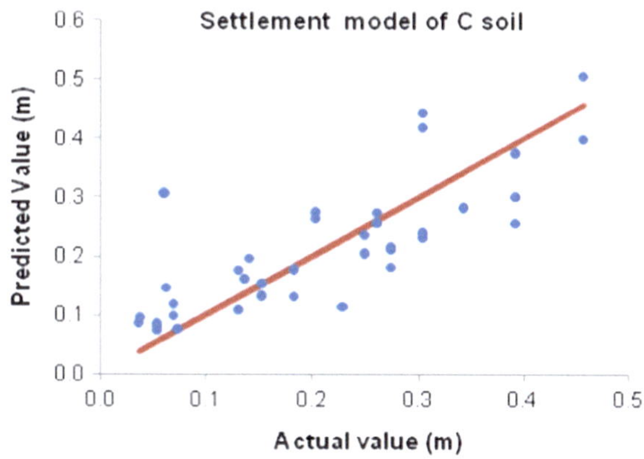

Fig. 4 Settlement model of C-soil with testing data

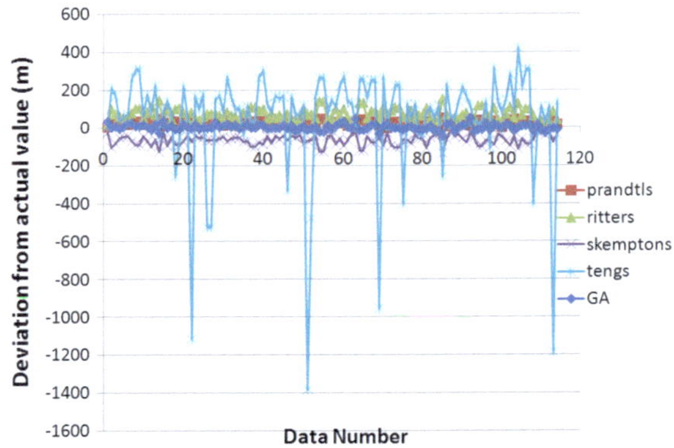

Fig. 5 Comparison of bearing capacity model of C-soil

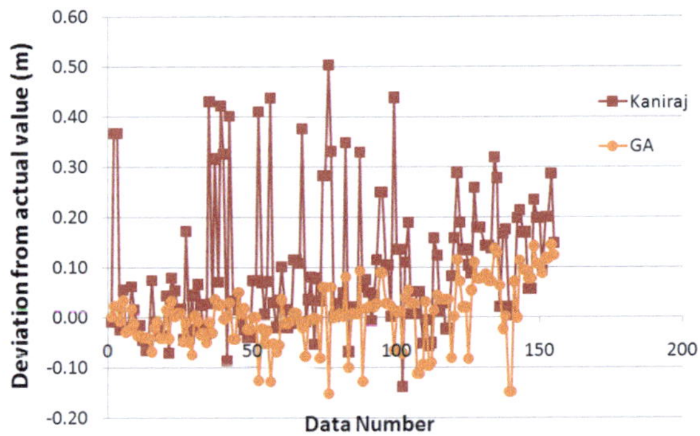

Fig. 6 Comparison of settlement model of C-soil

comparison results of bearing capacity and settlement models for C-soil are shown in Figs. 5 and 6 respectively.

Parametric study was done to evaluate the predictive capability of obtained formulas as well as the effect of varyings each of the input parameters on the corresponding output (AbdulRazzaq et al. 2011). This was done by fixing all input parameters to their mean value, except one and varying it between its range of minimum and maximum values and then noting down effect of the variation on the output.

For bearing capacity models, it is observed that the bearing capacity values increases as the values of SPT, water table depth, bulk density, cohesion and angle of internal friction values are increased (Shariatmadari et al. 2008). But in case of water content, bearing capacity values reduces as the value of water content increases.

For settlement models, it is observed that the settlement values decreases as the values of SPT, bulk density, cohesion, angle of internal friction and bearing capacity values are increased.

4 Conclusions

The prediction and calculation of bearing capacity and settlement is a wide area of research and each method or empirical formula has its own limitations that it can be used under certain conditions. In this paper Genetic Algorithm technique was used to develop expressions for bearing capacity and settlement of C-soil, φ-soil, and C-φ soil with the help of 832 datasets dealing with almost all soil types of Kerala, India. And thus 6 GA models and corresponding equations were developed for bearing capacity and settlement of C-soil, φ-soil, and C-φ soil. The developed equations were checked for their performance using training dataset and were found that the models developed were able to predict the output (either bearing capacity or settlement) near to the actual value. The ability and accuracy of the models while dealing with unseen data was also checked using testing or validation dataset. The models were then compared in performance with some conventional methods of computing bearing capacity and settlement, separately for C-soil, φ-soil and C-φ soil and models were found to be superior. The sensitivity analysis was also carried out to check the predictive capability of the models and to study the influence of each of the input parameters on the output, and is also checked and satisfied. These expressions developed in this study are hence suitable for predicting bearing capacity and settlement for either C-soil, φ-soil or C-φ soil, and will be mostly suitable for soils in Kerala, India.

The suitability of the said method for other states in India as well as outside India was not checked in this study. The developed equations are based on the soil database solely from Kerala, and that can be said as one of the limitation of the present study. The results of this study may require more refinement before they are used for any comparison purpose or for practical problems.

References

AbdulRazzaq, K. S., Hussein, W. A., & Hameed, A. H. (2011). Bearing capacity based on SPT-computer interpolation. *Diyala Journal of Engineering Sciences, 4,* 118–129.

Aggour, M. S. (2002). *Updating bearing capacity—SPT graphs.* Report Submitted to Maryland State Highway Administration Office of Policy and Research.

Akpila, S. B. (2013). Comparison of standard penetration test methods on bearing capacity of shallow foundations on sand. *Scientific Journal of Pure and Applied Sciences, 2,* 72–78.

Akpila, S. B. (2014). Bearing capacity and settlement response of raft foundation on sand using standard penetration test method. *Canadian Journal of Pure and Applied Sciences, 8,* 2769–2774.

Chan, C. M., Zhang, L. M., & Jenny, T. N. (2009). Optimization of pile groups using hybrid genetic algorithms. *Journal of Geotechnical and Geoenvironmental Engineering, 135*(4), 497–505.

Fischer, D. (2008). *Evolution with a teleology: The genetic programming heuristic approach to modeling.* Kuala Lumpur: Universiti Malaya.

Kaniraj, S. R. (1988). *Design aids in soil mechanics and foundation engineering.* New York: Tata Mcgraw Hill.

Koza, J. R. *Genetic programming: On the programming of computers by means of natural selection.* A Bradford Book The MIT Press Cambridge, Massachusetts London, England.

Momeni, E., Nazir, R., Jahed Armaghani, D., & Maizir, H. (2014). Prediction of pile bearing capacity using a hybrid genetic algorithm-based ANN. *Measurement, 57,* 122–131.

Porta, J., Parapar, J., Doallo, R., Rivera, F. F., Sante, I., & Crecente, R. (2013). High performance genetic algorithm for land use planning. *Computers, Environment and Urban Systems, 37,* 45–58.

Rezania, M., & Javadi, A. A. (2007). A new genetic programming model for predicting settlement of shallow foundations. *Canadian Geotechnical Journal, 44,* 1462–1473.

Shahnazari, H., Shahin, M. A., & Utunchian, M. A. (2013). Evolutionary-based approaches for settlement prediction of shallow foundations on cohesionless soils. *International Journal of Civil Engineering, 12,* 55–64.

Shariatmadari, N., Eslami, A., & Karimpour-fard, M. (2008). Bearing capacity of driven piles in sands from SPT–applied to 60 case histories. *Iranian Journal of Science and Technology, 32,* 125–140.

Influence of Configuration of Stone Columns on Combined Footings Resting on Reinforced Earth Beds

Priti Maheshwari

Abstract This work deals with the deformation analysis of combined footings resting on geosynthetic and stone column improved ground and subjected to symmetrically placed column loads. Simple mathematical model has been proposed which idealizes a combined footing as a beam. Just below the footing, there is the presence of a granular fill layer which has been modeled as a Pasternak shear layer. Geosynthetic layer, idealized as an elastic extensible membrane, has been placed in between this granular soil layer. There is an existence of a soft soil layer below geosynthetic-reinforced granular fill layer improved by stone columns. Stone columns and the soft soil layer have been represented by Winkler springs and Kelvin-Voigt body respectively. Nonlinear behavior of granular fill, stone columns and the soft soil has been considered in the analysis employing hyperbolic constitutive relationships. Governing differential equations have been derived and solution of these has been obtained with the help of appropriate boundary and continuity conditions. Finite difference method has been adopted for solving the model of footing-geosynthetic-reinforced granular fill-stone column reinforced soft soil system. Parametric study has been conducted to study the influence of configuration of stone columns. Five column loads have been considered. These results have been presented in the form of non-dimensional charts which can be used for deciding the optimum configuration of the stone columns in such soil-footing system. It has been seen that both these parameters, i.e., diameter and spacing of stone columns significantly influence the response of combined footing. The deflection of footing has been found to reduce with reduction in spacing to diameter ratio (*s*/*d*) of stone columns. The ratio, *s*/*d* has been found to possess an optimum value of 2.5–3.

Keywords Combined footings · Extensible geosynthetic · Stone columns Nonlinearity

---

P. Maheshwari (✉)
Department of Civil Engineering, Indian Institute of Technology Roorkee, Roorkee, India
e-mail: priti_mahesh2001@yahoo.com

A. I. V. and V. B. Maji (eds.), *Geotechnical Applications*, Lecture Notes in Civil Engineering 13, https://doi.org/10.1007/978-981-13-0368-5_3

1 Introduction

Nowadays, ground improvement by means of stone columns and the geosynthetics has been widely adopted. It helps in improving the bearing capacity of foundations and reducing total and the differential settlements. Many studies have been conducted for the analysis of stone column treated soft soil system (Balaam and Booker 1981; Alamgir et al. 1996; Shahu et al. 2000 etc.). In all these studies, geosynthetic has not been used as an inclusion. Some of the studies pertaining to the analysis of geosynthetic-reinforced earth beds include Madhav and Poorooshasb (1988), Ghosh and Madhav (1994), Shukla and Chandra (1994), Yin (1997), Maheshwari et al. (2004) etc. In all these studies, the natural soil was not reinforced with stone columns. Deb et al. (2007, 2010) presented models for analysis of geosynthetic-reinforced granular fill soft soil with stone columns system by considering inextensible and extensible nature of the geosynthetic layer respectively. However, the flexural rigidity of the footing was not considered in the analysis. Maheshwari and Khatri (2012) proposed generalized model for combined footings on Geosynthetic-Reinforced Granular Fill Stone Column Improved Soft Soil System. The geosynthetic layer was assumed to behave as inextensible rough elastic membrane.

The critical review of literature suggests that there is no study available dealing with the deformation analysis of footings resting on extensible geosynthetic-reinforced granular fill soft soil with stone columns system. In view of this, a mechanical model has been proposed for the purpose of such an analysis. The footing has been assumed to behave as a beam of finite length and flexural rigidity. The footing is resting on extensible geosynthetic-reinforced granular fill which lies above the stone column treated soft foundation soil (Fig. 1). The governing differential equations along with appropriate boundary conditions have been derived and solved employing finite difference method. Iterative Gauss Siedel method has

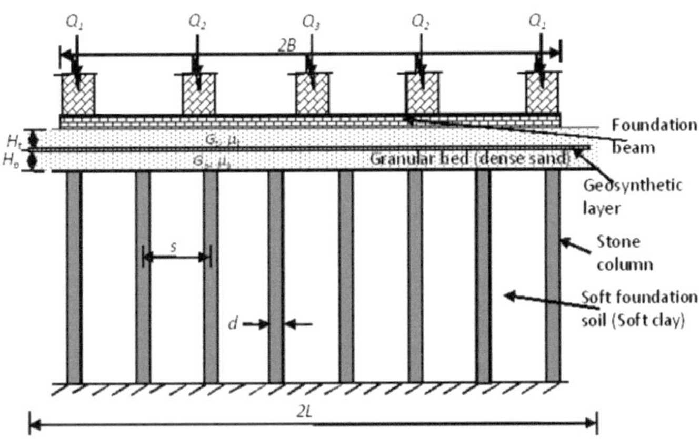

Fig. 1 Footing—geosynthetic reinforced granular bed—stone column reinforced soft soil system

been adopted for this purpose. Influence of configuration of stone columns, i.e., spacing and diameter of stone columns has been studied by means of detailed parametric study.

2 Analysis

A combined footing subjected to five concentrated loads and has been assumed to be loaded symmetrically (Q_1, Q_2, and Q_3). The footing has been assumed to have finite flexural rigidity (EI) and therefore idealized as a beam of finite length (=2B). The footing is resting on the geosynthetic-reinforced granular fill of width 2L over soft soil which has been treated by stone columns. Diameter and the spacing of stone columns are d and s respectively. The geosynthetic layer has been provided in the granular fill with tensile modulus as E_g as shown in the figure. The top and bottom thickness of granular fill layer is H_t and H_b respectively. The respective shear moduli of top and bottom layers are G_t and G_b. This soil-foundation system has been modeled as shown in Fig. 2. Stone columns have been represented by series of stiffer Winkler springs and saturated soil, by Kelvin–Voigt body. The granular fill layer has been idealized as Pasternak shear layer. Granular fill, soft soil and the stone columns has been assumed to exhibit nonlinear hyperbolic stress–strain behavior. Geosynthetic layer in between the granular fill layer has been idealized as extensible, linear, rough, and elastic membrane.

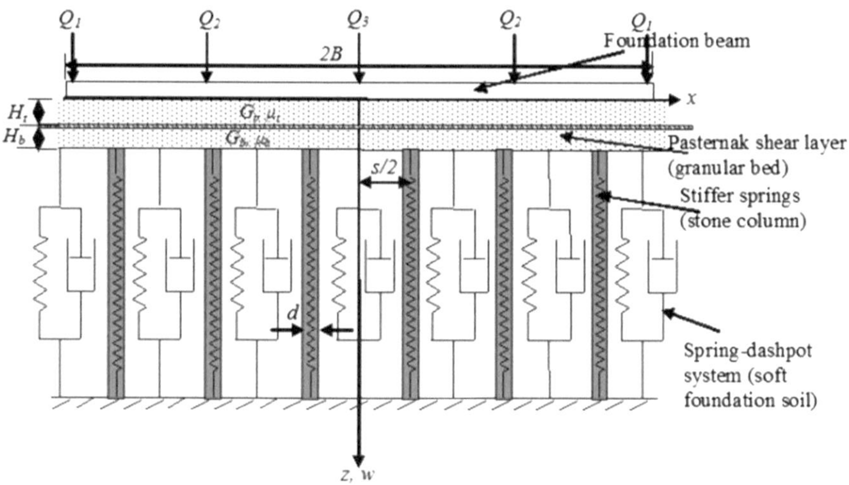

Fig. 2 Idealization of the problem

2.1 Governing Differential Equations

The governing differential equations have been derived from the free body diagram of the granular fill and the geosynthetic layer (Yin 1997). Considering the force equilibrium, the following equations have been derived:

Below the foundation beam and within soft soil

$$q = \frac{k_{\mathrm{so}} w}{U\left[1 + \frac{k_{\mathrm{so}} w}{q_u}\right]} - T\cos^3\theta \frac{\mathrm{d}^2 w}{\mathrm{d}x^2} - \sin\theta \frac{\mathrm{d}T}{\mathrm{d}x} - \left(G_t' H_t + G_b' H_b\right)\frac{\mathrm{d}^2 w}{\mathrm{d}x^2} \tag{1a}$$

Below the foundation beam and within stone column region

$$\begin{aligned} q = {} & \frac{k_{\mathrm{co}} w}{\left[1 + \frac{k_{\mathrm{co}} w}{q_{\mathrm{cu}}}\right]} - T\cos^3\theta \frac{\mathrm{d}^2 w}{\mathrm{d}x^2} \\ & - \sin\theta \frac{\mathrm{d}T}{\mathrm{d}x} - \left(G_t' H_t + G_b' H_b\right)\frac{\mathrm{d}^2 w}{\mathrm{d}x^2} \end{aligned} \tag{1b}$$

$$\begin{aligned} \frac{\mathrm{d}^2 T}{\mathrm{d}x^2} = {} & \sin\theta\cos\theta \frac{\mathrm{d}^2 w}{\mathrm{d}x^2}\frac{\mathrm{d}T}{\mathrm{d}x} \\ & + \frac{1}{\cos\theta}\left(\frac{G_t^o}{H_t} + \frac{G_b^o}{H_b}\right)\left[\sqrt{\left(\frac{T}{E_g} + 1\right)^2 - \left(\frac{\mathrm{d}w}{\mathrm{d}x}\right)^2} - 1\right] \\ & + \frac{H_b}{H_t}\left(\frac{G_t^o}{\tau_{ub}} - \frac{G_b^o}{\tau_{ut}}\frac{G_{to}}{G_{bo}}\frac{G_t^1}{G_b^1}\right)\frac{G_b^o H_t}{\left(G_t^o H_b + G_b^o H_t\right)}\frac{\mathrm{d}^2 w}{\mathrm{d}x^2}\frac{\mathrm{d}T}{\mathrm{d}x}, \end{aligned} \tag{2}$$

where, q is the reaction of the soil on beam; T, the mobilized tension in geosynthetic layer; θ, the slope of geosynthetic layer element (Yin 1997); w, the vertical displacement of beam; $\mathrm{d}x$, the projected element length in x direction; E_g, the tensile stiffness (N/m) of geosynthetic layer; G_{to} and G_{bo}, the initial shear modulus of top and bottom shear layers respectively; τ_{ut} and τ_{ub}, the ultimate shear resistance of top and the bottom shear layers respectively. k_{so} and k_{co}, are the initial modulus of subgrade reaction for saturated soft soil and the stone columns respectively; q_u and q_{cu}, the ultimate bearing resistance of soft soil and the stone columns respectively and U, is the average degree of consolidation at any time $t > 0$. G_t^o, G_b^o, G_t^1, and G_b^1 are defined as follows:

$$\begin{aligned} G_t^o &= \frac{G_{\mathrm{to}}}{1 + \frac{G_{\mathrm{to}}}{\tau_{\mathrm{ut}}}\left|\frac{\mathrm{d}w}{\mathrm{d}x}\right|}, \quad G_b^o = \frac{G_{\mathrm{bo}}}{1 + \frac{G_{\mathrm{bo}}}{\tau_{\mathrm{ub}}}\left|\frac{\mathrm{d}w}{\mathrm{d}x}\right|}, \\ G_t^1 &= \frac{G_{\mathrm{to}}}{\left[1 + \frac{G_{\mathrm{to}}}{\tau_{\mathrm{ut}}}\left|\frac{\mathrm{d}w}{\mathrm{d}x}\right|\right]^2}, \quad G_b^1 = \frac{G_{bo}}{\left[1 + \frac{G_{bo}}{\tau_{ub}}\left|\frac{\mathrm{d}w}{\mathrm{d}x}\right|\right]^2} \end{aligned} \tag{3}$$

The differential equation for bending of a beam with uniform cross section under external load intensity, p is written as

$$\mathrm{EI}\frac{\mathrm{d}^4 w}{\mathrm{d}x^4} + q = p \tag{4}$$

Combining Eqs. (1a, 1b) and (4) results into the governing differential equations along with Eqs. (2) and (3).

2.2 Solution Methodology

The equations have been written in nondimensional form employing the nondimensional parameters:

$$\begin{aligned} &W = w/B, H_t^* = H_t/B, H_b^* = H_b/B, E_g^* = E_g//k_{so}L^2,\\ &q^* = q/k_{so}B,\ q_u^* = q_u/k_{so}B,\ q_{cu}^* = q_{cu}/k_{co}B = q_{cu}/\alpha k_{so}B,\\ &\tau_{ut}^* = \tau_{ut}H_t/k_{so}B^2, \tau_{ub}^* = \tau_{ub}H_b/k_{so}B^2,\ T^* = T/k_{so}B^2,\\ &I^* = \mathrm{EI}/k_{so}L^4,\ Q_1^* = Q_2^* = Q_3^* = Q^* = Q/k_{so}B^2 \text{ and } \alpha = k_{co}/k_{so}, \end{aligned}$$

Q is the applied load and B is the half length of the beam considered.

The resulting nondimensional equations have been written in finite difference form and solved by iterative Gauss Siedel scheme employing the following boundary conditions:

At $x = 0$:

$$\frac{\mathrm{d}w}{\mathrm{d}x} = 0; -\mathrm{EI}\frac{\mathrm{d}^3 w}{\mathrm{d}x^3} = \frac{Q}{2} \text{ and } \frac{\mathrm{d}T}{\mathrm{d}x} = 0$$

At $x = B$:

$$\begin{aligned} &\tfrac{\mathrm{d}^2 w}{\mathrm{d}x^2} = 0; -\mathrm{EI}\tfrac{\mathrm{d}^3 w}{\mathrm{d}x^3} = -Q - \left(T\cos^3\theta + G_t'H_t + G_b'H_b\right)\\ &\tfrac{\mathrm{d}w}{\mathrm{d}x} - T\sin\theta \end{aligned}$$

At $x = L$:

$$\frac{\mathrm{d}w}{\mathrm{d}x} = 0 \text{ and } T = 0$$

3 Results and Discussion

First of all validation of the proposed methodology has been done by comparing the results from a particular case of the present study with that available in the literature (Deb et al. 2010). These have been found to be in good agreement with each other.

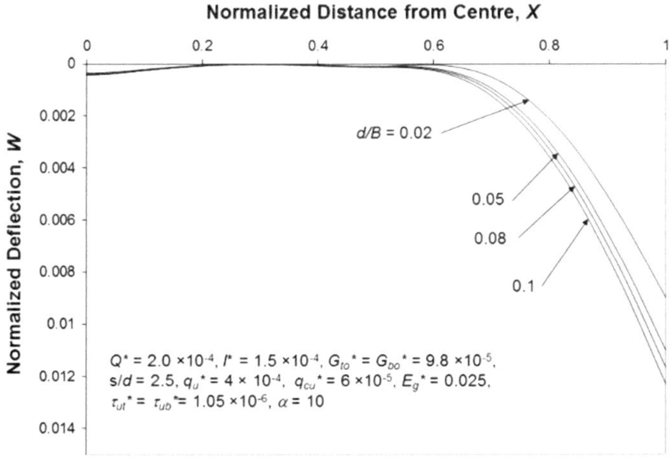

Fig. 3 Effect of diameter of stone columns on normalized deflection of foundation beam

An attempt has been made to study the influence of configuration of the stone columns on response of the soil-foundation system. The normalized diameter, i.e., *d*/*B*, has been varied from 0.02 to 0.1 while the spacing to diameter ratio, *s*/*d* has been varied from 2 to 4.

Figure 3 shows the effect of variation of normalized diameter on the deflection pattern of the foundation beam for the input parameters as mentioned in the figure. As the diameter of stone columns increases, the number of stone columns reduces and hence increase in maximum deflection is observed. This increase has been found to be about 37% as the normalized diameter of the stone columns increase from 0.02 to 0.1. Variation of maximum normalized tension mobilized in geosynthetic layer due to variation in diameter of stone columns has been presented in Fig. 4. As expected, the tension mobilized has been found to increase by about 64% with an increase in normalized diameter of the stone columns from 0.02 to 0.1.

Figures 5 and 6 exhibit the influence of spacing to diameter ratio, *s*/*d*, on normalized deflection and bending moment in the foundation beam for the input parameters as mentioned in the figures. It can be observed that as the parameter, *s*/*d* increases, the normalized deflection of the beam also increases. However, for *s*/*d* = 2.5–3, the variation is very small and therefore, it can be considered as optimum value of the parameter *s*/*d*. The normalized tension mobilized has been found to increase by about 83% as the parameter *s*/*d* increases from 2 to 4.

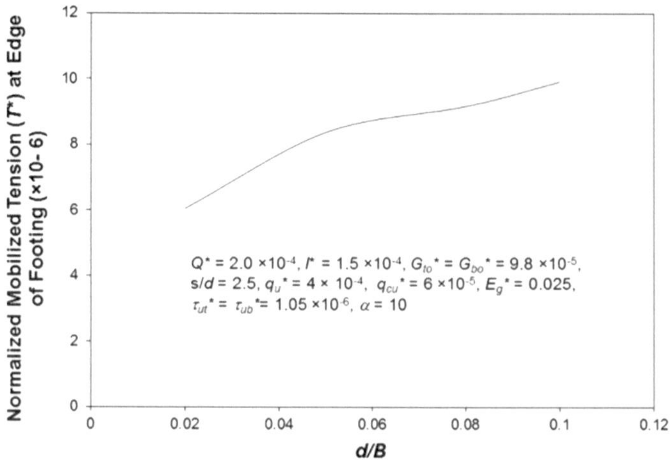

Fig. 4 Effect of diameter of stone columns on normalized tension mobilized in geosynthetic layer

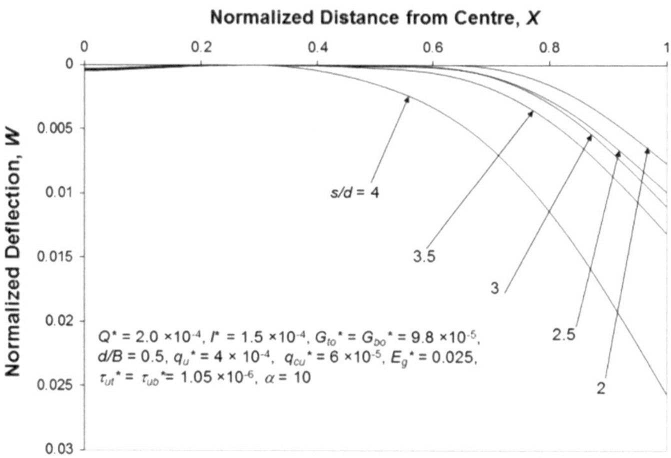

Fig. 5 Effect of spacing to diameter ratio of stone columns on normalized deflection of foundation beam

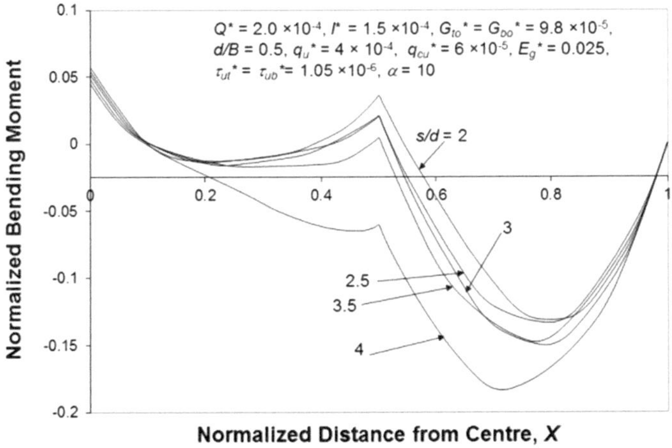

Fig. 6 Effect of spacing to diameter ratio of stone columns on normalized beaning moment in foundation beam

4 Concluding Remarks

The present work proposes a simple nonlinear model for the analysis of combined footings resting on geosynthetic reinforced granular fill-soft soil with stone column inclusions. Governing differential equations have been derived and solved by finite difference method with the help of appropriate boundary conditions. Iterative Gauss Siedel method has been adopted as solution algorithm. A parametric study has been conducted to study the effect of configuration of stone columns. It has been observed that spacing as well as diameter of the stone columns have significant influence on the response of soil-foundation system. For the range of physical input parameters, a value of spacing to diameter ratio of stone columns as 2.5–3 has been found to be optimum.

References

Alamgir, M., Miura, N., Poorooshasb, H. B., & Madhav, M. R. (1996). Deformation analysis of soft ground reinforced by columnar inclusions. *Computers and Geotechnics, 18*(4), 267–290.

Balaam, N. P., & Booker, J. R. (1981). Analysis of rigid rafts supported by granular piles. *International Journal for Numerical and Analytical Methods in Geomechanics, 5*, 379–403.

Deb, Kousik, Chandra, S., & Basudhar, P. K. (2010). Analysis of extensible geosynthetics and stone column-reinforced soil. *Ground Improvement, 163*(4), 231–236.

Deb, Kousik, Sivakugan, N., Chandra, S., & Basudhar, P. K. (2007). Generalized model for geosynthetic-reinforced granular fill-soft soil with stone columns. *International Journal of Geomechanics, ASCE, 7*(4), 266–276.

Ghosh, C., & Madhav, M. R. (1994). Settlement response of a reinforced shallow earth bed. *Geotextiles and Geomembranes, 13*, 643–656.

Madhav, M. R., & Poorooshasb, H. B. (1988). A new model for geosynthetic reinforced soil. *Computers and Geotechnics, 6*(4), 277–290.

Maheshwari, Priti, Basudhar, P. K., & Chandra, S. (2004). Analysis of beams on reinforced granular beds. *Geosynthetics International, 11*(6), 470–480.

Maheshwari, Priti, & Khatri, Shubha. (2012). Generalized model for footings on geosynthetic—reinforced granular fill—stone column improved soft soil system. *International Journal of Geotechnical Engineering, 6*(4), 403–414.

Shahu, J. T., Madhav, M. R., & Hayashi, S. (2000). Analysis of soft ground-granular pile-granular mat system. *Computers and Geotechnics, 27*(1), 45–62.

Shukla, S. K., & Chandra, S. (1994). A generalized mechanical model for geosynthetic-reinforced foundation soil. *Geotextiles and Geomembranes, 13,* 813–825.

Yin, J. H. (1997). Modelling geosynthetic-reinforced granular fills over soft soil. *Geosynthetics International, 4*(2), 165–185.

Dynamic Response Characteristics of Hollow Steel Single Pile Under Vertical Excitations

D. Srivastava, S. S. Choudhary, S. Biswas and B. Manna

Abstract In this present study, the nonlinear behavior of a single hollow steel pile having an outer diameter of 0.114 m and length of 3 m was investigated under vertical excitations of rotating machine. Forced vibration tests were performed in the field to determine the dynamic responses of the single pile for four different eccentric moments. First time-acceleration responses of the single pile were measured for different frequencies and finally from that the frequency-amplitude response curves have been obtained for different excitation forces. It is observed from the field test results that the measured frequency-amplitude response curves exhibit nonlinear behavior of the soil-pile system by showing the decrement in resonant frequencies and disproportional increment in resonant amplitudes with the increase of excitation intensities. Additionally, the inverse analytical method of Novak is used to quantify the variation of stiffness, damping, and effective mass of the pile for different excitation intensities using the measured response curves. It is observed that the calculated damping values are increased with the increase of excitation forces. However, the effective mass and average stiffness values are decreased with the increase of excitation forces. The frequency-amplitude response curves are also back-calculated using the theory of vertical vibration with the calculated parameters of the soil-pile system. The back-calculated response curves are compared with the field dynamic test results and it can be seen from the

D. Srivastava (✉) · S. S. Choudhary
Department of Civil Engineering, National Institute of Technology Patna,
Patna 800005, India
e-mail: devsri27@gmail.com

S. S. Choudhary
e-mail: shv.snkr@gmail.com

S. Biswas · B. Manna
Department of Civil Engineering, Indian Institute of Technology Delhi,
New Delhi 110016, India
e-mail: sanjit.jal@gmail.com

B. Manna
e-mail: bmanna@civil.iitd.ac.in

A. I. V. and V. B. Maji (eds.), *Geotechnical Applications*, Lecture Notes in Civil Engineering 13, https://doi.org/10.1007/978-981-13-0368-5_4

comparison curves that the dynamic nonlinear response obtained from the theoretical analysis have a very close match with the field test response curves.

Keywords Nonlinear response · Soil-pile system · Vertical excitation

1 Introduction

Many experimental and theoretical studies have been done to investigate the dynamic nonlinear characteristics of pile foundation. When these loads are dynamic in nature special considerations need to be adopted in the design and analysis of the pile foundation. One of the primary sources of the dynamic load on the pile foundation is machine-induced vibrations. The aim of pile foundations for supporting machinery equipment is to limit the vibration amplitude within an acceptable range. Therefore, the response of pile foundation under machine vibrations should be accurately determined to achieve this criterion.

Many researchers (Novak and Grigg 1976; EI Sharnouby and Novak 1984; EI Marsafawi et al. 1992) conducted the field test on single pile and group pile to obtain the dynamic responses under machine induced vibrations. Many theoretical studies (Novak 1971, 1974; Novak and Aboul-Ella 1978) have been done in order to determine the dynamic nonlinear response and impedance function (stiffness and damping) of the soil-pile system. More dynamic field tests and theoretical study (Manna and Baidya 2010; Elkasabgy and Naggar 2013) were carried out to study dynamic nonlinear characteristic of the soil-pile systems.

In this current study, dynamic field tests have been conducted on a single pile to measure the frequency-amplitude responses. The effective mass with stiffness and damping are evaluated using the theoretical approach proposed by Novak (1971) from the measured dynamic response curves. The dynamic responses are also back-calculated by theoretical method and compared with the tests results.

2 Description of Test Site

The testing site was located at Indian Institute of Technology Delhi campus, New Delhi, India. To obtain the soil properties of test site, SPT tests as well as different laboratory tests were performed. From all the in situ and laboratory tests, it was found that the subsurface site consists of clayey silt soil. The soil properties of test site with depth are presented in Fig. 1 (Table 1).

Fig. 1 Field vibration test setup of the single pile

Table 1 Soil properties in-situ

Layers	Type of soil	Depth (m)	Dry density (kN/m^3)	N_{corr}—value	Shear modulus (kN/m^2)
1	Clayey silt	0.00–0.75	18.20	18	16585
		0.75–1.50	16.12	12	11046
		1.50–2.50	11.67	15	13815
2	Clayey silt with gravel	2.50–3.50	14.77	26	23938

3 Dimensions of Pile and Its Installation

A hollow steel pile with an outer diameter of 0.114 m and thickness of 0.002 m was used as single pile. A single pile of length 3 m was driven into a borehole by a SPT hammer. To make the borehole, auger boring was done with an auger bit of 0.1 m so that the resistance over driving of pile can be minimized. A steel plate of dimension 0.9 m × 0.9 m × 0.037 m was used for pile cap. The pile cap was tightly joined with the pile by pile cap connectors.

4 Experimental Investigation

Forced vertical vibration tests were performed on the single pile under a static load (W_s) of 10 kN for four different excitation moments ($W.e$ = 0.868, 1.269, 1.631, and 1.944 Nm). A mechanical oscillator with two counter rotating shafts was used to produce the dynamic excitation forces. The W and e are weight and eccentric distance of the rotating masses.

First the mechanical oscillator was placed on the pile cap in such a way that it could generate the sinusoidal force in vertical direction. The mechanical oscillator was connected with a DC motor through a flexible shaft to run the oscillator. One accelerometer was attached on the top of the pile cap-loading system and frequency measuring sensor was attached to the DC motor to determine the acceleration and operating frequency during dynamic testing. The field test setup of the soil-pile system is presented in Fig. 1.

5 Theoretical Investigation

In the present study, the methodology involved to solve the inverse problem (Novak 1971) has been used to determine various parameters of the soil-pile system using dynamic test response curves. The response curves with nonlinear restoring force and linear damping have many geometric properties which can alleviate the data reduction from such response curves. The reduction of resonant frequency is acknowledged by the backbone curve which can be established by intersecting the measured response curve by a pencil of straight lines passing through the origin. The backbonc curve shows the undamped variation of resonant frequencies $\Omega(A)$ with amplitude. A simple relation of Ω (Novak 1971) is given below

$$\Omega = \sqrt{\omega_1 \omega_2} \tag{1}$$

where, ω_1 and ω_2 are the frequencies related to the points of interaction where the response curve and a line passing through the origin of coordinates. Using the Eq. (1) on each straight line, the points of curves of the natural frequencies (Ω) are obtained and by extrapolation of frequency axis, ω_0 is obtained. Where, ω_0 is the initial frequency of the back bone curve at which amplitude is zero.

In general, where the restoring force $F(A)$ is nonlinear for each steady state amplitude 'A' and damping force is linear the stiffness characteristics can be presented as

$$K_e(A) = \frac{F(A)}{A}, \tag{2}$$

where, $K_e(A)$ is the equivalent linear stiffness which is a function of amplitude A. The restoring force can be expressed by the series

$$F(A) = k_1 A + k_3 A^3 + k_5 A^5 + \cdots + k_n A^n, \tag{3}$$

where, k_i = coefficients and i = 1, 3, 5, … n (odd integers). In elastic nonlinear system, the steady vibrations are considered linear for the variable spring constant with a function of steady amplitudes. If $F(A)$ is considered as a function of the steady amplitudes, then the square of nature frequencies of linear theory is expressed as

$$\Omega^2(A) = \frac{F(A)}{Am_{\text{eff}}}, \tag{4}$$

where, $\Omega^2(A)$ is the variation of undamped natural frequencies with amplitudes and m_{eff} is the effective mass of the soil-pile system.

The damping and effective mass are calculated using the geometric properties of the dynamic responses. In the case of dynamic nonlinear response curves, the effective mass (m_{eff}) is obtained much higher than the mass of pile cap-loading system (m_s), where m_s is the total static load on the pile which includes pile cap, steel plates, and oscillator. The apparent additional mass can be defined in terms of the mass coefficient by

$$\zeta = \frac{m_{\text{eff}} - m_s}{m_s} \tag{5}$$

When effective mass and the undamped natural frequency is known from the backbone curve Ω, the restoring force $F(A)$ is calculated without any further use of the field response curve by

$$F(A) = Am_{\text{eff}}\Omega^2 \tag{6}$$

6 Theory Versus Experiment

The dynamic nonlinear response curves are back-calculated by the theory of vibration with the use of calculated parameters (Novak 1971) of the single pile. These back-calculated frequency-amplitude response curves are compared with the test results as shown in Fig. 2. In this theoretical plot, the backbone curves Ω is also plotted for different response curves. From the comparison of response curves, it is found that the field vibration test results and the measured frequency-amplitude response curves exhibit nonlinear behavior of the soil-pile system by showing the decrement in resonant frequencies and disproportional increment in resonant amplitudes with the increase of excitation forces. It is well observed that the theoretical resonant frequencies and amplitudes are very close to test results for all excitation intensities. It can be found from the nature of backbone curve that the stiffness characteristic of the soil-pile system changes with the excitation forces.

The variation of stiffness, damping, and effective mass for different excitation forces are presented Table 2. It can be also found from tabular results that the damping is increased but the stiffness and effective mass is decreased with increase of excitation forces. The higher value of additional effective mass at lower

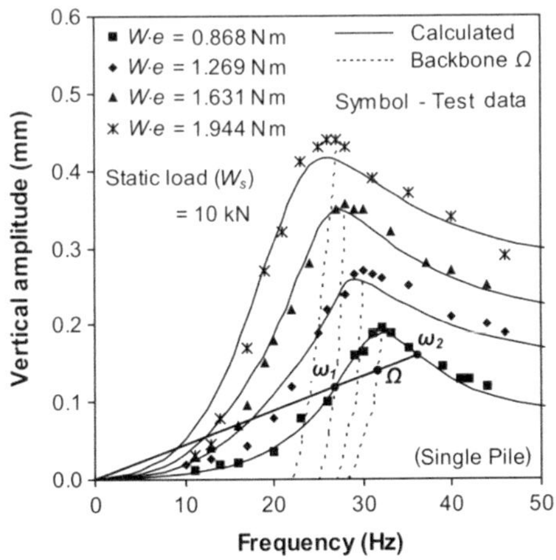

Fig. 2 Test and back-calculated response of single pile under field vibration test

Table 2 Calculated parameters of the single pile responses

Ecc. moment (Nm)	Effective mass		Damp.	Stiff. (kN/mm)
	Mass m_{eff} (kg)	Mass coeff. (ξ)		
0.868	9997	9.00	0.16	370.93
1.269	8677	7.68	0.17	245.65
1.631	7863	6.86	0.18	207.48
1.944	6369	5.37	0.19	144.10

excitation forces indicate that the partial separation between pile and soil may occur for higher excitation force.

From the theoretical back calculation of the effective mass and stiffness, the values of restoring forces and vertical displacements are evaluated from the field test results. The computed results are plotted in Fig. 3 for different eccentric moments. It can be observed from the figure that the values of the vertical amplitudes are increased with the increase of eccentric forces. These patterns show that the soil-pile stiffness is decreased with the increase of excitation forces.

In this paper, single pile of length (L) = 3 m and diameter (d) = 0.114 m (L/d = 26) is used. Many research works (Manna and Baidya 2010; Elkasabgy and Naggar 2013) have been carried out under dynamic loading with different pile length and L/d ration to understand the complex nature of soil-pile system. From the different research work it was found that, the natural frequencies are increased and vertical resonant amplitudes are decreased with increasing length of piles under dynamic loading. However, different researchers used 1 g-scaling law (Wood 2004) to understand the dynamic behavior of varying dimensions of model piles.

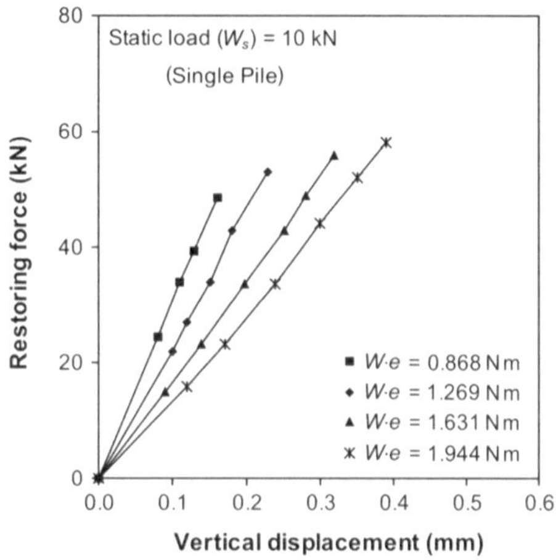

Fig. 3 Pile restoring force characteristic versus pile displacement of single pile

7 Conclusions

In this study, the field pile responses are measured and the frequency-amplitude response characteristics (Novak 1971) of the single pile are evaluated for machine induced vertical vibration. It is observed from the both test and analytical results that the dynamic response of vibration tests shows nonlinear behavior of the soil-pile system. It can be also observed from the comparison of test and theoretical results that the theoretical nonlinear response curves approximately matches with the field responses. Hence, a comprehensive analytical approach is possible for the soil-pile system with the evaluation of back-calculated parameters of the single pile responses under vertical vibrations. The dynamic test results of piles presented in this study are found nonlinear. Therefore, equivalent liner continuum model or nonlinear FEM could be used for dynamic analysis of piles under machine vibrations

References

El Marsafawi, H., Han, Y. C., & Novak, M. (1992). Dynamic experiments on two pile groups. *Journal of Geotechnical Engineering, ASCE, 118*(4), 576–592.

Elkasabgy, M., & El-Naggar, M. H. (2013). Dynamic response of vertically loaded helical and driven steel piles. *Canadian Geotechnical Journal, 50*(5), 521–535.

El Sharnouby, B., & Novak, M. (1984). Dynamic experiments with group of piles. *Journal of Geotechnical Engineering, ASCE, 110*(6), 719–737.

Manna, B., & Baidya, D. K. (2010). Dynamic nonlinear response of pile foundations under vertical vibration-theory versus experiment. *Soil Dynamics and Earthquake Engineering, Elsevier, 30*, 456–469.

Novak, M. (1971). Data reduction from nonlinear response curves. *Journal of Engineering Mechanics, ASCE, 97*(EM4), 1187–1204.

Novak, M. (1974). Dynamic stiffness and damping of piles. *Canadian Geotechnical Journal, 11,* 574–598.

Novak, M., & Aboul-Ella, F. (1978). Impedance functions for piles embedded in layered medium. *Journal of Engineering Mechanics, ASCE, 104*(3), 643–661.

Novak, M., & Grigg, R. F. (1976). Dynamic experiments with small pile foundations. *Canadian Geotechnical Journal, 13,* 372–385.

Wood, D. M. (2004). *Geotechnical modeling* (pp. 1–488). Boca Raton: CRC Press.

Analysis of Laterally Loaded Fixed-Headed Single Pile in Multilayered Soil Using P-Y Approach

Somenath Mukherjee and Arindam Dey

Abstract Structures resting on pile foundation are generally subjected to lateral loads and moments acting on the pile head in addition to vertical loads. Horizontal forces originate due to wind, traffic, earth pressure, water wave, and seismic forces or their combination. Analysis and design of piles subjected to lateral forces and moments is very important for ensuring the stability and serviceability of structure. In designing laterally loaded piles, the pile head deflection is very important which depends on soil type, pile installation, pile flexibility (or pile stiffness), loading condition, and type of fixity of pile with pile cap. The flexural behavior of a pile is a function of the interaction between the soil and the pile and is governed by the properties of both. In the present study, lateral load behavior of single piles in layered soils, i.e., alternate layer of clay and sand is conducted. The analysis is carried out considering fixed-headed pile and floating tip at the base. Since soil response is complex idealized models are used for the analysis. Analysis based on generated nonlinear *P-Y* curves representing the soil behavior using OASYS ALP v19.2 has been elucidated. Flexural response of the pile along with lateral pile capacity, moment and length of fixity is estimated considering all the analysis with different pile diameters.

Keywords Single fixed-headed pile · Laterally loaded pile · ALP
Multilayered substrata · Flexural response

S. Mukherjee
C. E. Testing Co. Pvt. Ltd., Kolkata, India
e-mail: somenath2004@gmail.com

A. Dey (✉)
Department of Civil Engineering, Indian Institute of Technology Guwahati,
Guwahati 781039, India
e-mail: arindamdeyiitk@gmail.com

A. I. V. and V. B. Maji (eds.), *Geotechnical Applications*, Lecture Notes in Civil Engineering 13, https://doi.org/10.1007/978-981-13-0368-5_5

1 Introduction

For piles supporting bridge structures, it is very important to estimate the ultimate lateral load capacity of the pile in order to have a priori idea about its failure. Conventionally, such piles are estimated for their ultimate load based on specific deflection criterion. This article reports the findings of a study to estimate the lateral load capacity of a single pile embedded in stratified soil deposit in river bed, based on the permissible deflection of 10% of the pile diameter at the cut-off. Flexural response of the pile (in terms of deflection, bending moment and shear force, and contact stress profiles) has been illustrated using OASYS ALP v19.2 incorporating nonlinear P-Y analysis. Analysis of Laterally Loaded Piles (ALP) has been found to be efficient since it has the ability to consider the effect of water table and nonlinear earth pressure generation in the soil surrounding the pile. Based on the results, the article reports the lateral load capacity of the piles embedded in stratified deposit.

2 Working Principle of ALP

ALP models the pile as a series of elastic beam elements and the soil as a series of non-interactive, nonlinear Winkler springs. The outcome conforms to the prediction of contact pressures, horizontal displacements, shear forces and bending moments induced in a pile when subjected to lateral loads. The load-deflection behavior of soil is modeled either assuming an Elastic-Plastic behavior, or by specifying or generating load-deflection (P-Y) data (depending on whether the soil is clayey or sandy). The pile is discretized into several nodes, and two separate stiffness matrices (one for the pile in bending and one for the adjacent soil) relating nodal forces to displacements are developed. The pile stiffness is provide at each node, and remains constant between successive nodes. The software allows for incremental load application which aids to monitor the progressive change in flexural response. ALP is capable of considering the presence of water table. Figure 1 describes the general schematic of a soil-pile interaction model in ALP.

3 Problem Statement

Figure 2 depicts a single pile embedded in a stratified soil. ALP has been used to determine the lateral load capacity for a specified deflection at its cut-off level (maximum allowable deflection criterion considered as 10% of the pile diameter), the length of fixity of the pile and the corresponding flexural response envelopes at the lateral load capacity of pile. The pile had been analyzed for three different diameters (1000, 1200 and 1500 mm). The pile is considered to be made of M25

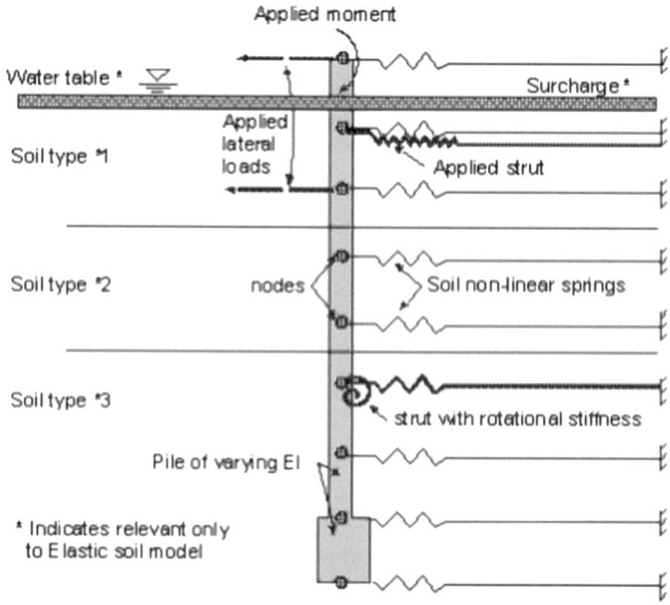

Fig. 1 General soil-pile interaction model in ALP

grade concrete, and is assumed to be fixed-headed. Similar exercise have been reported earlier for free-headed piles (Yadav 2013; Yadav and Dey 2013a, b).

4 Oasys ALP Nonlinear P-Y Analysis

ALP is capable of considering the effect of water table. The software also facilitates to determine the flexural response through automatic load increment, and hence has the capability to estimate the failure load. ALP also considers the passive resistance of the soil and indicates whenever the passive resistance of the soil is exceeded. The methodology taking into account of the generated *P-Y* curves has been considered in the present study. Figure 3 depicts a typical sample output as obtained from ALP. In order to maintain the pile head as fixed, a very high rotational restraint has been implied at the topmost node of the pile.

4.1 Estimation of Model Parameters

Both elastic and plastic soil properties are required to be provided in the ALP input model. Apart from the modulus of subgrade reaction, depending on the soil type,

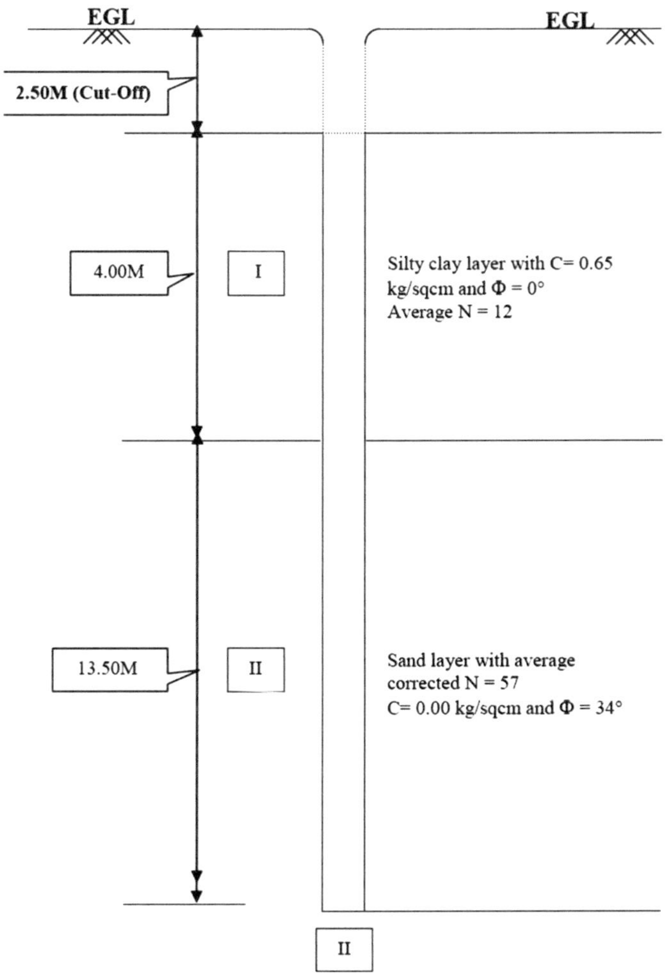

Fig. 2 Soil-pile configuration of the present study

the other parameters required as input are: (a) Unit weight, percentage strain, and cohesion for clayey soils and (b) unit weight, angle of internal friction and coefficient of earth pressure at rest for sandy soil. The unit weight of soil for all the problems is considered to be 18 kN/m^3, and the strain is considered to be of magnitude 5% for all soils. The coefficient of earth pressure at rest is computed using Jaky's expression (Alpan 1967) as follows:

$$K_0 = 1 - \sin \phi \tag{1}$$

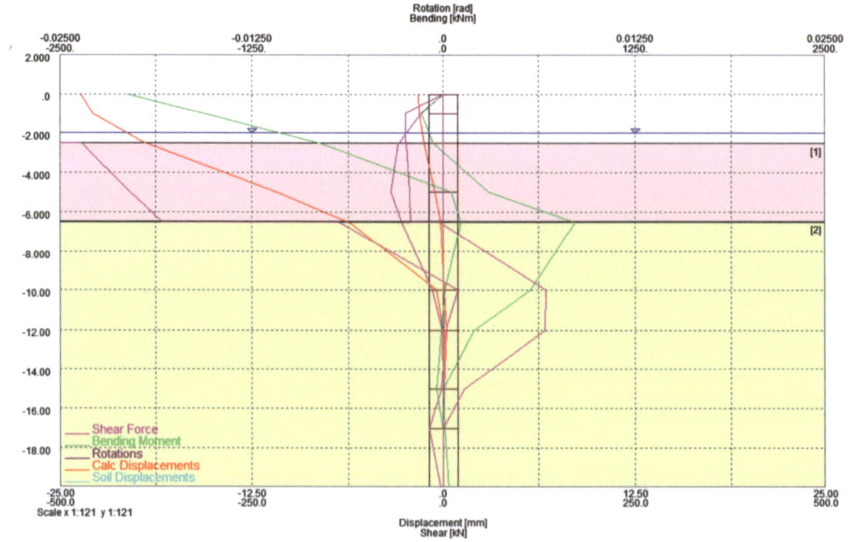

Fig. 3 Typical flexural response envelopes as obtained from ALP 19.2

The pile has been discretized into nodes with a uniform spacing of 0.5 m, with required refinements being made in the intersection with the water table and soil strata interface. The elastic modulus of concrete is calculated as per IS: 456 (2000).

5 Results and Discussions

The advanced feature of automatic load increment available in the software OASYS ALP 19.2 has been used to estimate the lateral load capacity of the pile. For the present problem, the load acting at the pile head has been set to a reasonable value (e.g., 500 kN) and the number of increments to reach the load has been provided (10 increments). This enables to obtain the cumulative flexural response of the pile with the load increments. Figure 4a–c shows a typical load increment procedure in terms of the flexural response of the pile (1500 mm diameter).

From the above observation, considering the deflection criterion as 10% of the pile diameter at cut-off level, the lateral load capacity of the pile has been estimated. Figure 5 depicts the lateral load capacity as estimated from the ALP software.

Using the estimated lateral load capacity, the flexural response of the pile has been determined for various diameters (1, 1.2 and 1.5 m) and is represented in Figs. 6a–d. It is observed that with the increase in the pile diameter, the maximum

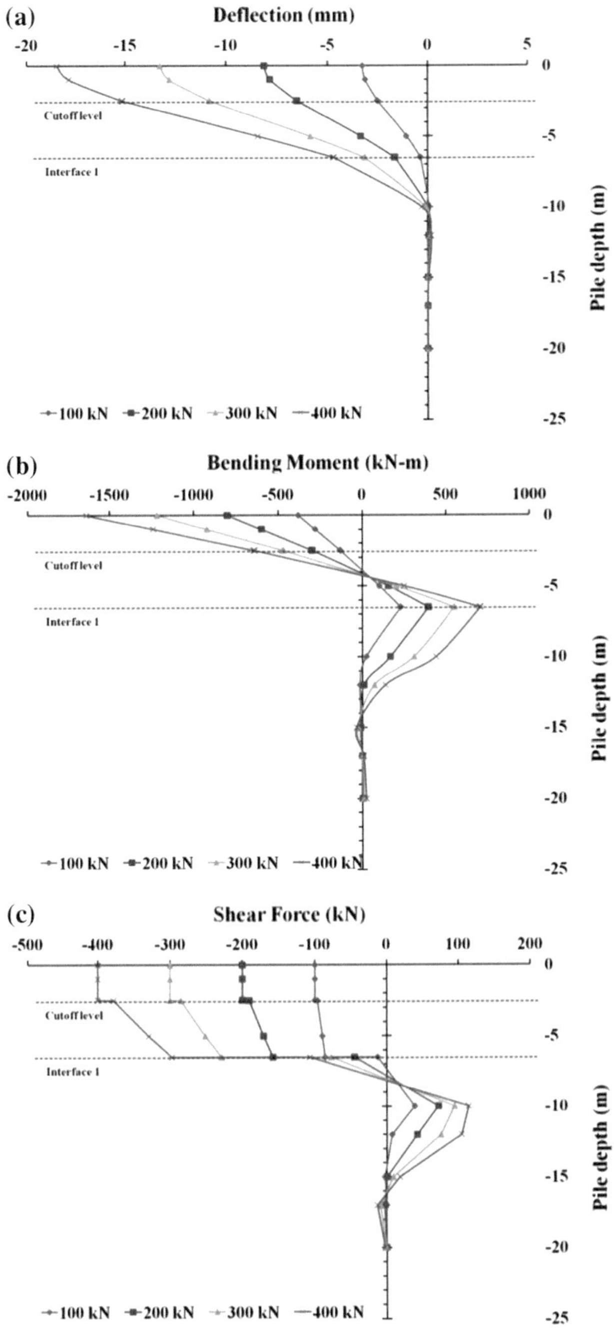

Fig. 4 Deflection, bending moment and shear response envelopes for load increments (pile diameter = 1500 mm)

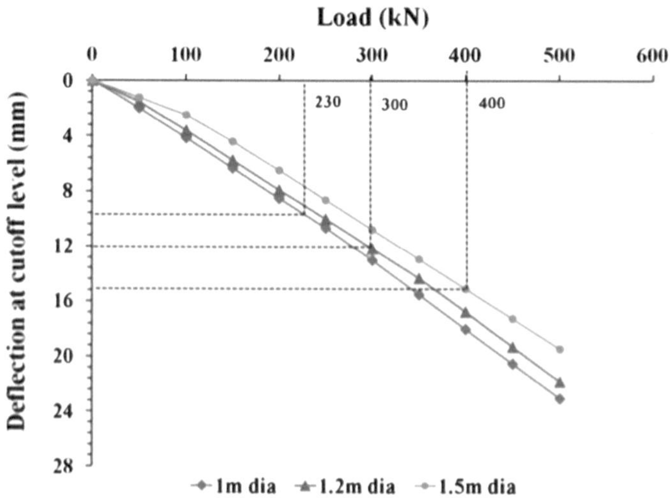

Fig. 5 Load-deformation curves and lateral load capacity of piles using OASYS ALP 19.2

responses (deflection, rotation, bending moment, shear force, and contact stresses) also reveal increment in their magnitudes. It is worth noticing that the point of fixity, represented by the position of maximum bending moment, does not reveal significant change with the change in the diameter of the pile. This is attributed to the fact that ALP considers the passive resistance offered by the soil. Moreover, the response of soil is generally nonlinear under high load and especially when the structures reach the verge of failure. The nonlinear *P-Y* curves provide realistic behavior of the soil under such condition. Moreover, the load-deformation behavior of the alternatively placed sandy and clayey stratum is significantly different. All this factors might actually result in maintaining the point of fixity of the pile to be same despite change in pile diameter. The results are enumerated in Table 1.

Figures 7a, b depicts the *P-Y* curves that are generated at the top and bottom of each soil layer. It is noticeable that the clayey layers (which are more plastic than the sandy layers) show more nonlinear behavior, and reaches the plastic limit condition. The sandy layers show mostly bilinear behavior and, as indicated in the figure, needs to be subjected to large load values to reach their ultimate state. Hence, the linear/nonlinear flexural behavior of the pile will be governed by the thickness and location of the clayey or sandy stratum.

Fig. 6 Flexural response of pile of different diameters

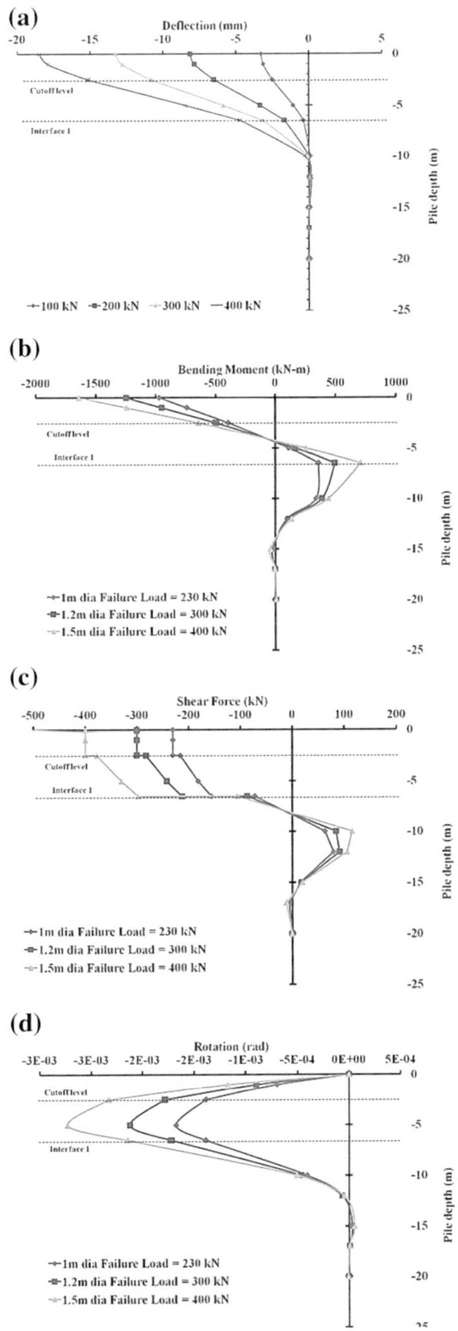

Table 1 Response of the pile at lateral load capacity as obtained from ALP

D (mm)	$y_{\text{cut-off}}$ (mm)	P_u (kN)	M_{max} (kNm)	L_{fix} (m)
1000	10	230	355.46	6.5–7
1200	12	300	493.92	6.5–7
1500	15	400	708.39	6.5–7

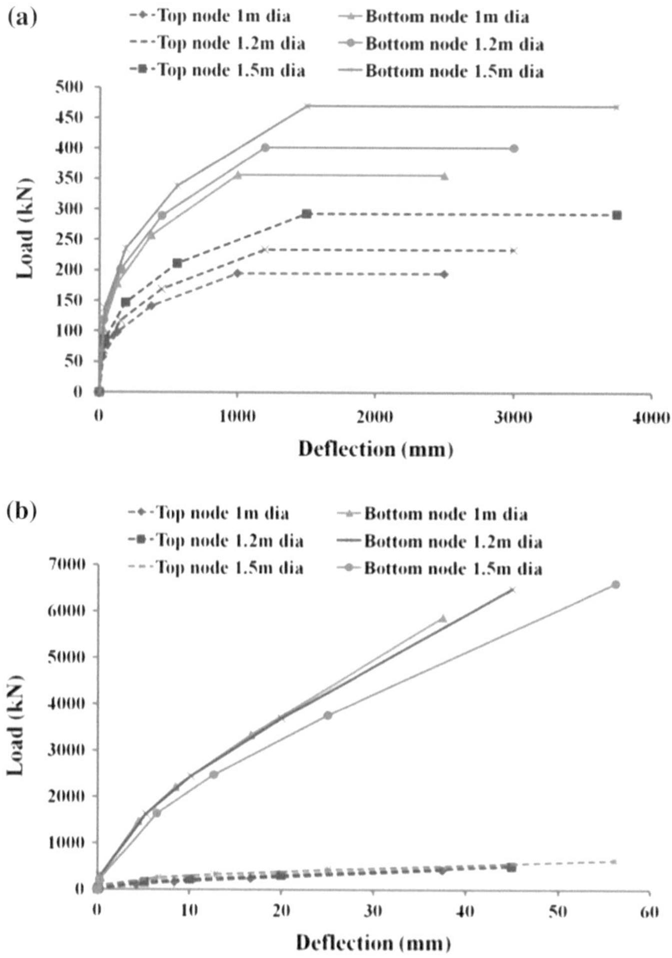

Fig. 7 Generated *P-Y* curves for the **a** clayey **b** sandy soils

6 Conclusions

Based on the conducted study, the following important conclusions can be stated:

- Owing to enhanced flexural rigidity, lateral load capacity of a pile increases with the increase in the diameter of the pile.
- An increase in the diameter of the embedded piles results in the increase in the maximums of the flexural responses of the pile.
- ALP results revealed that the length of fixity is not significantly affected by the change in the pile diameter. This is attributed to the fact that the length of fixity is not governed alone by the flexural rigidity of the pile, but by the relative flexural rigidity of the pile and the surrounding soil.
- *P-Y* curves generated reveal that the clay layers show prominence in nonlinear behavior and reaches the plastic limit under failure load condition, while sandy layers, commonly behaving as elastic, show at most bilinear behavior and indicates that this kind of soil require high loads to reach into their plastic limit.

References

Alpan, I. (1967). The empirical evaluation of coefficient of K_0 and K_{0R}. *Soils and Foundations, 7*(1), 31–40.

IS 456. (2000). *Plain and reinforced concrete*. New Delhi, India: Bureau of Indian Standards.

Yadav, A. K. (2013). *Lateral load capacity of piles in stratified soil deposit subjected to scour: Linear elastic and nonlinear P-Y analysis*. IIT Guwahati: BTech Project Report.

Yadav, A. K., & Dey, A. (2013) Using ALP to estimate the impact of scouring on the lateral load capacity of piles (pp. 1–5). http://www.oasys-software.com/casestudies?id=89/.

Yadav, A. K., & Dey, A. (2013) Lateral load capacity of piles in stratified deposits subjected to scour. *Indian Geotechnical Conference: IGC GANGA 2013* (pp. 1–10). Roorkee, India.

Application of Artificial Neural Network to Predict the Settlement of Shallow Foundations on Cohesionless Soils

T. Gnananandarao, R. K. Dutta and V. N. Khatri

Abstract The present study tries to predict the settlement of shallow foundation on granular soil using a mathematical model. The application of feed-forward neural networks with back propagated algorithm is followed for the same. For the development of ANN model, 193 in situ tests data were collected from the literature. The inputs required for the development of model were the foundation pressure, width of footing and the standard penetration number. The predicted settlement using this model was found to compare favourably with the measured settlement. Further the results of sensitivity analysis indicated that the width of foundation has highest impact on the predicted settlement in comparison to other input variables. The present study confirms the ability of ANN models to predict a complex relationship between the nonlinear data as in present case.

Keywords Shallow foundations · Settlement · Artificial neural networks Sandy soil

1 Introduction

Design of shallow foundations on cohesionless soils are depends on the two major criteria's; those are bearing capacity and settlement of the foundation. However, settlement plays a major role than the bearing capacity when the width of footing exceeds one meter (Maugeri et al. 1998). Further in case of foundations on cohessionless soil the settlements are likely to occur in very short span of time after

T. Gnananandarao · R. K. Dutta
National Institute of Technology, Hamirpur, Hamirpur, India
e-mail: anandrcwing@gmail.com

R. K. Dutta
e-mail: rakeshkdutta@gmail.com

V. N. Khatri (✉)
Indian Institute of Technology, Dhanbad, Dhanbad, India
e-mail: vishuiisc@gmail.com

A. I. V. and V. B. Maji (eds.), *Geotechnical Applications*, Lecture Notes in Civil Engineering 13, https://doi.org/10.1007/978-981-13-0368-5_6

load application. Such settlements origins rapid deformation of structures and even excessive settlement and may lead to a structural failure. So prediction of settlement on cohesionless soil is major concern in design of shallow foundations.

Forecasting of the settlement of shallow foundations on cohesionless soils is very difficult and not yet entirely well defined. At present many theoretical, experimental methods are available in the literature to predict the settlement of shallow foundations on cohesionless soils. Further the empirical correlations based on in situ tests such as cone penetration test, standard penetration test (SPT), dilatometer test, plate load test, pressuremeter test and screw plate load test are also available which may preclude the need to pick undisturbed sample of cohesionless soils. However most of these mentioned methods simplify the problem by incorporating several assumptions associated with the factors that affect the settlement of shallow foundations. Hence to predict the reliable estimate of settlement of shallow foundations on granular soils, mathematical model is an alternative approach wherein a relationship is developed between the experimental results and major effected calibrated parameters. In the present study, the use of Artificial Neural Networks (ANN) is sought out for the prediction of settlement of shallow foundation on the granular soil. The ANN models are followed since they are having an advantage of predicting the complex relationship between the nonlinear data sets. Its use however for geotechnical applications have been explored very recently (Dutta et.al. 2015). In the present study the ANN model is developed by using 193 in situ tests data collected from the literature (Boger and Guterman 1997; Bowles 1987; Papadopoulos 1992; Stone 1974). It should be noted that the developed ANN model is based on field data related to standard penetration test (N) only. After the development of model, a sensitivity analysis was also carried out to know the contribution of various input parameters and their effect on the output. Finally a nonlinear equation is proposed for the prediction of settlement on the basis of present study.

2 Overview of Artificial Neural Networks

Artificial neural networks are part of artificial intelligence, it means that their architecture try to mimic the function of human brains and nervous systems. They have the strength to relate the given input data and corresponding output data, that may be single or multiple parameters for solving linear or nonlinear problems.

A typical structure of artificial neural networks consists of a number of processing elements or nodes which are typically arranged in different layers namely: an input layer, an output layer and one or more hidden layers (Fig. 1).

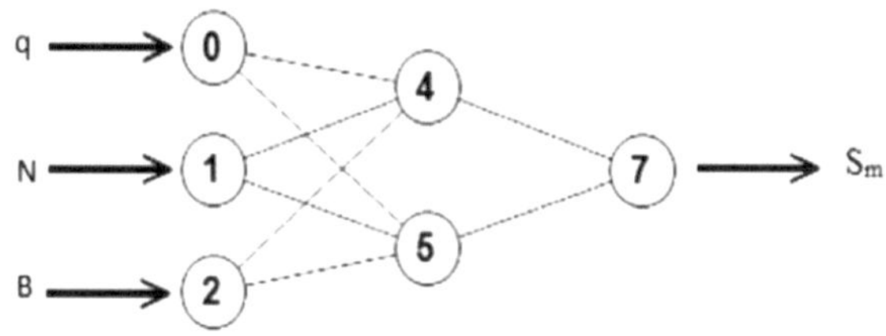

Fig. 1 Neural networks architecture for settlement of shallow foundations

3 Data Set

The data used in this study were collected from the literature (Boger and Guterman 1997; Bowles 1987; Papadopoulos 1992; Stone 1974). This data includes record of field measurements of settlements of shallow foundations along with the corresponding soil and footing information. The data comprised of total 193 cases which includes: (i) 119 individual data sets from (Boger and Guterman 1997), (ii) 2 cases from (Bowles 1987), (iii) 11 cases from (Papadopoulos 1992) and (iv) 61 cases from (Stone 1974). This data contains information on net applied pressure (q), standard penetration number (N), width of the footing (B), Depth of the foundation (D), and settlement of footing (S_m). In the present study, net applied pressure (q), standard penetration number (N), width of the footing (B) were taken as input parameters and settlement of footing as output parameter for the development of model. Whereas, the depth of footing was neglected on the basis of sensitivity analysis study reported by Shahin et al. (2002). Similar manner plan shape of footing is also a variable and later it is discarded on the basis of sensitivity analysis. On similar lines with Shahin et al. (2002) a Feed-forward network with back-propagation algorithms was used for both training and testing stages to predict the required output. Based on the recommendations of Stone (1974) the collected data was randomly divided into two separate data sets, one is for training (70% of total data set, i.e. 135) and another one is testing (30% of total data set, i.e. 58). A summary of input and output parameters belonging to chosen data set along with their respective maximum and minimum values are provided in Table 1.

Table 1 Data ranges used for artificial neural network model variables

Parameters	Minimum value	Maximum value
Footing net applied pressure, q_{net} (kPa)	2	697
Average SPT blow count, N	4	120
Footing width, B (m)	0.5	37.8
Measured settlement, S_m (mm)	0.6	269

4 Optimal Neural Networks Selection

The selection of neural networks architecture is most difficult task because determining the optimum number of layers and neurons in the hidden layer was generally based on the trial and error process. There is no effective method to get an optimal neural networks (NN) architecture and parameters setting. Moreover, the performance of neural networks well dependent on the initial weights and given input parameters. However, a time-consuming trial and error method still remains valid. In this study, the authors have fixed the two neurons in the hidden layer based on the thumb rule as suggested by Boger and Guterman (1997).

After the determination of hidden layer neurons, it was necessary to decide the number of iterations required for training. Because excessive training of the neural network leads to noise, however insufficient training of the neural network leads to poor predictions. Hence, using a trial-and-error method the number of iterations for training and testing data sets was finalized. Also after computing the mean square error between the actual and predicted values corresponding to different iterations, the least mean square error is chosen for finding the neural network structure. On the basis of above mentioned criterions about 300 iterations were selected. Therefore, the neural network model picked for our experimentation has the structure of 3-2-1 (Fig. 1) for the constitutive modelling.

5 Activation Function Selection

ANNs make use of various transfer functions to create a relationship between the input and output variables at each neuron layer. A huge number of activation functions are followed in artificial neural network however, sigmoid, linear, Gaussian, sin, cos, etc. are commonly used activation factions (Dutta et.al. 2015). In the present study, the initial trials were performed with the different activation functions and on the basis of obtained results, the sigmoid function was selected since its use resulted in the reliable predictions.

6 Performance Measures

After settlement prediction of footings, by using the above-mentioned procedure, the accuracy of the predicted settlement with respective to the targeted settlement was checked. However, it may be achieved after minimizing the error. Therefore, to check the accuracy of the predicted data some accuracy measures are selected to compute the magnitude of the predicted error. The various used statistical parameters and error models are presented in Table 2 with their respective mathematical

Table 2 Performance parameters and their mathematical expressions

Statistical coefficient	Mathematical expression
Correlation coefficient (r)	$r = \frac{\sum \sigma_{ht_i} \times \sigma_{hp_i} - \bar{\sigma}_{ht} \times \bar{\sigma}_{hp}}{(n-1) S_{\sigma_{ht}} S_{\sigma_{hp}}}$
Coefficient of determination (R^2)	$R^2 = 1 - \frac{\sum_i \left(\sigma_{hp_i} - \sigma_{ht_i}\right)^2}{\sum_i \left(\sigma_{hp_i} - \bar{\sigma}_{hp_i}\right)^2}$
Mean square error (MSE)	$\text{MSE} = \frac{1}{n} \sum_{i=1}^{n} \left(\sigma_{ht_i} - \sigma_{hp_i}\right)^2$
Root mean square error (RMSE)	$\text{RMSE} = \sqrt{\frac{1}{n} \sum_{i=1}^{n} \left(\sigma_{ht_i} - \sigma_{hp_i}\right)^2}$
Mean absolute error (MAE)	$\text{MAE} = \frac{1}{n} \sum_{i=1}^{n} \left\lvert \sigma_{ht_i} - \sigma_{hp_i} \right\rvert$
Mean absolute percentage error (MAPE)	$\text{MAPE} = \left[\frac{1}{n} \sum_{i=1}^{n} \left\lvert \frac{\sigma_{ht_i} - \sigma_{hp_i}}{\sigma_{ht_i}} \right\rvert \right] \times 100$

Note σ_{ht}, σ_{hp} target and predicted settlement respectively, $\bar{\sigma}_{ht}$, $\bar{\sigma}_{hp}$ mean of the target and predicted settlement respectively, $S_{\sigma_{ht}}$, $S_{\sigma_{hp}}$ standard deviation of the target and predicted settlement respectively, n number of observations

expressions. From the Table 2 the MSE as its name indicates, contains quadratic loss function as it squares and after averages the various errors.

RMSE is an alternative way of expressing the MSE, which was simple square root of the MSE. Both MSE and RMSE measure the uncertainty of the prediction. The MAE is also an absolute measure similar to MSE and its value ranges from 0 to + ∞. It gives the average size of error prediction when negative signs are ignored. The MAPE is a relative measure, it represents the errors as a percentage of the original data and it is useful for comparing the accuracy of more than one methods. In addition the range of MAPE is (i) less than 10%, (ii) in between 10 and 20%, (iii) between 20 and 50% and (iv) more than 50%. It implies that an excellent accurate prediction, good prediction, acceptable prediction and inaccurate prediction respectively. The MSE, RMSE, MAE and MAPE provide the measurement of the level of forecast errors. Smaller the values of above statistics will represent the better models. RMSE will always be more or equal to the MAE. If the difference is more between the RMSE and MAE, it indicates the variation is more in the individual error in the data set. Further the accuracy of the predicted data generally evaluated in terms of correlation coefficient (r) and coefficient of determination (R^2). Correlation coefficient (r) measures the direction of a linear relationship between two variables. The range of r value is in between $-1 \leq r \leq +1$, if it close to zero means that there is weak linear correlation. Coefficient of determination (R^2) is a number that shows the proportion of the variance in the dependent variable that is foreseeable from the independent variable. The coefficient of determination ranges from 0 to 1. To check the accuracy of the predicted settlement, the coefficient of determination (R^2), correlation coefficient (r), MSE, RMSE, MAE and MAPE were calculated for the training and testing data. Further, based on the best fit statistical results the appropriate activation function was chosen among the available activation functions. As mentioned previously in the present study the sigmoid function was selected based on the above mentioned facts.

7 Results and Discussions

After carrying out the process of Feed-forward network with back-propagation algorithms the results of statistical values are presented in Table 3 and weights between the input neuron and hidden neuron as well as hidden neuron and output neuron with biases are presented in the Table 4. The observation of Table 2 indicates that, the values of the performance parameters obtained from the present study are within the mentioned range. It means the predicted output is reasonable in relation to the actual value (Fig. 2). Further the plot between the measured settlement and predicted settlement generates the coefficient of determination (R^2) of 0.85 which implies the predicted settlement are within permissible degree of accuracy within the selected settlement range of 0.6–269 mm. Hence the present model is an improvement over the model presented by Shahin et al. (2002) with the range of settlement between 0.6 and 121 mm. It should be noted that the present ANN model is similar yet an improvement over the model reported by Shahin et al. (2002) which is based on, Burland and Burbidge (1985), data whereas the present model is based on 193 data.

8 Sensitivity Analysis

Sensitivity analysis was carried out in order to study the individual contribution of input variables on the settlement using a method reported by Schmertmann (1970) which was based on weight formation (Table 4). In this analysis the relative importance of individual variables are measured. The results of this analysis revealed that the width of the footing is the most important factor for the prediction of settlement. Its impact on output value is about 79%. Whereas the net applied pressure (q) and standard penetration number (N) influence the output by 11 and 10% respectively.

Table 3 Statistical values for the training and testing data

Statistical values for the training data							
	Activation function	r	R^2	MSE	RMSE	MAE	MAPE (%)
Training	Sigmoid	0.89	0.85	7.32	2.71	5.73	2.31
Testing	Sigmoid	0.87	0.84	8.1	2.85	5.98	3.11

Table 4 Final weights between the input neuron and hidden neuron as well as hidden neuron and output neuron

	Weights (w_{ji})				Biases	
Hidden neurons	*q*net	*N*	*B*	Settlement	b_{hk}	b_0
Hidden 1	0.7486	−0.6152	−5.1196	−3.3517	0.7211	0.0275
Hidden 2	0.6030	−0.6383	−4.9623	−3.2458	0.6211	

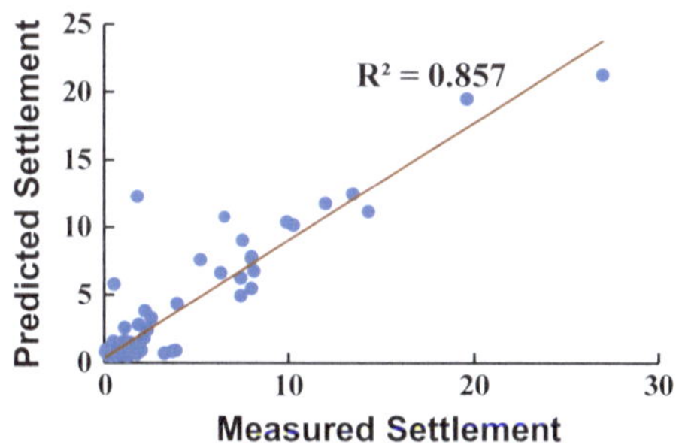

Fig. 2 Measured verses predicted settlement plot for the artificial neural networks

9 Model Equation

Finally the equation for the settlement prediction can be formulated based on the trained weights and biases of the neural network model. The model equation for the prediction of settlement of shallow foundations on granular soils was established using the values of the weights and biases shown in Table 4 as per the following expressions.

$$A = 0.72 + 0.75 * q - 0.62 * N - 5.12 * B$$

$$\mathrm{B} = 0.62 + 0.60 * q - 0.64 * N - 4.96 * B$$

$$\mathrm{E} = 0.027 + 0.72 * \frac{1}{1 + \mathrm{e}^{(-2*0.5*A)}} + 0.62 * \frac{1}{1 + \mathrm{e}^{(-2*0.5*B)}}$$

$$\text{Settlement (S)} = \frac{1}{1 + \mathrm{e}^{(-2*0.5*E)}}$$

10 Conclusions

Over the past decade, the applications of neural networks in geotechnical engineering are well established globally. A feed-forward neural networks with back-propagation algorithm are used to exhibit the capability of ANNs to predict the settlement of shallow foundations on granular soils. It contains 193 case records of field test data for settlement of shallow foundations on granular soils were used for model development and verification. Some activation functions are used in the input and output neurons of neural network among them sigmoid is the best one based on statistical parameters results. The statistical results (R^2 = 0.85, r = 0.89, RMSE = 2.71, MAE = 5.73, MSE = 7.32, and MAPE = 2.31%) indicated that the neural networks are able to predict the settlement of shallow foundations on cohesionless soils. The sensitivity analysis indicated that the width of the footing is the most important factor for the settlement of shallow foundations on cohesionless soils.

References

Boger, Z., & Guterman, H. (1997). Knowledge extraction from artificial neural network models. *IEEE International Conference on Computational Cybernetics and Simulation, 4,* 3030–3035.

Bowles, J. E. (1987). Elastic foundation settlements on sand deposits. *Journal of Geotechnical Engineering, 113*(8), 846–860.

Burland, J. B., & Burbidge, M. C. (1985). Settlement of foundations on sand and gravel. *Proceedings Institution of Civil Engineers Part I, 78*(6), 1325–1381.

Dutta, R. K., et al. (2015). Prediction of deviator stress of sand reinforced with waste plastic strips using neural network. *International Journal of Geosynthetics and Ground Engineering, 1*(11), 1–12.

Maugeri, M., et al. (1998). Observed and computed settlements of two shallow foundations on sand. *Journal of Geotechnical and Geoenvironmental Engineering, 124*(7), 595–605.

Papadopoulos, B. P. (1992). Settlements of shallow foundations on cohesionless soils. *Journal of Geotechnical Engineering, 118*(3), 337–393.

Schmertmann, J. H. (1970). Static cone to compute static settlement over sand. *Journal of Soil Mechanics & Foundations Div, American Society of Civil Engineers, 96*(SM3), 1011–1043.

Shahin, M. A., Maier, H. R. & Jaksa, M. B. (2002). Predicting settlement of shallow foundations using neural networks. *Journal of Geotechnical and Geoenvironmental Engineering, ASCE, 128*(9), 785–793.

Stone, M. (1974). Cross-validatory choice and assessment of statistical predictions. *Journal of Royal Statistical Society, B, 36,* 111–147.

Interference of Two Nearby Footings Resting on Clay Medium

Lohitkumar Nainegali and Anupkumar G. Ekbote

Abstract The study pertains the behaviour of two closely spaced strip footings resting on the surface of the semi-infinite clay soil medium. The effect of interference on characteristic behaviour like bearing pressure, settlement and tilt are observed using two dimensional plane strain finite element analysis. The foundation soil medium is modelled using Mohr–Coulomb soil model that follows linear-elastic perfectly plastic behaviour using finite element analysis software PLAXIS. The soil domain is discretized into 15-noded triangular elements with fine meshing in vicinity of the footings and coarser meshing towards boundary of the domain. Parametric study is performed by varying the clear spacing between the footings and depth of the footings; their effect on the load-settlement characteristic, variation of settlement and bearing pressure with clear spacing are studied. The results are presented in terms of non-dimensional efficiency factors defined as the ratio of settlement/bearing pressure of interfering footings to that of the isolated footing. It is found the interference affect the performance of isolated footing. The bearing pressure is found to decrease with decrease in clear spacing between the footings and in contrast, the settlement is found to increase compared to that of the isolated footing. The results may help in finding the minimum spacing between the footings for better performance.

Keywords Interference effect · Clay soil · Strip footings · Finite element analysis Mohr–coulomb model

---

L. Nainegali (✉) · A. G. Ekbote
Department of Civil Engineering, Indian Institute of Technology (Indian School of Mines), Dhanbad, Dhanbad 826004, India
e-mail: lohitkumarsn@gmail.com

A. G. Ekbote
e-mail: anupge@gmail.com

A. I. V. and V. B. Maji (eds.), *Geotechnical Applications*, Lecture Notes in Civil Engineering 13, https://doi.org/10.1007/978-981-13-0368-5_7

1 Introduction

Lack of construction site, rapid urbanization, structural restrictions, etc. may enforce the structure or the foundations or group of foundations to come up ever close to each other. Such situations may lead to interference phenomenon; wherein the stress zones below the foundations may overlap each other causing distraction in the failure mechanism, settlement and bearing capacity responses of the footings in comparison with that of the isolated footings. The observation of such phenomenon was first reported by Stuart (1962); the study was carried out using limit equilibrium method on ultimate bearing capacity (UBC) of two closely spaced strip footings resting on the surface of cohesion-less medium. However, the momentum gained only in recent years using both numerical/theoretical and experimental analysis and the same are highlighted by Nainegali (2013a). Thereafter, many researchers (Srinivasan and Ghosh 2013; Nainegali et al. 2013b; Eltohamy and Zidan 2013; Naderi and Hataf 2014; Noorzad and Manavirad 2014; Ghosh et al. 2015) have reported on different aspects of the problem. Subsequently, it has been observed that the UBC of the interfering footings increases with decrease in spacing and attains a peak at certain critical spacing and the same is true for settlement at failure. UBC and settlement of the interfering footings is observed to be greater than that of the isolated footing having similar properties. However, from the literature it has been noted that most of the studies have been carried out for two or more interfering footings resting on the surface of cohesion-less soil medium and very few attempts have been made for footings resting on the clay as foundation medium. Henceforth, the studies are carried out to observe the effects of interference on the two nearby strip footings resting on the surface and embedded in clayey soil medium.

2 Problem Definition

Figure1, shows two nearby symmetrical strip footings of width, B placed at a clear spacing, S loaded simultaneously with the uniform load intensity, q while embedded in semi-infinite, homogenous and dry clay soil medium. The analyses are performed by varying the clear spacing ratio (S/B) between the two footings resting on the surface ($D/B = 0$) and embedded ($D/B = 1$) in clay medium. The effect of interference on the bearing capacity, settlement and tilt of the footings against the clear spacing between the footings has been studied considering the parameters presented in Table 1.

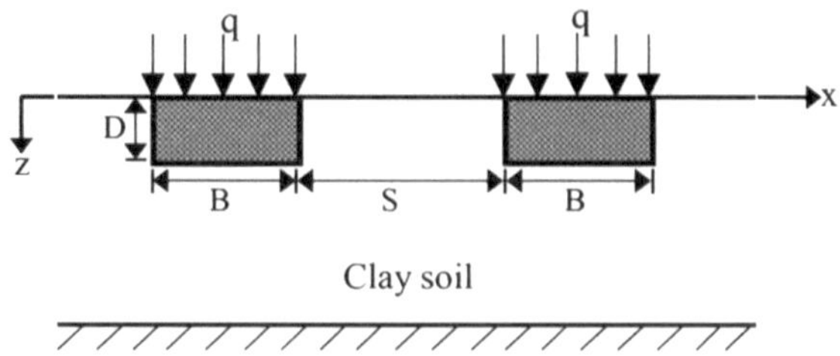

Fig. 1 Problem definition

Table 1 Soil properties and the varying parameters

Soil properties	
Parameter	Value
Young's modulus (*E*)	2000 kN/m^2
Poisson's ratio (υ)	0.3
Unit weight (γ)	16 kN/m^3
Cohesion, *c*	40 kN/m^2
Friction angle (ϕ)	0^0
Varying parameters	
D/*B* = 0.0 (surface footing), 1.0 (embedded footing)	
S/*B* = 0.5, 1.0, 1.5, 2.0, 2.5, 3.0, 3.5, 4.0, 5.0	

3 Numerical Modelling

3.1 Modelling

The analyses have been performed considering plane strain problem, since length of the footing is assumed long enough in comparison with the width of the footing. The commercial finite element software PLAXIS 2D 2016 is used for modelling two nearby strip footings resting on the homogenous semi-infinite clay deposit.

The foundation soil is considered to obey Mohr–Coulomb model that follows linear-elastic perfectly plastic behaviour. The surface footings are modelled using the plate elements which are based on the Mindlin's beam theory with the bending stiffness, EA = 2.35e8 kN/m and axial stiffness, EI = 1.958e7 kN m^2/m while the embedded footings are modelled using the linear-elastic non-porous type of material, with Young's modulus, E = 2.496e7 kN/m^2 and Poisson's ratio, υ = 0.2. The base of the footings is considered rough. The foundation soil domain is discretized using 15-noded triangular elements in association with the fine meshing near the vicinity of the footings. Since the interfering footings are symmetrical with respect to both geometry and loading and hence only half of the foundation domain is considered for the analyses assigning proper boundary conditions; both vertical and horizontal displacements are restricted for the bottom horizontal boundary whereas only the horizontal displacements are restricted for the vertical boundaries. Prior to the calculation, initial stresses of the soil due to its own weight have been generated using the K_0 procedure based on Jaky's formula.

3.2 Sensitivity Analysis

Sensitivity analyses (Nainegali et al. 2013a) are carried out for foundation domain size and mesh to optimize the computational time and effort without affecting the accuracy of the results. Considering $S/B = 0.5$, the far domain in the horizontal, x and vertical downward, z direction are varied between 8B to 12B and 10B to 12B, respectively. Thereafter, the farthest domain in x direction have been chosen to be 12B from the outer edge of the footing and that in z direction is chosen to be 12B below the base of the footing for both surface and embedded footings. As an example the load-settlement curves of the footing varying domain size in x direction with $S/B = 0.5$ and $z/B = 12B$ are shown in Fig. 2 and it can be observed that between 11B and 12B (x direction) the domain has very negligible effect. A fine meshing scheme of PLAXIS 2D 2016 have been adopted in the present analysis apart from finer meshing in vicinity of the interfering footing.

3.3 Validation

Prior to the analysis of the above said problem, the finite element model is validated for isolated surface footing, surface and embedded interfering footings. With the properties as mentioned in Table 1, the UBC of isolated surface footing obtained from the present finite element analysis is 215.6 kN/m^2 and that obtained from the Meyerhof (1963) theory is 208.3 kN/m^2, the solution of present analysis is on higher side by 3.5%. Next, the load-settlement curve of interfering surface footings placed at $S/B = 0.5$ is compared with that of Nainegali et al. (2013b), considering the same properties of soil, footing and domain of Nainegali et al. (2013b) and the same is shown in Fig. 3. It can be observed (Fig. 3) that the load-settlement curve

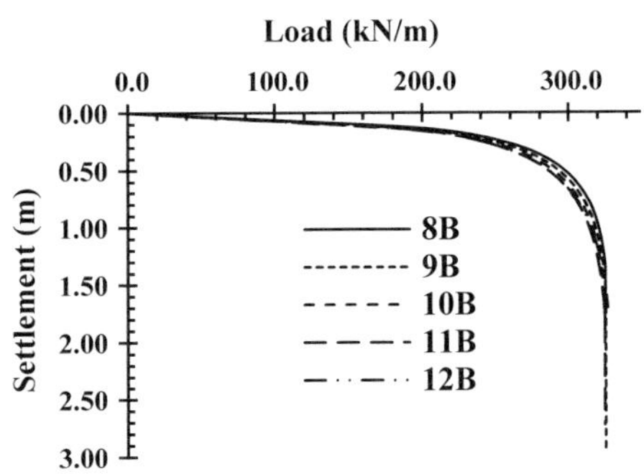

Fig. 2 Load-settlement curve varying domain in x direction

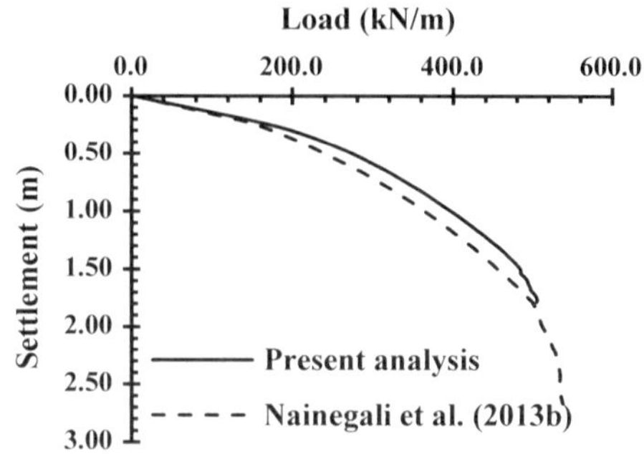

Fig. 3 Validation with Nainegali et al. (2013b)

obtained from the present analysis match well with that of Nainegali et al. (2013b) and the small difference may be due to choice of finite element type and meshing scheme; however the UBC obtained from the present analysis is 6.08% lesser. Similarly, validation has been carried out with that of Vivek (2011) for embedded ($D/B = 1.0$) interfering footings placed at $S/B = 0.5$ using their properties and it has been observed that the load-settlement curve of present analysis and that given by Vivek (2011) has reasonably matched and the UBC obtained differed by 4%. However, due to space and brevity the validation of the same has not been presented.

4 Results and Discussions

4.1 Bearing Pressure Variation with the Spacing

Varying the clear spacing between the two footings, the load-settlement curves obtained for the interfering surface ($D/B = 0.0$) and embedded ($D/B = 1.0$) footings are shown in Figs. 4 and 5, respectively for different S/B ratio. It can be observed that, the load-settlement curves of surface interfering footings (Fig. 4) do not show clear distinction with the isolated footing as that reported in the literature for footings on cohesion-less soil medium. The same is true with the embedded interfering footings (Fig. 5) with small distinction observed for $S/B = 0.5$ compared with isolated footing.

The UBC of interfering footings compared to that of isolated footing is not significant for footings on clay soil medium. At $S/B = 0.5$, a difference of 2.8 and 9.12% in UBC with that of isolated footing is observed for $D/B = 0.0$ and 1.0, respectively. However, for footings on cohesion-less soil medium, the variation of UBC is quite significant as observed from the literature. As an example the study conducted for two closely spaced strip footings on sand by Kumar and Bhoi (2009) showed that, at $S/B = 0.14$ the efficiency factor observed is 1.63 which is approximately 60% predominant as compared with isolated footing. Also, as seen from the results of Kumar and Saran (2003) for two nearby strip footings on sand,

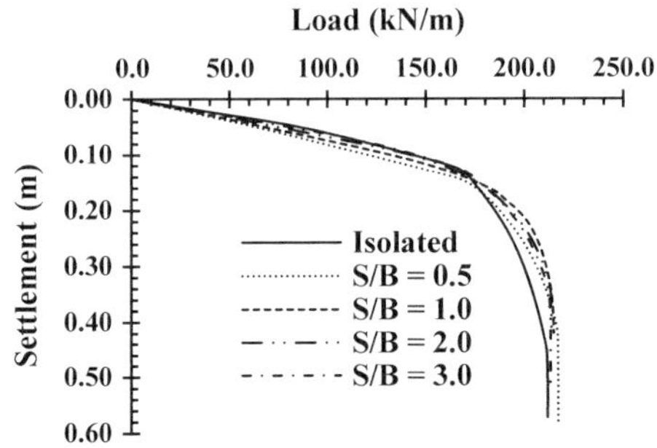

Fig. 4 Load-settlement curves for $D/B = 0.0$ footings

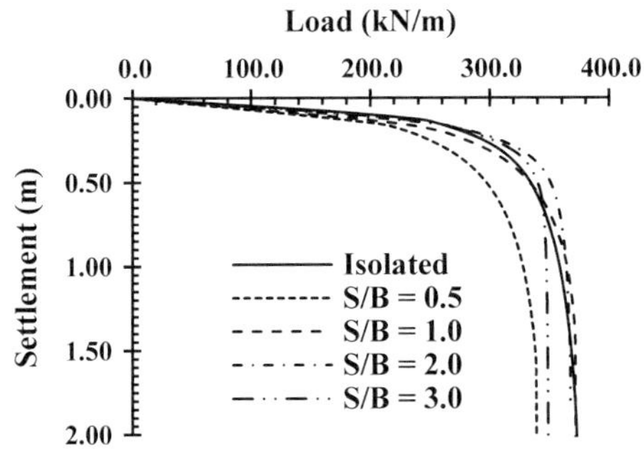

Fig. 5 Load-settlement curves for $D/B = 1.0$ footings

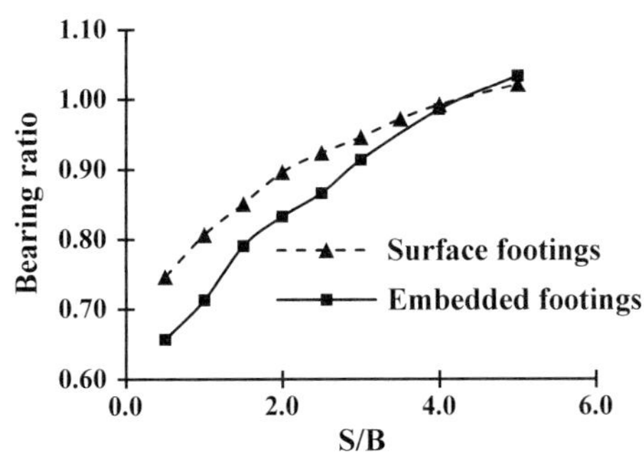

Fig. 6 Variation of bearing ratio with S/B ratio

the efficiency factor for $S/B = 0.5$ is about 1.75 which is approximately 75% predominant. Further, the effect of interference on bearing pressure corresponding to allowable settlement of 50 mm (IS 1904-1986) is studied. The variation of the same in non-dimensional factor 'bearing ratio' (defined as the ratio of bearing pressure of interfering footing to that of isolated footing both measured corresponding to allowable settlement of 50 mm) against S/B ratio is presented in Fig. 6 for both surface and embedded footings. It can be seen that at $S/B = 0.5$, the bearing ratio is less than one. It indicates that the bearing pressure of interfering footing is less than isolated footing, measured corresponding to allowable settlement and increases with increase in *S*/*B* ratio attaining one at *S*/*B* = 5 and more. It is observed that decrease in bearing pressure at *S*/*B* = 0.5 is 25.4 and 36% for *D*/*B* = 0.0 and 1.0, respectively.

4.2 Settlement Variation with the Spacing

Figure 7, presents the variation of settlement ratio (defined as the ratio of the settlement of interfering footing to that of isolated footing, measured corresponding to the pressure of isolated footing against allowable settlement) with respect to spacing ratio for both surface and embedded footings. It can be seen that, the settlement ratio is greater than one (signifying settlement of interfering footing is greater than isolated footing) and decreases with increase in *S*/*B* ratio and reaches a value of one at *S*/*B* = 5 and more, wherein footings can specified to be acting individually. It is noted that increase in settlement at *S*/*B* = 0.5 is about 36 and 62% in case of surface and embedded footings, respectively, which is significant.

4.3 Tilt of the Footings

It is observed that, the settlement at base of the footing is non-uniform, which influences the tilt of the footing; however, uniform settlement is observed for isolated footing. Maximum and minimum settlement is respectively observed at inner and outer edge (tilt in footing occurs towards each other) of the interfering footings and is illustrated in Fig. 8. For the analysis, 'tilt ratio' is defined as the ratio of settlement difference at inner and outer edge to the width of the footing is considered. The variation of tilt ratio with *S*/*B* ratio is presented in Fig. 9 for case of *D*/*B* = 0.0 and 1.0 and can be observed that the tilt decreases with increase in *S*/*B* ratio. At *S*/*B* = 0.5 significant tilt occurs in footing which is noteworthy for practical considerations.

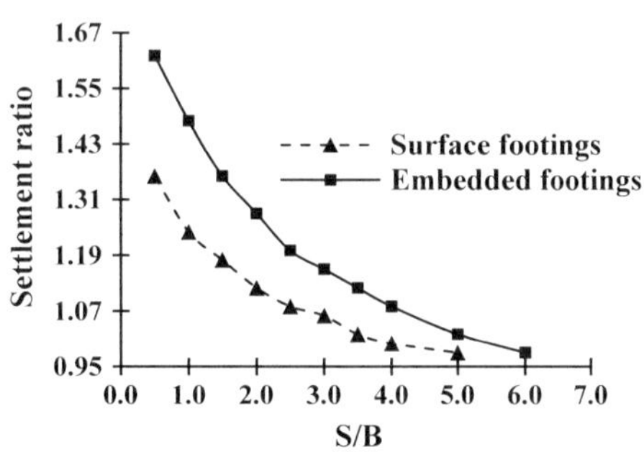

Fig. 7 Variation of settlement ratio with *S*/*B* ratio

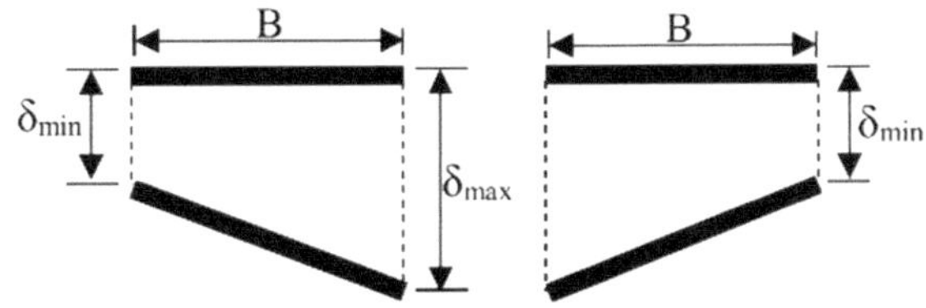

Fig. 8 Tilt of the footing

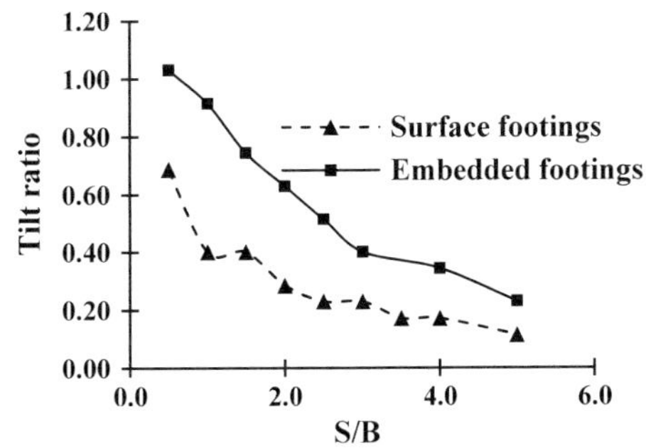

Fig. 9 Variation of tilt ratio with *S/B* ratio

5 Conclusions

The studies on two nearby surface and embedded strip footings are carried out on homogenous clay medium. The effect of interference for footings on pure clay medium is very negligible with respect to UBC, while the effect is noticeable for the bearing pressure and the settlement measured in the allowable range. Moreover, it is noticed that significant tilt occurs for the footings placed very close to each other. In the zone of major interference ($S/B = 0.5$) decrease in bearing pressure and increase in settlement is noticed compared to that of isolated footing, measured at allowable settlement and pressure, respectively.

References

Eltohamy, A. M., & Zidan, A. F. (2013). Performance of interfering strip footings resting on reinforced sand under uniform and non-uniform load-experimental and numerical study. *Journal of American Science., 9*(1), 421–430.

Ghosh, P., Basudhar, P. K., Srinivasan, V., & Kunal, K. (2015). Experimental studies on interference of two angular footings resting on surface of two-layer cohesionless soil deposit. *International Journal of Geotechnical Engineering, 9*(4), 422–433.

IS 1904-1986 (reaffirmed 2006) Indian standard code of practice for design and construction of foundations in soils: general requirements. BIS, New Delhi.

Kumar, J., & Bhoi, M. K. (2009). Interference of two closely spaced strip footings on sand using model tests. *Journal of geotechnical and geoenvironmental engineering., 135*(4), 595–604.

Kumar, A., & Saran, S. (2003). Closely spaced footings on geogrid-reinforced sand. *Journal of geotechnical and geoenvironmental engineering., 129*(7), 660–664.

Nainegali, L. S. Ghosh, P., & Basudhar, P. K. (2013b). Interaction of nearby strip footings under inclined loading. In *Proceedings of the 18th International Conference on Soil Mechanics and Geotechnical Engineering, Paris*. pp. 58–66.
Naderi, E., & Hataf, N. (2014). Model testing and numerical investigation of interference effect of closely spaced ring and circular footings on reinforced sand. *Geotextiles and Geomembranes, 42*(3), 191–200.
Nainegali, L. S., Basudhar, P. K., & Ghosh, P. (2013). Interference of two asymmetric closely spaced strip footings resting on nonhomogeneous and linearly elastic soil bed. *International Journal of Geomechanics, 13*(6), 840–851.
Noorzad, R., & Manavirad, E. (2014). Bearing capacity of two close strip footings on soft clay reinforced with geotextile. *Arabian Journal of Geosciences, 7*(2), 623–639.
PLAXIS 2D 2016. Delft, Netherlands, Plaxis bv.
Srinivasan, V., & Ghosh, P. (2013). Experimental investigation on interaction problem of two nearby circular footings on layered cohesionless soil. *Geomechanics and Geoengineering., 8* (2), 97–106.
Stuart, J. G. (1962). Interference between foundations with special reference to surface footings on sand. *Geotechnique, 12*(1), 15–23.
Vivek, P. (2011). *Static and dynamic interference of strip footings in layered soil*. M.Tech. Thesis, Department of Civil Engineering, Indian Institute of Technology, Kanpur.

Static and Incremental Cyclic Loading of Ring and Circular Footings on Coir Geocell-Reinforced Sand

Afi R. Sudhakar and M. N. Sandeep

Abstract Ring footings are widely used as foundation for water tanks, television antennas, silos, chimneys, oil storage tanks, etc. This paper presents an experimental study to investigate the cyclic as well as static behavior of model ring footing and circular footing resting on coir geocell-reinforced sand. The parameters studied are coir geocell width, and depth of embedment of geocell. The studies have shown that, with the provision of geocell-reinforced sand cushion, there is substantial reduction in settlement of both ring and circular footings due to modified stress distribution. The beneficial effect in terms of increased load carrying capacity and reduced settlement is related to the width and depth of placing geocell mattress.

Keywords Ring footing · Circular footing · Cyclic loading · Coir geocell

1 Introduction

In case of axisymmetric structures such as water towers, transmission towers, television antennas, silos, chimneys and oil storage, ring footings are generally used to support columns or walls. The use of ring footing decreases the amount of materials used and it is more economical.

Introducing reinforcement inclusions within the soil is an effective and reliable method in improving the bearing capacity of soil. The more recent advancement of reinforced soil is to provide three-dimensional confinements to soil by using geocells. Geocell foundation mattress consists of a series of interlocking cells, constructed from polymer geogrids, which contains and confines the soil within its pockets. Nowadays natural material is introduced as reinforcement to improve the engineering properties of soil. Coir is a natural fiber extracted from husk of coconut.

A. R. Sudhakar (✉) · M. N. Sandeep
IES College of Engineering, Thrissur, India
e-mail: afirsudhakar@gmail.com

M. N. Sandeep
e-mail: sandheepmn@gmail.com

A. I. V. and V. B. Maji (eds.), *Geotechnical Applications*, Lecture Notes in Civil Engineering 13, https://doi.org/10.1007/978-981-13-0368-5_8

Several studies have been reported on stabilization of soil using coir fiber. This study presents a series of laboratory results on the behavior of ring and circular footings placed on sand reinforced with coir geocells.

From the literatures, it was found that for ring footings the improvement in bearing capacity ratio was maximum for an inner to outer diameter ratio of 0.39 (El Sawwaf and Nazir 2012; Hashim et al. 2013). Laboratory model test of bearing pressure of circular and ring footings on sand shows that ring footing with an optimum inner to outer diameter ratio of 0.4 has nearly same load carrying capacity as that of circular footing. Several investigations have also been reported on the use of synthetic geocells to improve the bearing capacity of weak soil. It has been found that geocell enhances the footing performance on sand and it was also found that optimum width of geocell mattress is around four times the width of footing. Studies by Nikhil John & Asha (2013) also shows that in order to obtain the maximum benefit, the top of geocell mattress should be at a depth of 0.1 times the diameter of footing. Cyclic loading on geocell-reinforced foundations exhibit a fourfold increase in bearing capacity of footing compared to unreinforced case (Tafreshi and Dawson 2015; Choobbasti et al. 2010).

This paper focuses on the effect of geocell reinforcement on load carrying capacity of ring footing and circular footings under vertical and cyclic loading using experimental approach. The specific objectives include investigating the effect of geocell width, depth of placing reinforcement, and geocell pocket size on bearing capacity of both footings.

2 Experimental Setup

2.1 Test Setup

The test facility developed in this study consists of a loading frame, a steel box, controlled tamping device, model footings, and loading arrangement loading arrangement. The vertical compressive load was applied to model footing through a hydraulic jack of 100 kN capacity, which was supported centrally at the bottom flange of the steel girder made of channel sections. It was operated manually by a hydraulic pump. The load was transferred to the model through a proving ring which was fixed on the bottom of jack and the model footing was connected to the proving ring through a loading platform. Settlement of model footing was measured using dial gauges having travel of 25 mm and least count of 0.01 mm. The schematic view of the experimental setup is given in Fig. 1.

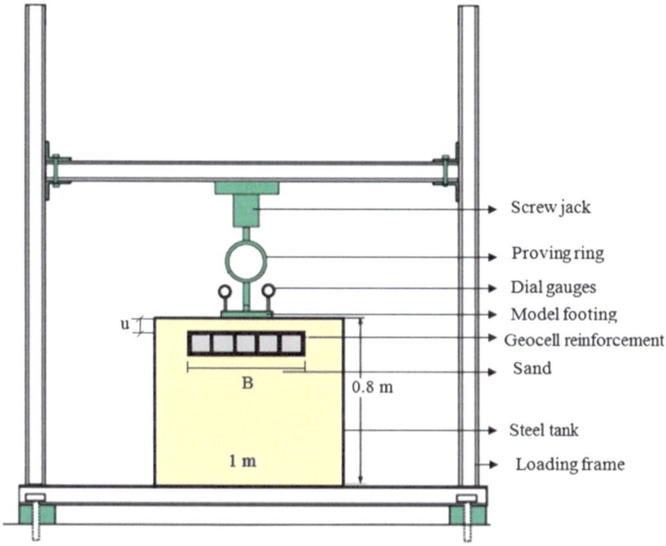

Fig. 1 General arrangement of experimental set up

2.2 *Test Medium*

Clean river sand was used in all the experiments as a test medium. The index properties obtained are presented in Table 1 and the sand is classified as poorly graded sand (SP).

Table 1 Index properties of sand

Description	Value
Specific gravity, G	2.67
Effective size, D_{10} (mm)	0.27
D_{30} (mm)	0.43
D_{60} (mm)	0.78
Coefficient of uniformity (C_u)	2.87
Coefficient of curvature (C_C)	0.86
Type of soil	SP
Maximum dry density (kN/m^3)	18.4
Minimum dry density (kN/m^3)	15.3
Specimen	Dry sand
Relative density (%)	50
Undrained cohesion	0
Angle of internal friction	40°

2.3 Test Box

A rectangular rigid steel test box of inner dimensions 1 m × 1 m × 0.8 m with 10 mm wall thickness was fabricated and used for conducting the experiments on the model footings in the laboratory. With a wall thickness of 10 mm and several stiffeners on the outer side, the box is considered to be rigid. The inside face of the tank was graduated at every 100 mm depth intervals to aid preparation of sand bed in layers by fixed density method.

2.4 Model Footings

Model ring footing made of mild steel with an outer diameter of 150 mm and inner diameter of 60 mm. Model circular footing having diameter same as the outer diameter of ring footing was made from mild steel. The inner to outer diameter ratio of the ring was 0.4. The thickness of model ring and circular footing was 15 mm. The ring was placed on the correct middle of the test tank on the soil surface. Four steel columns of height 100 mm welded to a circular plate of 15 mm thick placed over the surface of footings in order to simulate the over head structures on footing.

2.5 Soil Reinforcement

Geocell was used as the soil reinforcement. Geocell was made from coir belt. The properties of coir geocell are given in Table 2.

3 Methodology

A total of five series of experiment were carried out on model footings. Series 1 consists of static and cyclic load test on ring footing and circular footings without reinforcement. Series 2 consist of static load test on ring and circular footings with coir geocell of varying width. Series 3 consists of bearing capacity test on ring

Table 2 Properties of geocell

Description	Value
Thickness (mm)	8.88
Mass per unit area (gsm)	1267
Pocket size (cm × cm)	8 × 8
Tensile strength (kN/m)	15.8
Geocell height (cm)	10

footing and circular footings with coir geocell placed at different depth under static loading. Series 4 consists of static and cyclic test on ring and circular footing with optimum width and depth of placing geocell. A total of 30 tests were carried out on the model ring and circular footings supported on unreinforced and reinforced soil.

4 Result Analysis and Discussion

The performance improvement of footings due to geocell is expressed using a non-dimensional term called improvement factor (IF). IF is defined as the ratio of bearing load with geocell reinforcement at a given settlement to the corresponding load on unreinforced soil at the same settlement. When this ratio is calculated at settlements beyond the ultimate load of unreinforced soil, the ultimate load of the footing (q_{ult}) is used instead of q_0. The footing settlement "*s*" were expressed as a non-dimensional form in terms of the footing width as *s*/*B* (%).

4.1 *Static and Cyclic Test in Unreinforced Soil*

It was clear Figs. 2 and 3, that ring footing with inner to outer diameter ratio of 0.4 is having almost same load carrying capacity compared to that of circular footing. However the settlement values are lesser for ring footing compared to circular footing. This may be due to the presence of additional shear failure surface starting from the internal edge of ring footing and finished at the end of wedge zone of shear failure surface. This additional failure surface increases the friction area of active zone.

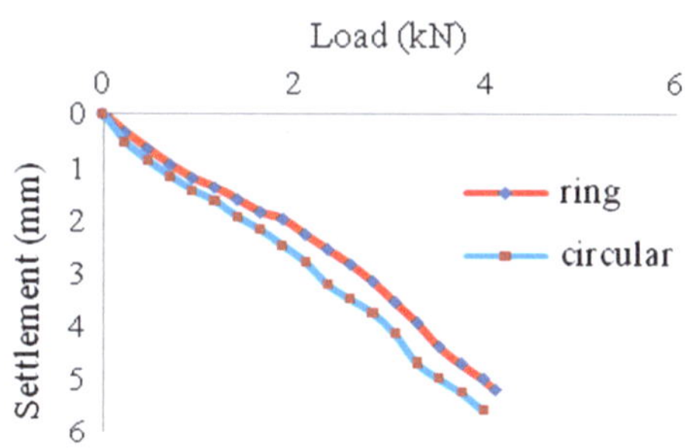

Fig. 2 Load-settlement of footings on unreinforced soil under static loading

Fig. 3 Load-settlement of footings on unreinforced soil under cyclic loading

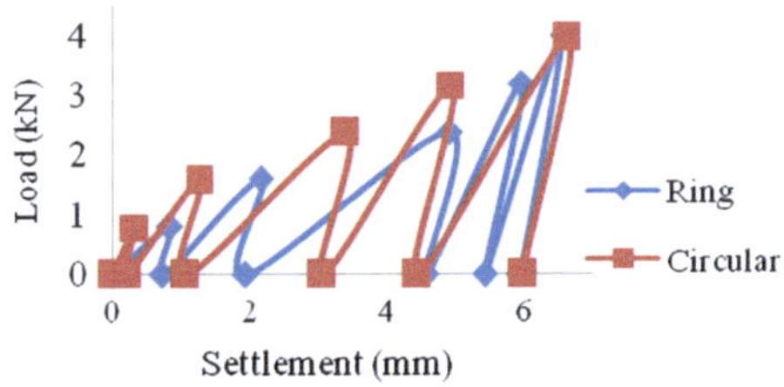

4.2 *Effect of Width of Geocell*

The tests were conducted on both the footings with single layer of geocell reinforcement and with a pocket size of 0.53B placed at a depth of 0.1B. The width of reinforcing layer (b) was varied as 1B, 2B, 3B, 4B, and 5B. It is clear that no much improvement is obtained with geocell mattress of size same as that of loading plate. The performance increases with increase in size of geocell mattress from 1B to 4B. Maximum improvement is obtained when the width of reinforcement is 4B. Beyond 4B, there is no much significant improvement (Figs. 4 and 5).

Geocell reinforcement of width 4B gives an improvement of 1.7 times in case of ring footing and 2.09 times in case of circular footing.

4.3 *Effect of Depth of Geocell Embedment*

Geocell placed at 0.1 times footing width gives greater improvement at smaller loads, but geocell at a depth of 0.25 B gives much improvement at higher loads also. In this particular study, the optimum depth to the top of geocell is taken as 0.25 times the width of footing which provides an average of 1.7 times improvement in bearing load of ring footing (Figs. 6 and 7).

In case circular footing, the depth of reinforcement when increased from 0B to 0.1B, load carrying capacity increased and settlement decreased. The optimum depth is obtained as 0.1B and all other tests were conducted at this particular depth.

Fig. 4 Improvement factor of ring footing with varying geocell width under static loading

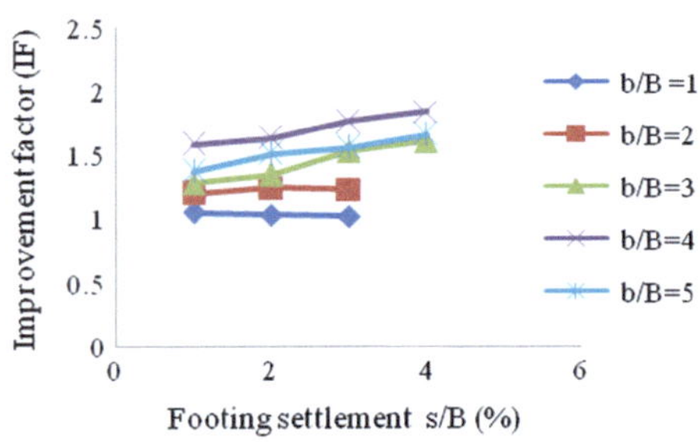

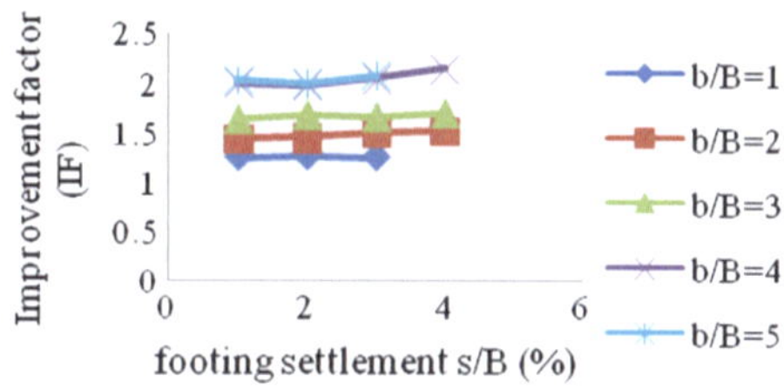

Fig. 5 Improvement factor of circular footing with varying geocell width under static loading

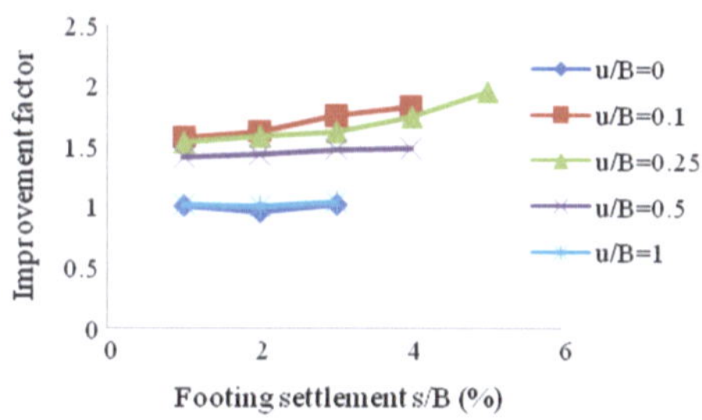

Fig. 6 Improvement factor of ring footing with varying geocell depth under static loading

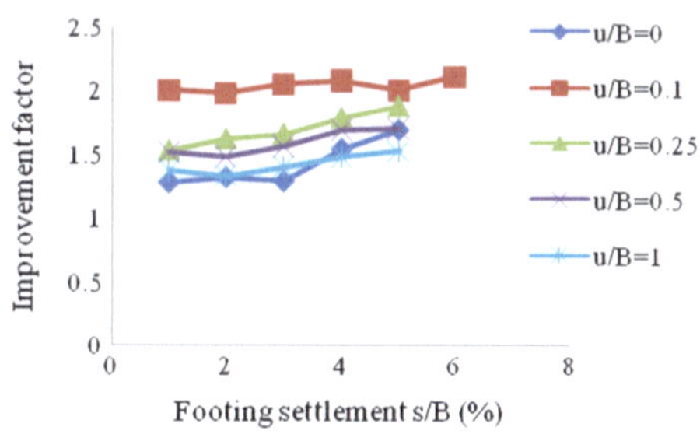

Fig. 7 Improvement factor of circular footing with varying geocell depth under static loading

The embedment of geocell at this depth provides an average of 2.05 times improvement in load carrying capacity of circular footing.

4.4 *Comparison of Reinforced and Unreinforced Footings*

In static case, provision of geocell reinforcement of optimum width, depth and pocket size, increases the load by 77% in case of ring footing and 88% in case of circular footing compared to unreinforced case (Fig. 8).

Cyclic load test was also conducted on both ring and circular footing with geocell reinforcement provided at optimum width, and depth. In case of ring footing, geocell reinforcement provides 60% reduction in settlement, while in circular footing, 68% decrease is obtained when compared with unreinforced case (Fig. 9).

Fig. 8 Load-settlement of unreinforced and reinforced footings under static loading

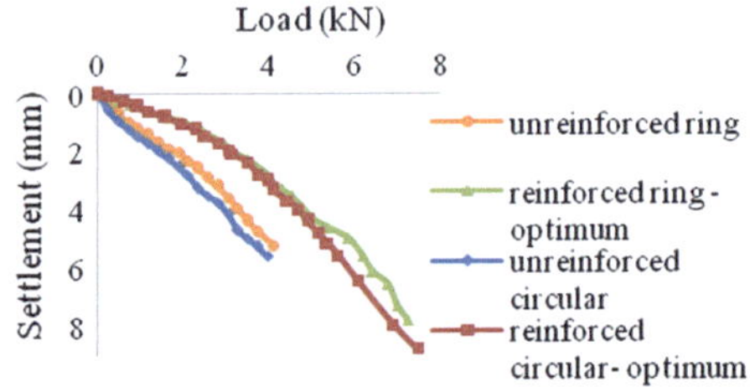

Fig. 9 Load-settlement of unreinforced and reinforced footings under cyclic loading

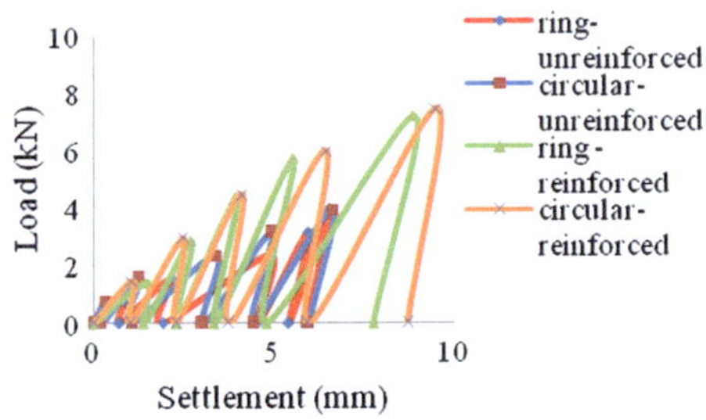

5 Conclusions

From the experimental program, the following conclusions were made:

The bearing capacity of ring footing with inner to outer diameter ratio of 0.4 is nearly same as that of circular footing in static and cyclic case. The optimum width of geocell mattress is equal to four times the diameter of footing in case of both ring and circular footing. Surface heaving and settlement reduces with increase in width of geocell. The optimum depth of placing coir geocell is obtained as 0.1 times the diameter of footing in case of circular and 0.25 times the outer diameter in case of ring footing. The results show that footing performance due to cyclic loading is better for geocell-reinforced soil than that of unreinforced soil.

References

Choobbasti, A. J., Hesami, S., Najafi, A. et al. (2010). Numerical evaluation of bearing capacity and settlement of ring footing; case study of Kazeroon cooling towers. *International Journal of Research and Reviews in Applied Sciences,* 263–271.

El Sawwaf, M., & Nazir, A. (2012). Behavior of eccentrically loaded small-scale ring footings resting on reinforced layered soil. *Journal of Geotechnical and Geoenvironmental Engineering, ASCE, 138*(3), 376–384.

Hashim et al. (2013). Experimental Investigation of the Bearing Pressure for Circular and Ring Footings on Sand. *Tikrit Journal of Engineering Sciences, 20*(3), 64–74.

Nikhil John, K., & Asha, N. P. (2013). Behaviour of small scale ring footing resting on geotextile reinforced soil. *Proceedings of Indian Geotechnical Conference, 2013* (pp. 1–8).

Tafreshi, S. N., & Dawson, A. R. (2015). *Repeated load response of square footings on geocell reinforced soil: Comparing use of single and multiple layered geocells*. World Congress on Engineering Education 2015.

Experimental Study on Uplift Capacity of Horizontal Circular and Strip Anchor Plates in Two-Layered Cohesionless Soil

Paramita Bhattacharya

Abstract Load-displacement behaviour and vertical uplift capacity of horizontal anchor plate embedded in layered sand deposits have been investigated by conducting laboratory model tests. Strip and circular plate anchors are used for this purpose. The layered sand deposits have been prepared with local sand having relative densities 25% (loose sand upper layer) and 65% (medium dense sand with bottom layer), respectively. The medium dense sand layer is always kept closer to the anchor plate and the loose sand layer is kept close to the top surface in the layered sand system. The experiments have been conducted for different embedment ratios with different relative thickness of the dense sand layer (H_{dense}/H). The maximum displacement of the anchor plate experienced at failure and the uplift capacity of the anchor plate have been found to increase with an increase in the thickness of the dense sand layer (H_{dense}) at any particular embedment depth (H). The present uplift capacity of the anchor plate determined experimentally is compared well with the available numerical solution.

Keywords Circular anchor · Strip anchor · Layered sand
Load-displacement behaviour · Uplift capacity

1 Introduction

Anchors are commonly used as foundation of transmission towers, utility poles, aircraft moorings, buried pipelines to resist the uplift forces. A large number of research investigations were carried out to study the vertical uplift capacity of horizontal anchor plate in homogeneous cohesionless soil by different researchers such as: Balla (1961), Meyerhof and Adams (1968), Das and Seeley (1975), Ovesen (1981), Rowe and Davis (1982), Murray and Geddes (1987), Sakai and

P. Bhattacharya (✉)
Department of Civil Engineering, Indian Institute of Technology Kharagpur,
Kharagpur 721302, India
e-mail: paramita@civil.iitkgp.ernet.in

A. I. V. and V. B. Maji (eds.), *Geotechnical Applications*, Lecture Notes in Civil Engineering 13, https://doi.org/10.1007/978-981-13-0368-5_9

Tanaka (1998), Ilamparuthi et al. (2002), Merifield and Sloan (2006). Only a few investigations are performed to study the uplift capacity of anchor plate in layered sand system found commonly in practical case (Bouazza and Finlay 1990; Kumar 2003; Sakai and Tanaka 2007). Bouazza and Finlay (1990) conducted laboratory model tests with circular anchor plates embedded in a dense sand stratum underlying below (i) loose sand or (ii) medium sand. The model tests revealed the fact that the pullout capacity of circular plate in layered sand system depends on the relative strength of the two layers, the thickness of the upper layer and the total embedment ratio. Kumar (2003) studied the vertical uplift resistance of shallow strip and circular anchor plates embedded in two-layered sand system by using upper bound limit analysis with an assumption of a linear collapse mechanism. Kumar (2003) concluded that the vertical pullout capacity of horizontal anchor plate depends on the relative position of the dense sand layer with respect to the position of anchor plate along with the thickness of the dense sand stratum and embedment ratio. Sakai and Tanaka (2007) found from small scale mode tests and elasto-plastic finite element analysis that the maximum uplift resistance offered by a shallow circular anchor plate embedded in two layered sand system increases with an increase in thickness of the dense lower layer placed immediately above the anchor plate. Bhattacharya and Kumar (2016) performed lower bound finite element limit analysis on the uplift capacity of both strip and circular anchor plate embedded in two layered sand system to study the effects of (i) the relative thickness and position of the dense sand stratum (H_{dense}/H) with respect to the anchor plate, (ii) dilatancy (ψ) of the sand layers and (iii) embedment ratios on the vertical uplift capacity of horizontal anchor plate.

From the above literature it has been noticed that the experimental works on layered sand system only dealt the uplift capacity of circular plate anchor. No experiment on strip or rectangular anchor plate has been reported. The present paper attempts to study the load-displacement behaviour of both strip and circular plate anchor till failure and a comparative study has been done on the load-displacement behaviour of both anchor plates along with its maximum uplift resistance prior to failure.

2 Problem Statement

A horizontal anchor plate, either strip or circular in shape, is embedded in a two layers sand system comprised of a loose sand layer of thickness H_1 with 25% relative density (I_{D1}) near to the top surface and a medium dense sand layer of thickness H_2 with 65% relative density (I_{D2}) near to the anchor plate as shown in Fig. 1a. It is required to investigate the load-displacement behaviour of the anchor plate till it fails and to determine the maximum pullout load (P_u) per unit area (A) of the anchor plate. The uplift capacity factor (F) of the anchor plate due to the self-weight of the two layered sand system is defined as:

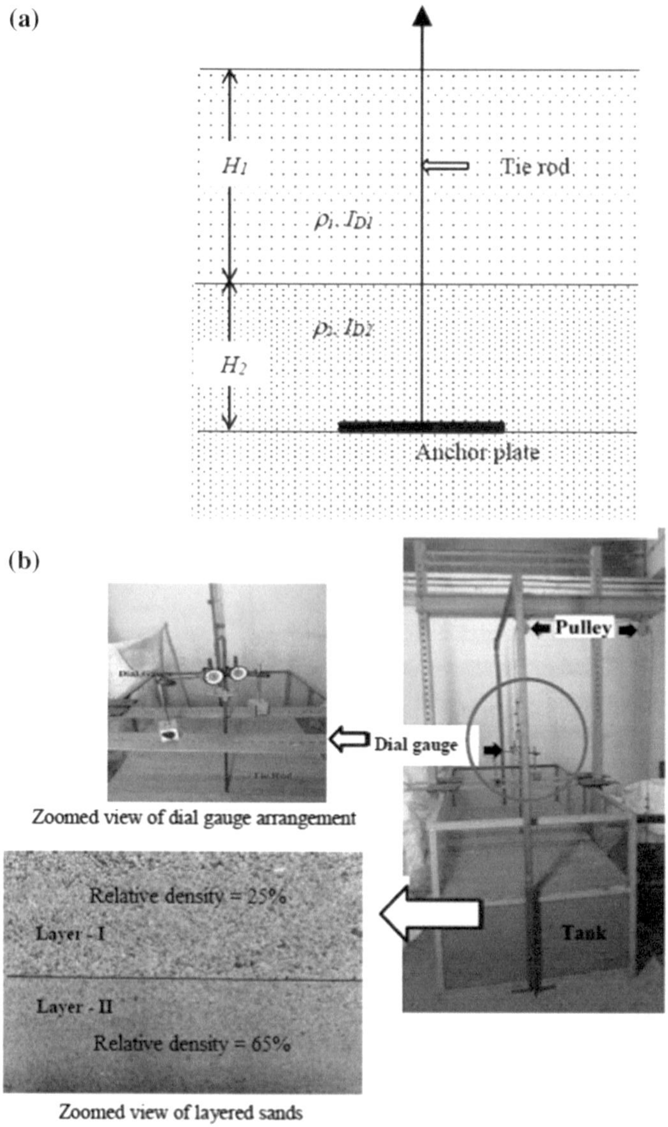

Fig. 1 **a** Schematic diagram of the problem, **b** experimental set up to conduct the laboratory model test

$$F_{\gamma} = \frac{P_u}{A(\rho_1 H_1 + \rho_2 H_2)g}, \tag{1}$$

where ρ_1 and ρ_2 are mass densities of upper loose and lower medium dense sand strata, respectively, and g is the acceleration due to gravity.

3 Experimental Set Up

A rectangular tank of size 1.8 m (length) × 0.9 m (width) × 0.8 m (height) was used to conduct the small scale laboratory tests. A glass sheet of 10 mm thickness was used to make the peripheral walls of the tank. Vertical and horizontal stiffeners were used along the sides and the bottom of the tank to strengthen it. The model tests were carried out by using (i) a circular steel anchor plate having 50 mm diameter and 8 mm thickness and (ii) a strip anchor plate of steel with 50 mm width, 550 mm length and 8 mm thickness. Two dial gauges, each having a displacement sensitivity of 0.01 mm, were placed normal to steel plate as shown in Fig. 1b to measure the displacements of the anchor plate. Average value of the two dial gauges' readings was considered as the vertical displacements. The loading frame consisting of two vertical channel sections was anchored into a concrete platform of size 1.5 m × 2.0 m. The vertical channels were placed at a distance of 1.0 m and provided with adequate lateral support at the bottom. In addition to the bottom support two lateral channels are provided at the top to stiffen the vertical channels. There were two pulley brackets mounted to conduct the pullout test in which one was mounted at the centre of the horizontal stiffeners and another one was at the end of horizontal stiffeners. The vertical uplift load was applied on the anchor plate by using a steel wire which moves freely over a coupling of two pulleys and connected with a gravity type loading arrangement. The loading arrangement was such that it pulls out the anchor plate vertically upward by placing the weights on its other end.

4 Sample Properties

Specific gravity, grain size distribution, mass density, relative density and direct shear tests were carried out as per Indian Standard Codes to characterize the sand sample. The grain size distribution curve is shown in Fig. 2 and found coefficient of uniformity C_u= 3 and coefficient of curvature C_c = 0.75. The sand used for the experiment has minimum and maximum densities equal to $\rho_{\min}$ = 1.48 g/cc and $\rho_{\max}$ = 1.71 g/cc, respectively. The estimated specific gravity (G) for the foundation material was 2.64, and maximum and minimum void ratios are equal to $e_{\max}$ = 0.78 and $e_{\min}$ = 0.54, respectively. For a vertical normal stress ranging from 50 to 150 kPa, the peak internal friction angles (ϕ_p) of the chosen sand were found to be 29.6° and 37.8° corresponding to ρ_d = 1.53 g/cc and ρ_d = 1.62 g/cc, respectively, from the direct shear test.

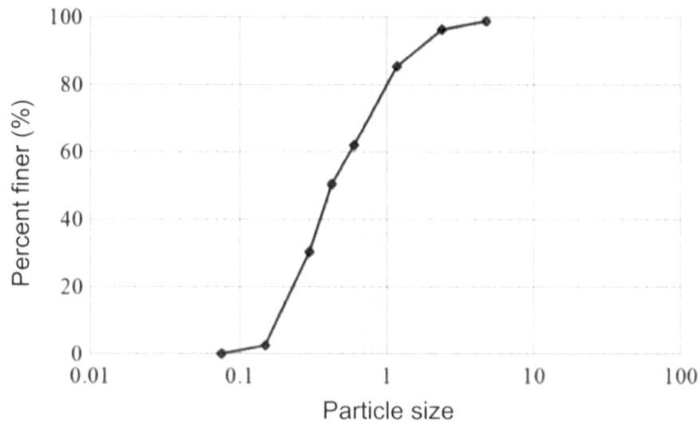

Fig. 2 Grain size distribution curve of the sand used in the experimental work

5 Results and Comparisons

5.1 *Variation of Uplift Capacity for Circular and Strip Anchor*

The load versus displacement curves of the strip and circular anchor plates embedded in two-layered sand system at embedment ratio (H/B) equal to 3, 4 and 5 have been presented in Figs. 3 and 4, respectively, for different values of H_{dense}/H. The load and displacement are expressed as non-dimensional quantity, say $P_u/(A_{av}B)$ and δ/B in percentage, respectively, where $\gamma_{av} = \frac{(H_1 + {}_2H_2)g}{H_1 + H_2}$, δ is the displacement of the anchor plate embedded in layered sand system.

The ultimate resistance capacity (F) of the anchor plate for different combination of H_{dense}/H and H/B has been determined from Figs. 3 and 4 by using the Eq. (1). The variation of the uplift capacity factors (F) of circular and strip anchor plates are shown in Fig. 5. F increases with an increase in H_{dense}/H. The rate of change of F is more in case of circular plate than the same in case of strip anchor plate keeping H, H_{dense}/H and relative densities of the layered sand system constant. However the rate of change of F has been found to decrease with an increase in H_{dense}/H for both circular and strip anchor plate. The anchor plate experiences greater displacement before failure with an increase in (i) the relative thickness of medium dense sand, and (ii) the embedment ratio of the anchor plate. Moreover keeping constant H/B, H_{dense}/H and relative densities of layered sands it is found that the circular anchor plate undergoes lesser displacement than the strip anchor plate.

5.2 *Comparison*

In Fig. 5 the experimentally obtained F values of both circular and strip anchor plates are compared with the numerical results provided by Kumar (2003) and Bhattacharya and Kumar (2016). The experimental results match well with the

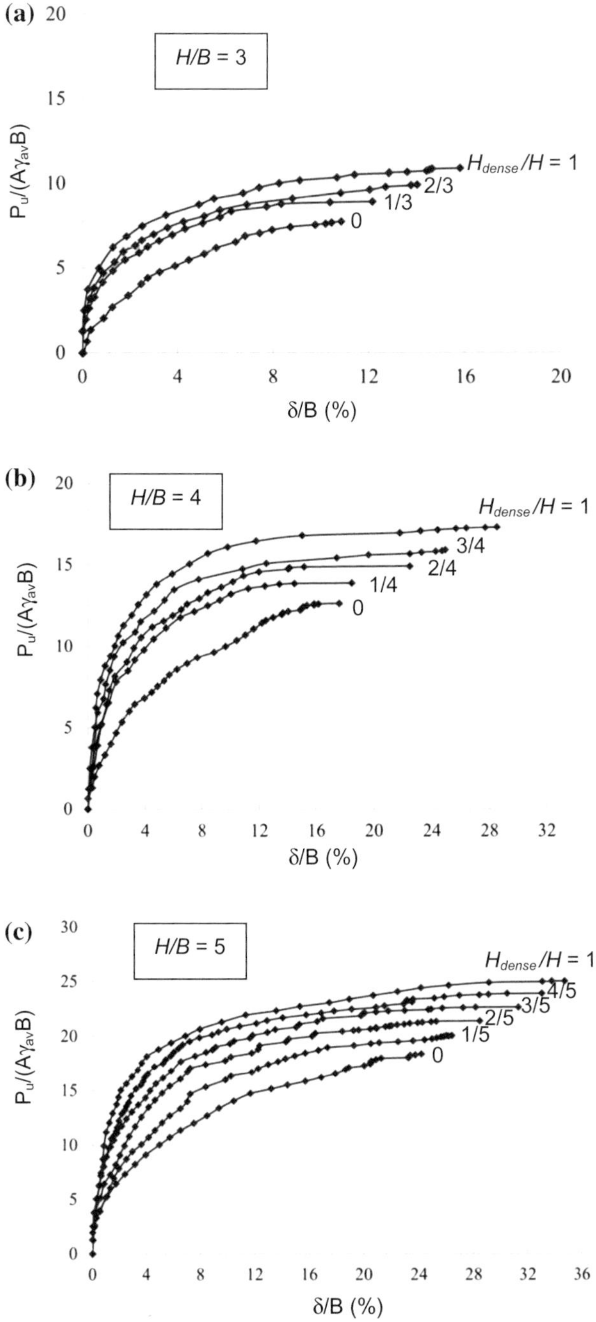

Fig. 3 Load-displacement response for strip anchor plate embedded in two layered sand system with: **a** *H/B* = 3; **b** *H/B* = 4; and **c** *H/B* = 5

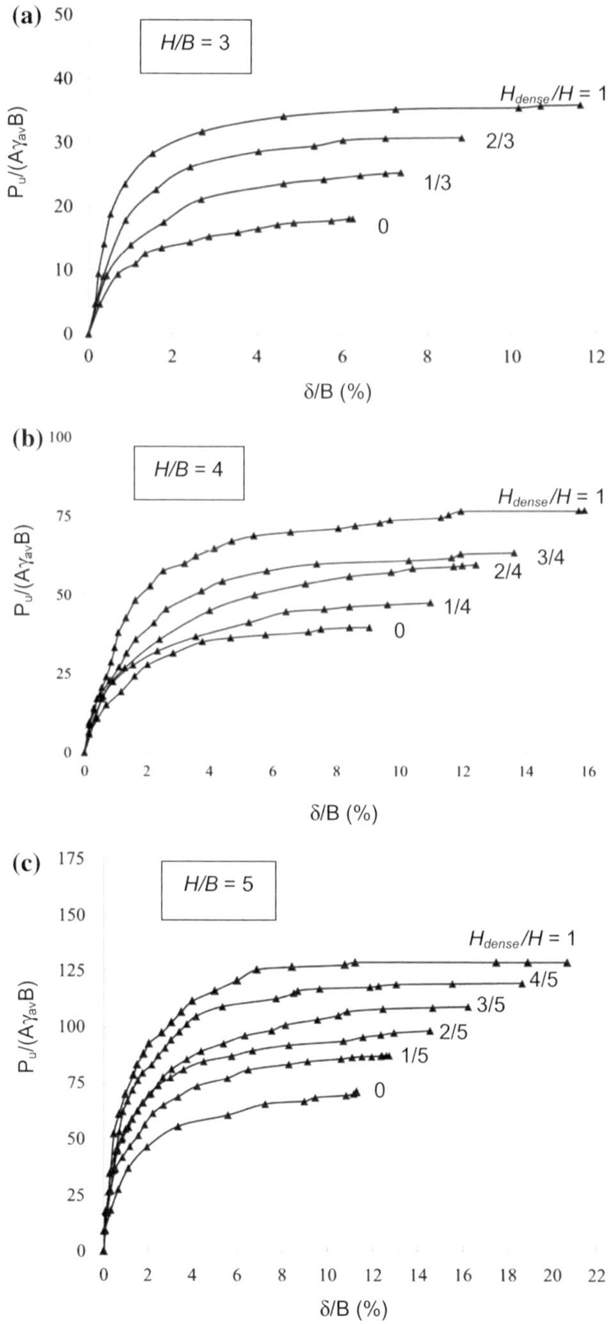

Fig. 4 Load-displacement response for circular anchor plate embedded in two layered sand system with: **a** $H/B = 3$; **b** $H/B = 4$; and **c** $H/B = 5$

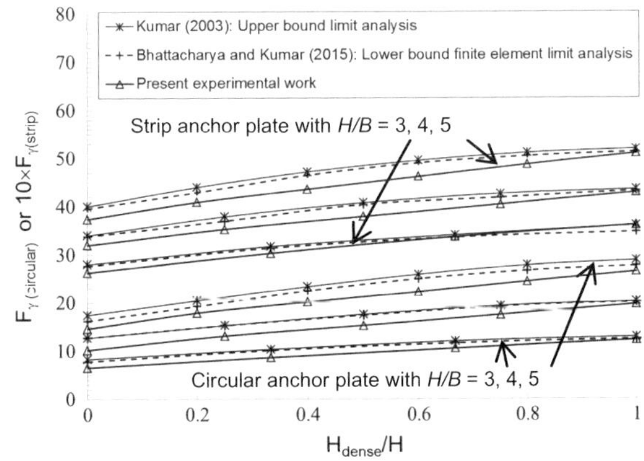

Fig. 5 A comparison of present experimental and available theoretical pull out factors for circular and strip anchor embedded in layered sand

numerical results with a maximum difference of around 8% for circular plate anchor and 4% for strip plate anchor. The dilatancy angle (ψ) measured from direct shear test with stress level 50–150 kPa varies from 2° to 15° which is higher than the average stress level (2–30 kPa) during model tests. The numerical solutions are for a material following an associated flow rule (i.e. $\phi = \psi$).

6 Conclusion

The present experimental investigations illustrate the vertical uplift resistance of horizontal circular and strip anchor plates embedded in two-layered sand system.

The uplift capacity of anchor plate and the magnitude of maximum displacement experienced by it prior to failure increases with an increase in H_{dense}/H keeping H/B constant. However the pullout capacity of circular anchor plate of diameter 50 mm is invariably greater than that of strip anchor plate of width 50 mm but the displacement undergoes by the strip anchor plate is always higher than that of the circular anchor plate keeping the relative thickness of the medium dense and loose sand layers and total embedment depth constant.

References

Balla, A. (1961). The resistance of breaking-out of mushroom foundations for pylon. In *Proceedings of 5th International Conference on Soil Mechanics and Foundation Engineering,* Vol. 1, pp. 569–576, Paris.

Bhattacharya, P., & Kumar, J. (2016). Uplift capacity of anchors in layered sand using finite element limit analysis: Formulation and results. *International Journal of Geomechanics, 16*(3). https://doi.org/10.1061/(asce)gm.1943-5622.0000560.

Bouazza, A., & Finlay, T. W. (1990). Uplift capacity of plate anchors buried in a two-layered sand. *Geotechnique, 40*(2), 293–297.

Das, B. M., & Seeley, G. R. (1975). Breakout resistance of shallow horizontal anchors. *Journal of Geotechnical Engineering, Division, ASCE, 101*(9), 999–1003.

Ilamparuthi, K., Dickin, E. A., & Muthukrisnaiah, K. (2002). Experimental investigation of the uplift behaviour of the circular plate anchors embedded in sand. *Canadian Geotechnical Journal, 39*(3), 648–664.

Kumar, J. (2003). Uplift resistance of strip and circular anchors in a two layered sand. *Soils and Foundations, 43*(1), 101–107.

Merifield, R. S., & Sloan, S. W. (2006). The ultimate pullout capacity of anchors in frictional soil. *Canadian Geotechnical Journal, 43,* 852–868.

Meyerhof, G. G., & Adams, J. I. (1968). The ultimate uplift capacity of foundations. *Canadian Geotechnical Journal, 5*(4), 225–244.

Murray, E. J., & Geddes, J. D. (1987). Uplift of anchor plates in sand. *Journal of Geotechnical Engineering, 113*(3), 202–215.

Ovesen, N. K. (1981). Centrifuge tests on the uplift capacity of anchors. In *Proceedings of 10th International Conference on Soil Mechanics and Foundation Engineering*, Vol. 1, pp. 717–722, Stockholm.

Rowe, R. K., & Davis, E. H. (1982). The behaviour of anchor plates in sand. *Geotechnique, 32*(1), 25–41.

Sakai, T., & Tanaka, T. (1998). Scale effect of shallow circular anchor in dense sand. *Soils and Foundations, 38*(2), 93–99.

Sakai, T., & Tanaka, T. (2007). Experimental and numerical study of uplift behavior of shallow circular anchor in two layered sand. *Journal of Geotechnical and Geoenvironmental Engineering., ASCE, 133*(4), 469–477.

Footing Load Tests on Sand: Validating Theoretical Predictions

Sanjay Gupta, Ravi Sundaram and Sorabh Gupta

Abstract As structures get taller and the load demand on the foundation soils increases, it has become imperative to reconsider conventional geotechnics, and calibrate and validate our theoretical predictions with ground conditions. A prototype footing load test was conducted on a 2 m × 2 m size concrete footing for a better assessment of the load-settlement behavior at a site in Noida. This paper presents the test results and compares them to results from conventional methods of analysis. A numerical simulation of the test conditions has also been performed using finite element method to validate the field results with mathematical simulation. It has been found the results of the field test are in good agreement with theoretical predictions, thus enhancing the factor of reliability of our foundation analysis.

Keywords Foundation design · Footing load test · Finite element analysis

1 Introduction

Delhi NCR has witnessed a spate of high-rise construction for residential and commercial projects in the past few years. For residential projects comprising of a group of high-rise towers, shallow foundations/raft foundations are typically considered as a practical and economic alternative to pile foundations.

For the geotechnical engineer, it is important to accurately characterize the ground, and reliably predict foundation settlements that will occur during the life span of the structure.

S. Gupta (✉) · R. Sundaram · S. Gupta
Cengrs Geotechnica Pvt. Ltd., New Delhi, India
e-mail: sanjay@cengrs.com

R. Sundaram
e-mail: ravi@cengrs.com

S. Gupta
e-mail: sorabh@cengrs.com

A. I. V. and V. B. Maji (eds.), *Geotechnical Applications*, Lecture Notes in Civil Engineering 13, https://doi.org/10.1007/978-981-13-0368-5_10

This paper presents a case study in which field results from a footing load test are compared to theoretical predictions of foundation settlement.

2 Background

A residential project comprising of several group housing towers (14- to 24-storeyed structures with single basement) is being constructed in Greater Noida (U.P.), about 1.5 km west of the Yamuna Expressway. The project shall also boast of a central tower (studio) which is planned to have 36 upper floors with two basements.

The authors carried out a detailed geotechnical investigation at the project site, consisting of 23 exploratory boreholes to 30 m depth, along with Standard Penetration Tests (SPT) at 1.5–3 m depth intervals and a comprehensive laboratory test program.

2.1 Regional Geology

The project site is in the Indo-Gangetic plains. Greater Noida forms the part of Ganga-Yamuna doab; the eastern boundary is marked by Ganga River and the Yamuna River defines the western boundary. The area represents an almost monotonous flat plain, dissected by drainage of different orders.

Regionally, the eastern half of Greater Noida forms part of Ganga alluvial plain, whereas it's western part is in close proximity of Hindon and Yamuna rivers, representing a marginal alluvial plain.

Tectonically, the alluvial plain of Ganga basin represents a structural trough (fore-deep) or down wrap of earth crust, the origin of which is correlated to plate tectonics and Himalayan uplift. The area is underlain by Quaternary sediments, with thickness increasing from west to east and also towards the northeast.

2.2 Site Stratigraphy and Groundwater

Geotechnical investigations at the site revealed that the soils at the site are alluvial in nature and primarily consist of alternating strata of silty sand and sandy silt to the maximum explored depth of 30 m.

The top 5 m soil is loose fine sand. This is underlain by low plastic sandy silt to 11 m depth. A thick stratum of medium dense to dense silty fine sand is encountered between 11 and 30 m depths, with an intervening layer of sandy silt (1–5 m thick) between 23–24 m and 26–29 m depths.

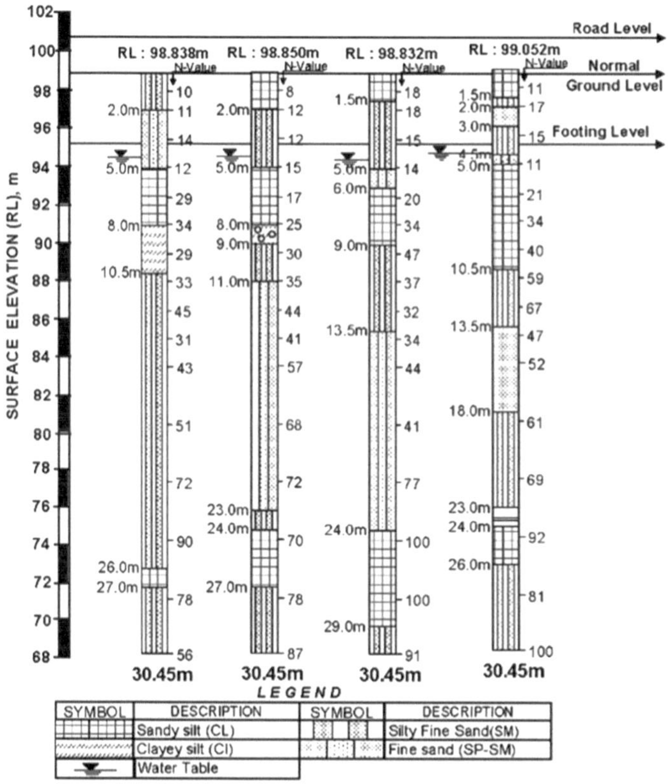

Fig. 1 Typical borehole data

Groundwater was met at 2.0–4.6 m depth below EGL (about RL 95 m) during the period of our field investigations (June, 2012).

A pictorial representation of typical borehole data has been shown on Fig. 1. The stratigraphy encountered at the site is typical of the region.

3 Foundation Analysis

Settlement analysis has been done for a 2 m × 2 m size foundation by the following methods:

1. Classical approach, computing the immediate settlement using elastic theory.
2. Hough's curves (Hough 1969).
3. Finite Element analysis using PLAXIS software.

Table 1 Soil Parameters used in the design

Depth below NGL (m)		Density (kN/m^3)	*N*	Cohesion intercept (kPa)	Angle of internal friction	Modulus of elasticity (MPa)
From	To					
0	2	17.0	15	70	0	5
2	6	17.0	15	0	34	8.7
6	10	18.5	25	90	0	10
10	15	19.0	25	90	35	19
15	24	19.5	30	0	35	28
24	30	20.0	35	120	0	29

4. The computed settlements have been compared with the results of the footing load test to make an assessment of the expected foundation behavior versus the predictions.

4 Conventional Settlement Analysis Methods

The immediate settlement has been computed using the theory of elasticity. Settlement analysis has also been done based on the methods proposed by Hough.

The following soil parameters were selected for the settlement analysis, based on the field and laboratory test results obtained from the borings done on site (Table 1).

The modulus of elasticity values selected for analysis were based on empirical correlations with SPT values, as well as our experience on nearby sites. The weighted average of the modulus of elasticity values within the footing influence zone works out as about 13.3 MPa, which is in very good agreement with the modulus of elasticity value of 17.8 MPa as interpreted from the footing load test data (See Fig. 4). However, the authors strongly recommend the use of pressuremeter and static cone penetration test data for a more realistic determination of the modulus of elasticity profile in situ.

5 Field Method: Footing Load Test

One (1) footing load test (FLT) was performed at the site at 4 m depth (RL 94.8 m) using a 2 m × 2 m size test RCC footing. M-35 grade of concrete was used in order to make the 1 m thick prototype footing. The test procedure was in general accordance with IS: 1888–1982.

Fig. 2 Footing load test in progress

A photograph and schematic sketch showing the test arrangement are presented on Figs. 2 and 3, respectively. A load frame arrangement was used to carry out the test. Eight (8) anchor piles (bored, cast-in situ, 500 mm diameter, 14 m length) were used to provide the reaction system. The top 6 m length of the reaction piles was filled with a mix of sand and cement (designed to approximate the strength characteristics of the surrounding sand) to avoid any interference within the footing influence zone. The anchors piles were concreted and reinforced between 6 and 14 m depths (i.e. 8 m effective length) to provide the uplift resistance.

The footing was loaded by pushing up against the anchor arrangement using two 3000 kN capacity hydraulic jacks. Four dial gauges measured the footing settlement with reference to a stable reference bar. The load was applied in 9 small increments of 282.5 kN each; up to a maximum loading intensity of 2545 kN. Considering the loaded area of 4 m^2, this is equivalent to a maximum bearing pressure of 636 kPa exerted on the plate. Each load was held until the time rate of settlement became negligible (less than 0.02 mm per minute).

The settlement of the footing at the maximum test load of 2545 kN was 59 mm. Figure 4 represents the load versus settlement graph obtained from the Footing Load Test.

6 Numerical Simulation

The footing load test was simulated using a commercially available finite element software (PLAXIS 2D).

The coordinate system used is the standard Cartesian coordinate system, where y-axis corresponds to the direction of ground surface and the x-axis corresponds to the horizontal stretch of the ground. For the vertical boundaries at the nodes, the vertical displacement u_y is left free and the horizontal displacement u_h is restrained,

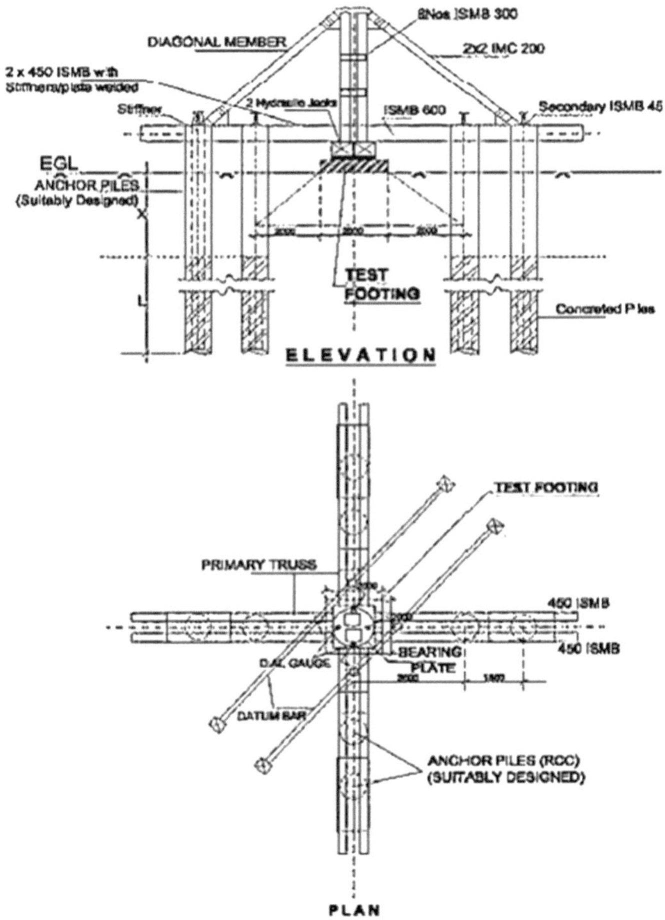

Fig. 3 Schematic sketch of footing load test arrangement

allowing only for a normal stress σ and no shear stress τ. The problem was modeled as an axi-symmetric model considering the footing to be symmetrical about y-axis. Figure 5 illustrates the basic model used.

Mohr–Coulomb constitutive model was used to model the soil layers. The problem is modeled in two phases. In Phase 1, excavation till the foundation level is carried out. In phase 2, the footing is given a prescribed displacement. The corresponding load that causes the prescribed displacement was then calculated.

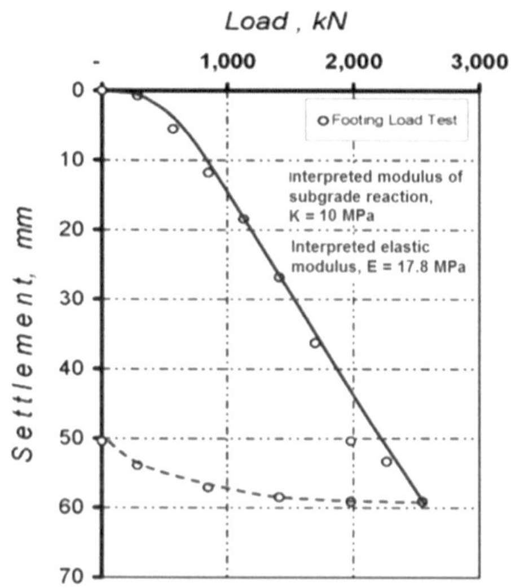

Fig. 4 Footing load test results

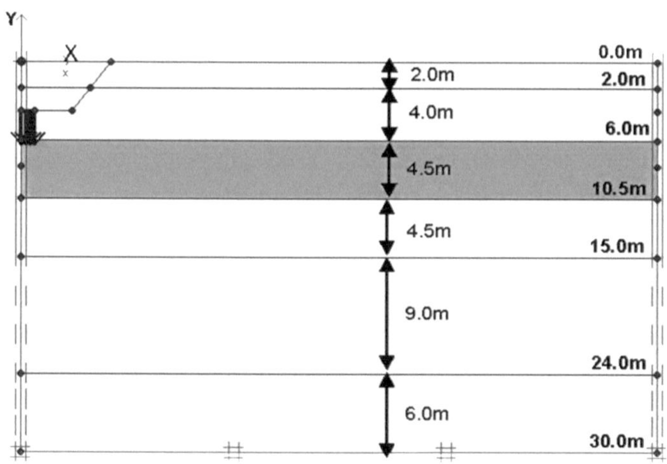

Fig. 5 Dimensions of the basic FEM model

7 Comparison

Figure 6 shows the combined graph of load versus settlement measurements obtained from the four methods discussed above, namely—elasticity theory, Hough's method, PLAXIS 2D modeling and footing load test.

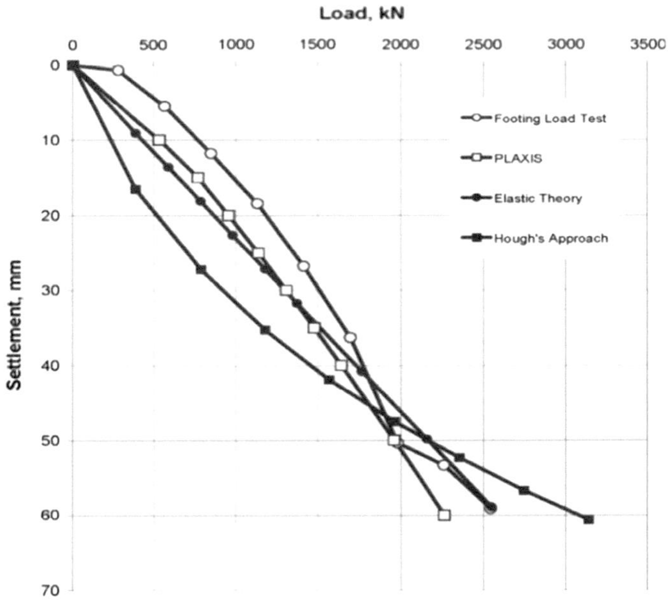

Fig. 6 Comparative load-settlement graph

The results of field test are in good agreement with theoretical predictions, with most theoretical methods over-estimating the settlements. The results differ with each other within tolerable limits. Table 2 shows the settlements corresponding to different loads from the various methods used for analysis. Table 3 summarizes the factors of safety (predicted settlement/actual settlement) for each prediction, with respect to the footing load test results.

Based on the above, it is clear that the predicted settlements are in good agreement with the actual settlements recorded during the footing load test at site. The best prediction was obtained from Elastic Analysis and PLAXIS; whereas Hough's method tends to over-predict settlements at lower load levels.

Table 2 Summary of results

Load (kN)	Bearing pressure (kPa)	Settlement (mm)			
		Elastic theory	Hough's	FEM	Footing load test
500	125	11.5	20.0	10.0	4.5
1000	250	23.0	32.0	21.0	15.0
1500	375	35.0	41.0	35.5	30.0
2000	500	46.0	48.0	50.0	51.0

Table 3 Factor of safety

Load (kN)	Bearing pressure (kPa)	Factor of safety (with respect to FLT)			
		Elastic theory	Hough's	FEM	Footing load test
500	125	2.6	4.4	2.2	1.0
1000	250	1.5	2.1	1.4	1.0
1500	375	1.2	1.4	1.2	1.0
2000	500	0.9	0.9	1.0	1.0

Interestingly, while the predicted settlements are somewhat conservative (FS $\sim$ 2.0–2.5) at lower load levels, the factor of safety reduces at higher loads to as much as 0.9–1.0. This is expected, since analytical tools are based on the assumption that the soil is linear elastic, which is naturally not the case.

8 Conclusions

Conventional geotechnical wisdom is limited by empiricism, heterogeneity and complexity of geo-materials and intrinsic scatter in in situ/laboratory test data. Our foundation settlement predictions are only as good as our test data, correlations used, and judgment exercised. The authors are of the opinion that footing load tests must be carried out where possible, in order to calibrate the theoretical design profiles against actual site performance. Footing load tests are particularly important for projects where the foundations are heavily loaded, or where the foundations are bearing on heterogeneous soils, weathered rock or boundary strata.

The innovative test setup for footing load tests developed by the authors helps reduce the size and cost of the footing load test arrangement. We hope that this will encourage greater adoption of this in situ tool for performance-based design of shallow foundations.

References

Hough, B. K. (1969). *Basic soils engineering*. New York: Ronald Press.

IS: 1888–1982. *Method of load test on soils* (Second Revision). New Delhi: Bureau of Indian Standards.

Experimental Study of Lateral Load Displacement Behavior of Piles in Sand

Amanpreet Kaur, Ashish Hans and Amit Kumar

Abstract Pile foundations are often subjected to lateral forces in addition to vertical loads. The lateral forces acting on piles may arise due to wind pressure, seismic waves, wave action in offshore structures, and lateral earth pressure in earth retaining structures. Lateral displacement at the pile head is an important criterion for a successful design of pile that supports lateral load. In this paper, results observed from model tests conducted on laterally loaded piles embedded in medium dense sand are presented. The effect of variation in pile embedment length to diameter (*L*/*D*) ratio and flexural stiffness on lateral load carrying capacity of single pile was studied by varying pile embedment length and pile diameter. Stainless steel pipes have been used as model piles. The piles were kept free headed with floating tips. Various value of pile length to diameter ratio (*L*/*D* = 10, 15, 20, 25, 30, and 35) were employed in the model testing. Pile group tests were also performed for comparison with single pile. The observed results are presented in the form of lateral load displacement curves and design charts are prepared for lateral load displacement response of piles in cohesionless soils.

Keywords Lateral displacement · Lateral load · Model pile · Pile group

A. Kaur (✉) · A. Hans · A. Kumar
Department of Civil Engineering, Rayat Bahra Institute of Engineering and Nano Technology, Hoshiarpur 146104, Punjab, India
e-mail: preetaman_79@yahoo.com

A. Hans
e-mail: ashishhans6@gmail.com

A. Kumar
e-mail: kumar124.ak@gmail.com

A. I. V. and V. B. Maji (eds.), *Geotechnical Applications*, Lecture Notes in Civil Engineering 13, https://doi.org/10.1007/978-981-13-0368-5_11

1 Introduction

Pile foundations are normally used to transfer loads at greater depths when provision of shallow foundations is not feasible. These foundations are often subjected to lateral loads in addition to axial loads. Lateral loads on the pile foundations may occur due to various reasons in different conditions like earth retaining structures, harbors, and offshore structures, high-rise buildings, etc. For laterally loaded piles, not only the ultimate loads but also the lateral deflection is required to be known to find the working loads for design as well as to check the serviceability limits. Several analytical methods have been developed for predicting the lateral resistance of pile foundations using subgrade reaction approach (Matlock and Reese 1960) or elastic continuum approach (Phanikanth and Choudhury 2013) including finite element method. Also lot of experimental testing has been performed to study the pile behavior in cohesionless soils (Uncuoğlu and Laman 2011). However most of the analytical methods require enormous computation time and resources. In the present work, a parametric study was carried out including model tests on laterally loaded piles in cohesionless soil in which the load displacement behavior of model piles of hollow steel circular sections embedded in medium dense sand was studied. The objectives of this study were to analyze the lateral load displacement behavior of a free head flexible pile in medium dense sand and to study the effect of variation in length to diameter (*L*/*D*) ratio on the lateral load carrying capacity of single piles. Effect of variation in flexural stiffness (EI) of the pile on lateral load capacity was also studied by changing the diameter of pile keeping other parameters constant. Comparison was also made between single pile and pile group of same *L*/*D* ratio to check the efficiency of a 2 × 2 pile group in medium dense sand. Design charts were prepared on basis of observed results.

2 Model Test and Instrumentation

2.1 Model Test

The schematic diagram of the model test arrangement is shown in Fig. 1. The model tests were conducted in a testing tank with internal dimensions of 0.62 m 0.62 m and 0.95 m in depth. The thickness of the soil below the pile tip was kept at least seven times the pile diameter to minimize the influence of base of the testing tank.

Steel pipes of external diameter 19.05 and 25.4 mm having wall thickness 0.5 mm were used as model piles. Bending test of model piles was conducted to find modulus of elasticity, *E* which was found to be 0.87×10^5 N/mm^2. Flexural rigidity (EI) of the piles was 109×10^6 N/mm^2 and 263.8×10^6 N/mm^2 for 19.05 mm and 25.4 mm diameter piles respectively. The dimensions of the model piles were determined by dimensional analysis using Buckingham's Pi theorem

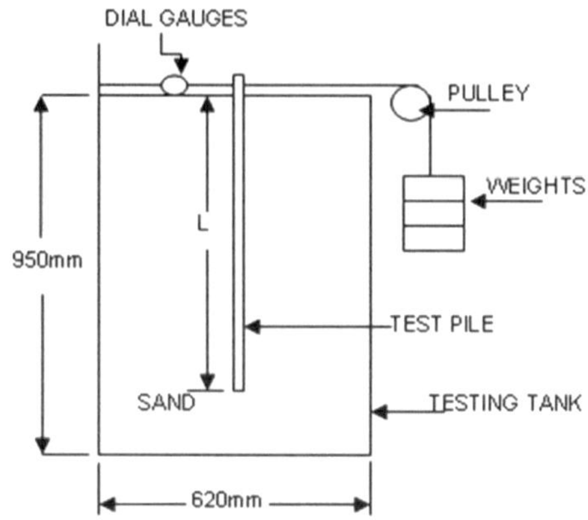

Fig. 1 Schematic diagram of model test arrangement

Table 1 Model test series detail

Pile diameter, *D* (mm)	*L*/*D* ratio	No. of piles
19.05	10, 15, 20, 25, 30, 35	1
19.05	35	4
25.4	10, 15, 20, 25, 30	1

including variables lateral displacement *y*, pile diameter *D*, pile length *L*, moment of inertia *I*, lateral load *P* and modulus of elasticity *E*. To satisfy the similarity between model testing and prototype, respective scaling factors used are $1/N$, $1/N$, $1/N$, $1/N^4$ and $1/N^2(E_p/E_m)$. For instance, steel pile of modulus of elasticity 0.87×10^5 N/mm^2 is subjected to lateral load of 10 N, then the corresponding lateral load on M25 concrete prototype pile will be 4597.7 N when a scale factor of 40 is used. The various series of model piles as used in study are given in Table 1.

Lateral load to the piles was applied through static loading with help of a pulley and string system. Lateral displacement at the top of the piles was measured with help of three magnetic base dial gauges fixed to a rigid plate on one side of the testing tank.

2.2 Sand Bed Preparation

Locally available poorly graded sand was used to prepare model ground in the test program. The index properties and particle size distribution curve of sand used are given in Table 2 and Fig. 2 respectively.

Model ground, i.e., sand bed was prepared by sand raining technique. Sand was spread with the help of a spreader from a height of 400 mm above the current surface of the sand. It helped to achieve a uniform relative density of 40%. To decide the height of free fall of sand, sand was spread from various heights into the cubical boxes of 150 mm size placed at the base of testing tank. After filling the

Table 2 Index properties of sand

Soil property	Value
Specific gravity, G	2.64
Uniformity coefficient, C_u	1.68
Coefficient of curvature, C_c	0.87
Mean particle size, d_{50} (mm)	0.3
Maximum density, γ_{max} (kN/m^3)	20.9
Minimum dry density, γ_{min} (kN/m^3)	15.1
Relative density, D_r (%)	40

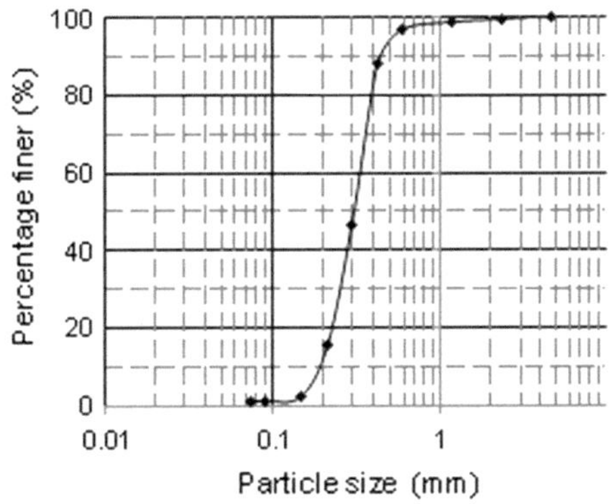

Fig. 2 Particle size distribution curve of sand

boxes, weight of each box was measured and corresponding relative density values were calculated. From the obtained data, height of free fall was decided to obtain required relative density.

Model piles were embedded in the center of testing tank. The verticality of piles was maintained with the help of guiding frame.

3 Results and Discussion

A series of model tests was conducted to investigate the effects of variation in pile parameters on response of laterally loaded piles embedded in medium dense sand. To check the effect of a particular parameter, that parameter was varied while other parameters were kept constant. Following sections consists of detailed discussion on the results observed.

3.1 Effect of Variation in Length of the Pile on Lateral Load Displacement Response

To study the effect of length on lateral load displacement response of piles, pile length to diameter (*L*/*D*) ratio was varied keeping the diameter constant and varying the length of the pile. Figures 3 and 4 present the comparison of load displacement response of piles with different *L*/*D* ratios for piles of diameter 19.05 and 25.4 mm respectively.

It was observed that for piles of 19.05 mm diameter, lateral load carrying capacity increases with increase in *L*/*D* ratio up to *L*/*D* = 25. However, with further increase in slenderness ratio, load deformation curves showed a decline in load carrying capacity of piles. This may be occurred because of medium denseness of sand (D_r = 40%) and free head piles behave like long columns subjected to lateral forces which start deflecting under lateral forces after certain limit of slenderness ratio. On the other hand, for piles of 25.4 mm diameter, the lateral load carrying capacity was found to be increasing with increase in *L*/*D* ratio. This difference in behavior of piles may be due to increased flexural stiffness of the piles as diameter of the piles increased.

To study the effect of variation of length on load deformation behavior of piles, comparison was made on the basis of fixed values of pile head displacement. Figures 5 and 6 show the load carried by piles at 2, 4, and 6 mm lateral displacement. It was observed that for 25.4 mm diameter pile, at 6 mm displacement, maximum increase in load is about 70% with increase in *L*/*D* ratio from 10 to 15. However, for longer piles, effect of variation in *L*/*D* ratio on load carried goes on decreasing. For the same diameter pile at same displacement, increase is only about 4% with increase in *L*/*D* ratio from 25 to 30. Similar trend was noted for 19.05 mm diameter pile with maximum increase of about 75% at 6 mm displacement with change in *L*/*D* ratio from 10 to 15. This remains only about 15% when *L*/*D* ratio changes from 20 to 25. With further increase in length, a decrease was observed in load.

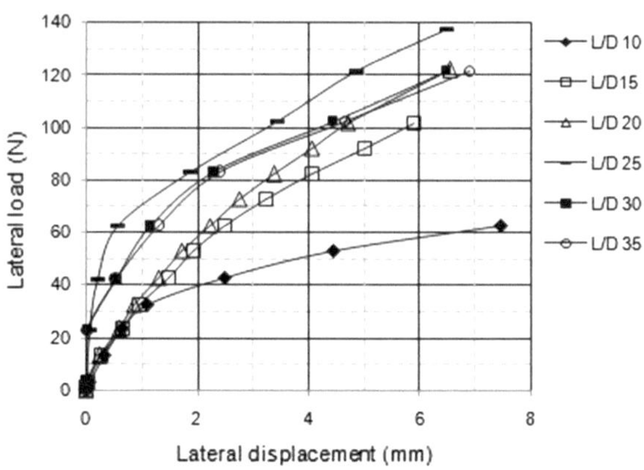

Fig. 3 Load displacement response of single piles of 19.05 mm diameter with varying *L*/*D* ratio

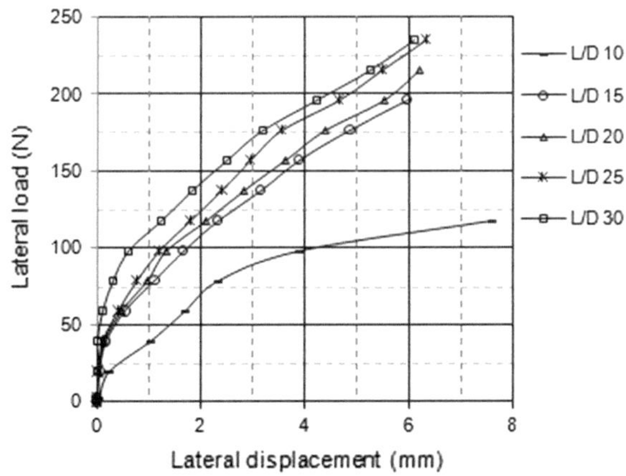

Fig. 4 Load displacement response of single piles of 25.4 mm diameter with varying *L/D* ratio

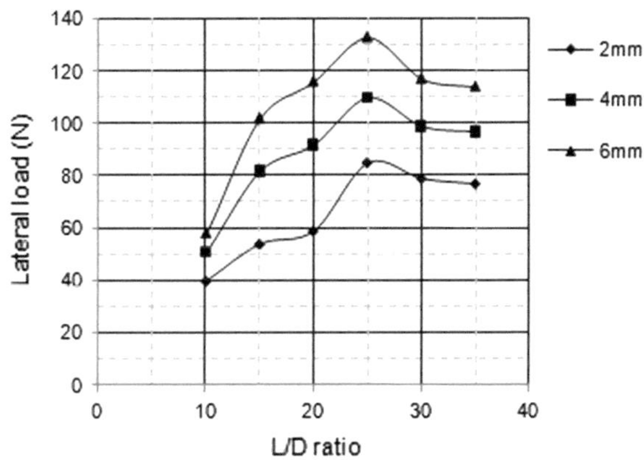

Fig. 5 Lateral load at 2, 4, 6 mm displacement for various *L/D* ratios of 19.05 mm diameter pile

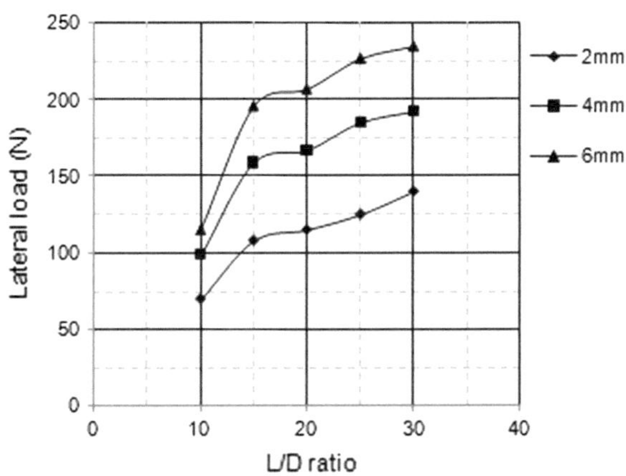

Fig. 6 Lateral load at 2, 4, 6 mm displacement at various *L/D* ratios of 25.4 mm diameter pile

3.2 Effect of Variation in Flexural Stiffness of the Pile on Lateral Load Carrying Capacity

The results observed from tests conducted on piles with different values of diameter were compared to study the effect of change in flexural stiffness on lateral load carrying capacity of single piles. It was observed that an increase in diameter results in increase in flexural stiffness which causes enhancement of lateral load carrying capacity of the piles with same value of *L/D* ratio. Observed values of load carried at 2, 4, and 6 mm displacement for various *L/D* ratios for both 19.05 and 25.4 mm diameter piles were compared to compute percentage increase in load with increase in flexural stiffness. It was noted that for different lengths of piles, percentage increase in load varies from 47 to 100% at 2 mm displacement, from 68 to 94% at 4 mm displacement and from 71 to 98% at 6 mm displacement with increase in flexural stiffness keeping *L/D* ratio same. Minimum values of percentage increase were found to be occurring in case of *L/D* 25. The increase in load in this case was 47% at 2 mm, 68% at 4 mm and 71% at 6 mm displacement. Figure 7 shows the comparison of load carried by 19.05 mm diameter and 25.4 mm diameter pile for various values of *L/D* ratio. More effect of increase in flexural stiffness was observed at 6 mm displacement.

3.3 Comparison of Single Pile and Pile Group

To check the efficiency of pile group, the results obtained from pile group test were compared with that of single pile test of same length pile. The test was performed on free head 2 × 2 pile groups with pile spacing of three times diameter of pile.

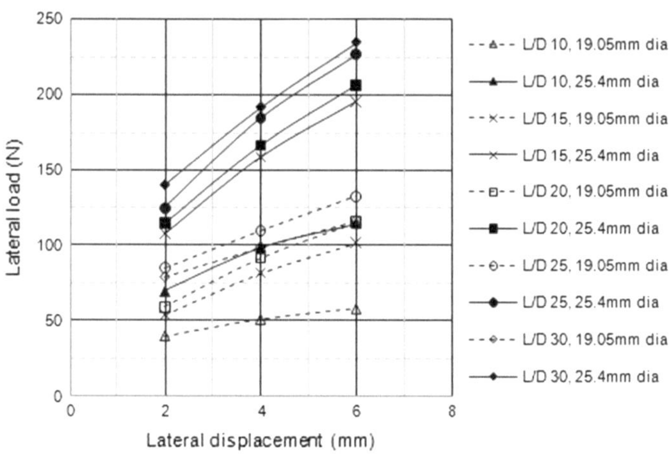

Fig. 7 Lateral load at various displacements for pile diameter 19.05 and 25.4 mm

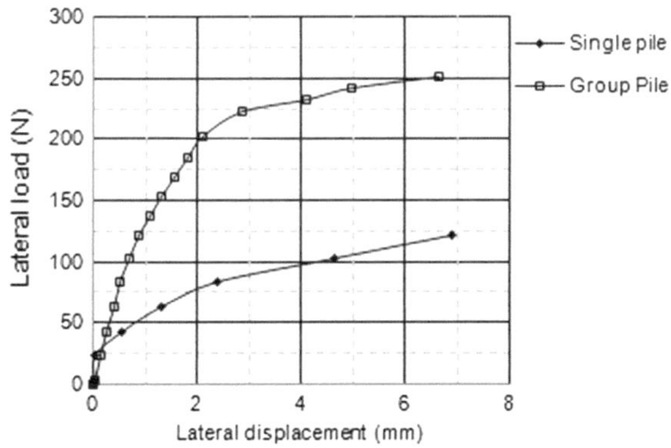

Fig. 8 Lateral load displacement response of single pile and pile group

Figure 8 illustrates the comparison between load displacement response of single pile and pile group of *L*/*D* 35 having pile diameter 19.05 mm.

Group efficiency η at given displacement is expressed in terms of ratio of lateral resistance of a pile group to that of a single pile.

$$\eta = \frac{Q_g}{Q_s n_1 n_2} \times 100,$$

where, Q_g and Q_s are lateral resistance of the pile group and of single pile respectively and n_1 and n_2 are number of rows and columns in pile group.

Group efficiency was found to be 56% for 6 mm displacement and 64% for 2 mm displacement for this case which is in accordance of results by Kim and Yoon (2011) who reported that for 2 × 2 pile group in medium dense sand, the group efficiency varies from 0.52 to 1.0 with lower values for pile spacing of three times pile diameter.

4 Conclusions

From the experimental test series conducted, following conclusions were drawn:

- The lateral load carrying capacity increases with increase in length to diameter ratio of the pile up to a particular optimum length, beyond which effect of increase in length on load carrying capacity starts decreasing with further increase in length.
- The lateral load carrying capacity of piles increases with increase in flexural stiffness of pile. Observed percentage increase in load with 2.4 times increase in flexural stiffness is 47–100% at 2 mm displacement, 68–94% at 4 mm displacement and 71–98% at 6 mm displacement for various values of *L*/*D* ratios.
- Efficiency of 2 × 2 pile group in medium dense sand with pile spacing three times pile diameter varies from 56 to 64%.

References

Kim, B. T., & Yoon, G. L. (2011). Laboratory modeling of laterally loaded pile groups in sand. *KSCE Journal of Civil Engineering, 15*(1), 65–75.

Matlock, H., & Reese, L. C. (1960). Generalized solutions for laterally loaded piles. *Journal of Geotechnical Engineering, 86,* 63–91.

Phanikanth, V. S., & Choudhury, D. (2013). Single piles in cohesionless soils under lateral loads using elastic continuum approach. *Indian Geotechnical Journal, 44*(3), 225–233.

Uncuoğlu, E., & Laman, M. (2011). Lateral resistance of a short rigid pile in a two-layer cohesionless soil. *Acta Geotechnica Slovenica, 2,* 19–43.

Effect of Excavation on the Response of Circular Footing in Sandy Soil

G. Sathiya and S. Karthigeyan

Abstract This paper presents the results of numerical studies carried out to investigate the effect of excavation on the adjacent circular footing. To obtain the response of the circular footing, series of numerical analyses were performed on the footing by incorporating varying parameters such as the footing location from an excavation, varying density of soil and depth of excavations. In the numerical analysis, the footing and sheet pile walls were idealized as a linear elastic materials and nonlinear behaviour of soils using a Mohr–Coulomb model. The paper mainly discusses on the effect of excavations to an adjacent circular footing embedded in medium and dense sandy soil. It is observed from the results that the settlement and bending moment of the footing increases with increase in depth of an excavation. The location at which the effect of excavation on the footing is less significant as termed a safe location or critical location of the footing from excavations. The critical location of the footing based on settlement criteria is 1.5 m from an excavation faces in case of the footing in medium sand and 0.5 m in dense sand.

Keywords Footing · Sandy soil · Excavation · Settlement and bending moment

1 Introduction

In urban areas due to heavy traffic and lack of adequate space has forced the civil engineers to build the new infrastructures very closer to an old buildings. Also, to create an additional floor space to meet the increasing space requirements for amenities, parking, and for housing. Hence, it is unavoidable to have an excavation for new construction within closer distances from old structures. This may cause

G. Sathiya (✉) · S. Karthigeyan
Department of Civil Engineering, Anna University, Guindy, Chennai 600025, India
e-mail: sathyaganesan119@gmail.com

S. Karthigeyan
e-mail: mahamaha2001@yahoo.com

A. I. V. and V. B. Maji (eds.), *Geotechnical Applications*, Lecture Notes in Civil Engineering 13, https://doi.org/10.1007/978-981-13-0368-5_12

severe soil movements in ground adjacent to an excavation and soil movements beneath an old footing behind excavations. Dinakar and Prasad (Dinakar and Prasad 2014), Karthigeyan and Samanta (2011), Ramadan et al. (2013) have reported that an excavation will cause both vertical and lateral soil movements but, lateral soil movement is very critical.

The performance of old structures supported by shallow foundations is severely affected due to a deep excavation for new construction adjacent to it. A major problem in this case is to prevent or minimize damages to the footing. Footings are usually designed to support the vertical loads which are applied directly from the super structure. However, old footings may also to carry additional loads from an adjacent excavation for new construction. These additional loads on the footing may induce a displacement, sliding, overturning and rotation. It may lead to distress of the structure or even failure in some cases. Therefore, it is essential to enumerate the effect of excavations on the response of the footing and assessments of the damage risk become an essential part of the planning, design, and construction of footings.

2 Numerical Analysis

The finite element based numerical analyses have been carried out to investigate the response of footings due to the effect of adjacent excavations using PLAXIS3D software. In the numerical analysis, the project geometry is modeled using a top view approach, i.e., model is extended in a vertical direction. A 3-D finite element mesh is generating from this project geometry. The input parameters consisting of soil strata, structures, construction stages, loads and boundary conditions are mainly based on the convenient graphical user interface, which allows a detailed and accurate modeling of major features. In the analysis, the circular footing and sheet pile wall was assumed to be a linear elastic material and the soil was treated as an elasto-plastic material, obeying Mohr–Coulomb failure criterion. The various material properties were used in the analyses is given as follow;

Properties of medium sand: Unit weight (γ_S)—17 kN/m^3, Elastic modulus of soils (E_s)—28000.0 kN/m^2, Poisson's ratio (μ_s)—0.30, Angle of internal friction ($\phi°$)—32°.

Dense sand: Unit weight (γ_S)—18.0 kN/m^3, Elastic modulus of soils (E_S)—50000.0 kN/m^2, Poisson's ratio (μ_s)—0.30 Angle of internal friction ($\phi°$)—36°.

Footing: Shape—Circular, (B_F)—1.5 m, Depth of the footing (D_F)—1 m, Unit weight (γ_F)—25.0 kN/m^3, Young's modulus of the footing (E_F) = 27.0 × 10^6 kN/m^2, Poisson's ratio (μ_F)—0.15.

Sheet Pile Wall: Unit weight—78.5 kN/m^3, Length—10.0 m, Thickness (t)—0.014 m, Elastic modulus—1.8 × 10^{11} kN/m^2.

3 Parametric Study

Parametric studies were carried out after validating the software using the published results reported in Mostafa and Ashraf (2011). Figures 1 and 2 show the schematic definition of the problem and the three-dimensional finite element mesh discretization respectively. A series of three-dimensional finite element analyses were carried out to study the effect of excavation on the behavior of an adjacent circular footing by incorporating various design parameters. The various parameters considered in the analysis that includes the footing location from an excavation, density of soil, and depth of excavation.

The numerical analyses were carried out in eight different phases. In the initial phase, the in situ stresses were calculated by performing a K_0 analysis. In the phase I, the footing installation was simulated in the model. In the phase II, vertical load was applied incrementally on the footing. In the phase III, the sheet pile was inserted and the phase IV to phase VIII, each step involves the removal of 1.0 m of soil layer and continued until the desired depth of an excavation was achieved. During the excavation stages, the vertical load applied in phase II was remains kept as constant.

The vertical displacement, horizontal displacement, and bending moment of the footing with respect to various design parameters were interpreted from numerical results. The main design concern of the footing is whether safe against an adjacent excavation or not. Therefore, numerical results were interpreted to investigate the safe or critical location of the footing from an excavation face by based on the allowable settlement and bending moment criteria of the footing.

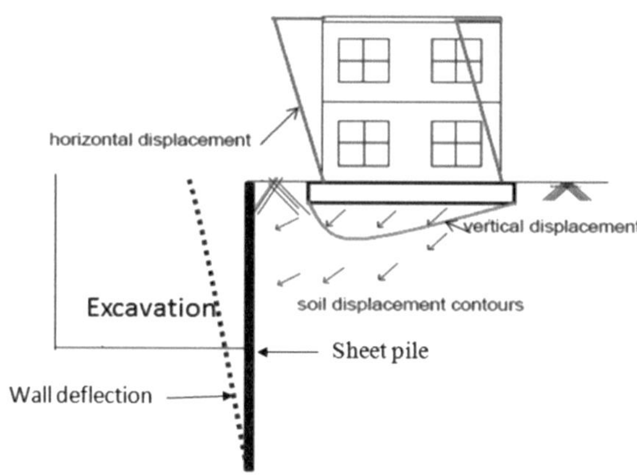

Fig. 1 Schematic definition of the problem

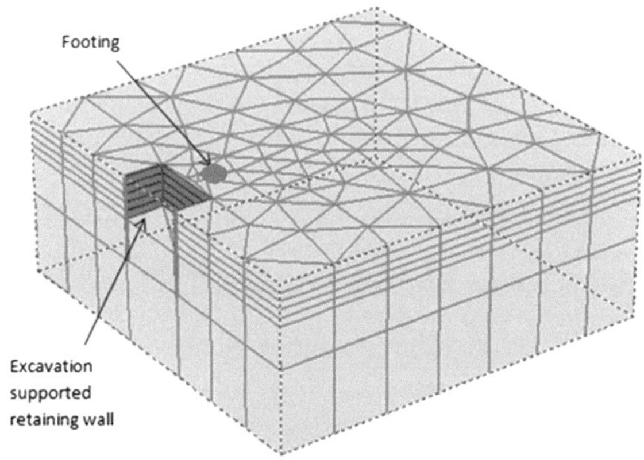

Fig. 2 Three-dimensional finite element mesh discretization

3.1 Depth of Excavation (d_e)

Figures 3, 4 and 5 show the vertical displacement, horizontal displacement, and bending moment of the circular footing having a diameter of 1.5 m, depth of the footing of 1.0 m and dimension of an excavation is 10.0 m × 10.0 m. The results presented in figures are pertaining to a maximum depth of an excavation of 5.0 m and the footing was placed at 0.5 m away from an excavation.

Figures 3 and 4 show the effect of excavations on the footing is increasing with increase in depth of an excavation. It can be seen from the figures that the vertical and horizontal displacement of the footing is increasing slightly up to a 2.0 m depth of excavation and sharply increased beyond this depth. Similar trend can be seen from Fig. 5 on the bending moment of the footing is increasing with increase in depth of excavations. This trend is observed to be common for the footing embedded in both medium and dense sandy soil.

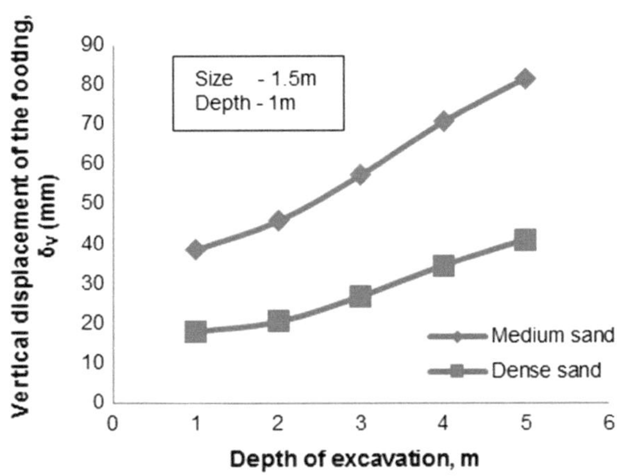

Fig. 3 Vertical displacement of the footing with respect to the depth of excavations

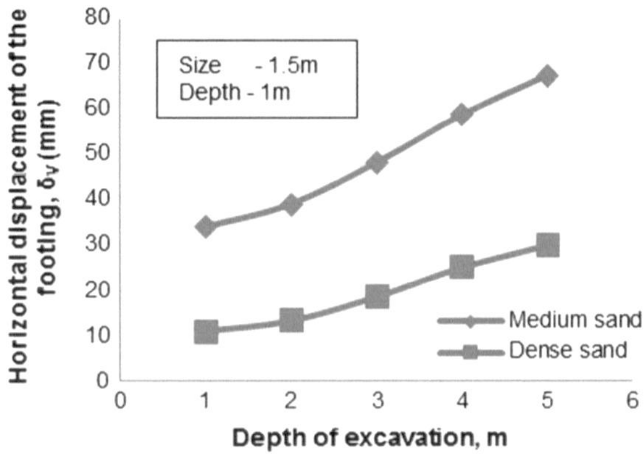

Fig. 4 Horizontal displacement of the footing with respect to the depth of excavations

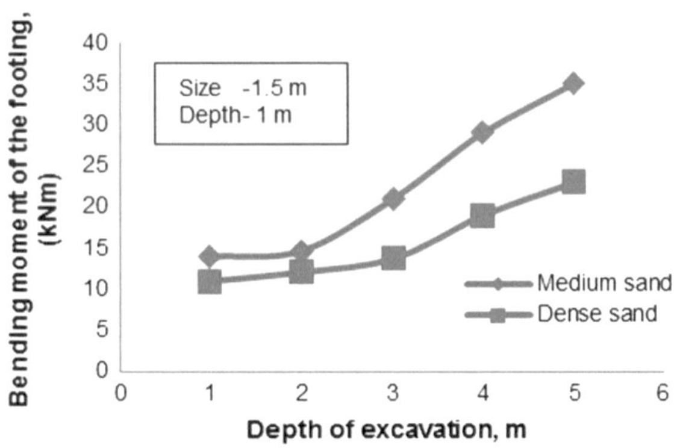

Fig. 5 Bending moment of the footing with respect to the depth of an excavation

Further, it can be seen from Figs. 3, 4 and 5 that the vertical displacement, horizontal displacement and bending moment of the footing are much higher for the footing embedded in the medium sand than in the dense sand. In general, it is observed that the effect of an excavation is very significant in case of a deeper depth of excavations as compared to a shallower depth of excavations.

3.2 *Location of the Footing (X_F) from an Excavation*

Figure 6 shows the view of an excavation adjacent to the footing with various locations away from an excavation. In the normal practices, footings are designed for vertical loads from the superstructure and checked for settlement of the footing to be within allowable limit or as prescribed by the structural design codes. In view of the same, the maximum vertical settlement and maximum bending moment induced on the footing due to the effect of an adjacent excavation were analyzed with respect to the footing located at different distances (X_F) away from an

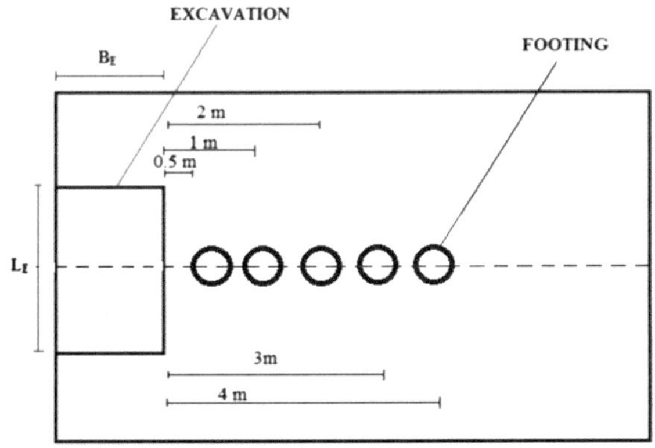

Fig. 6 Plan view of location of the footing from an excavation

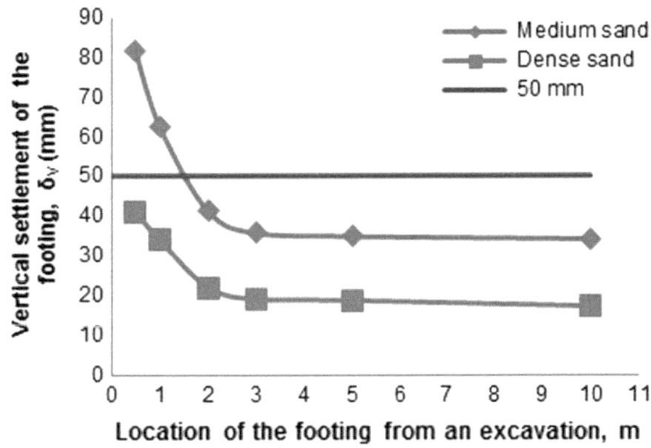

Fig. 7 Vertical settlement of the footing with respect to the location from an excavation

excavation faces. A series of three-dimensional finite element analyses were carried out on the footing located at varying distances away from an excavation. The numerical results were interpreted to plot the curves of vertical settlement and bending moment versus location of the footing from an excavation.

Figures 7 and 8 show the vertical settlement and bending moment of the footing with respect to varying locations of the footing from an excavation faces. The results presented in the figures were pertaining to a circular footing having a diameter of 1.5 m, depth of the footing of 1.0 m and excavation dimension of 10.0 m × 10.0 m with a maximum depth of an excavation is 5.0 m.

It can be observed from Fig. 7 that the vertical settlement of the footing is reducing with increase in distances of the footing from an excavation. It can also be seen that the influence of excavation on the footing is more significant in case of the footing in medium sand than in dense sand.

For the sake of understanding and to estimate safe location of the footing from an adjacent excavation, the maximum vertical settlement of the footing under vertical load is also plotted in the same figure with solid horizontal line. The interest here is mainly to investigate whether the footing designed for vertical load is safe

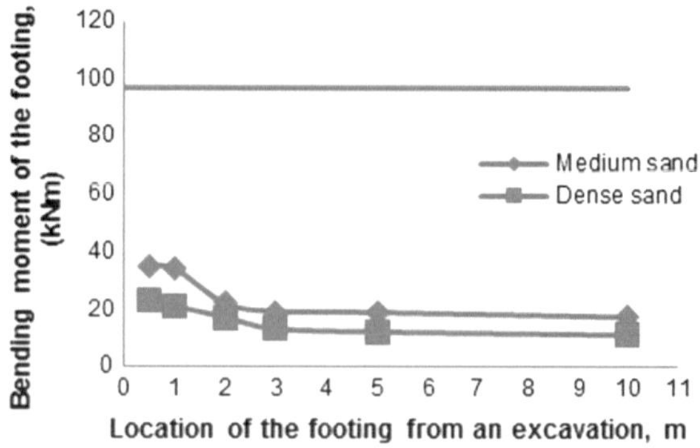

Fig. 8 Bending moment of the footing with respect to the location from an excavation

against an additional load due to the effect of excavations or not and also to arrive a safe location of the footing from the excavation face. A permissible settlement of 50.0 mm is considered for arriving the safe location of the footing in sand as prescribed by IS code 1904: 1986. It can be noticed that the footings which falls in above the horizontal line are unsafe and the footings that falls below the horizontal line are safe, i.e., the safe location of the circular footing in medium sand is 1.5 m and that of in dense sand is 0.5 m based on allowable settlement criteria.

It can be seen from Fig. 8 that the bending moment of the footing gets reducing with increase in location of the footing away from an excavation face. In the figure the straight horizontal line shows the maximum moment capacity of the footing. It is interesting to note that the bending moment of all footings fall below the horizontal line indicates the footing is safe against bending moment criteria.

4 Conclusions

This paper presented the numerical results to investigate the effect of excavations on the adjacent circular footing. The following conclusions can be drawn from the study:

- The effect of excavations to an adjacent circular footing has shown a significant influence on the settlement and bending moment of the footings. However, it depends mainly on the depth of an excavation.
- The vertical displacement, horizontal displacement, and bending moment of the footing are increases slightly up to 2.0 m depth of excavation and increase sharply beyond this depth.
- The effect of excavations on the footing is reducing with increase in location of the footing away from an excavation face.

- Safe location of the circular footing is found to be 1.5 m in medium sand and 0.5 m in dense sand based on allowable settlement criteria. Based on the bending moment criteria, the footing is safe in any location away from the excavation.
- In general, the effect of excavation on the adjacent circular footing is very significant for the footing embedded in medium sand than in dense sand.

References

Dinakar, K. N., & Prasad, S. K. (2014). Behavior of tie back sheet pile wall for deep excavation using plaxis. *International Journal of Research in Engineering and Technology, 3,* 97–103.

IS: 1904. (1986). *Code of practice for design and construction of foundations in soils: General requirements.* New Delhi, India: Bureau of Indian Standards.

Karthigeyan, S., & Samanta, M. (2011) Lateral response of pile under indirect loading due to an adjacent excavations. In *Proceedings of Indian Geotechnical Conference—2011* (pp. 883–886). December 15–17, 2011, Kochi.

Mostafa, E. I., & Ashraf, K. N. (2011). The effect of deep excavation induced lateral soil movements on the behavior of strip footing supported on reinforced sand. *Journal of Advanced Research, 3,* 337–344.

Ramadan, E. H., Ramadan, M., Khashila, M. M., & Kenawi, M. A. (2013) Analysis of piles supporting excavation adjacent to existing buildings. In *Conference of Soil Mechanics and Geotechnical Engineering* (pp. 2835–2838).

Part II
Theme 6: Stability of Earth and Earth Retaining Structures

Analysis of Slope Stability of Fly Ash Stabilized Soil Slope

Tarun Kumar Rajak, Laxmikant Yadu and Sujit Kumar Pal

Abstract Fly ash is a waste product of thermal power plants, can be used as embankment material, stabilization of soft soils, road sub-base constructions, and other geotechnical fields. In this study, fly ash is used as soil stabilizer for an embankment slope. This paper discusses the shear strength parameters of the soil stabilized with fly ash. The soil has been mixed with 10, 20, 30, and 40% fly ash by dry weight for conducting modified Proctor compaction test and direct shear test. The experimental results indicate that the dry density and cohesion value of soil decreases where as the angle of internal friction increases with increase in the percentage of fly ash. The analysis of slope stability of native soil and stabilized soil has been studied by Fast Lagrangian Analysis of Continua (FLAC) slope software. Parametric study has been done to calculate the factor of safety by considering different slope height at constant slope angle under summer and rainy season. The analysis revealed that the slope with native soil was stable up to 12.0 m height under both summer and rainy season. With further increase in height, the slope becomes critically stable and failed under rainy season at 14.0 m height. The addition of fly ash enhances the strength and provides resistance to slope instability under both the conditions up to 14.0 m height. It has been found from the analysis that the factor of safety increases with increase in percentage of fly ash at a particular height. 30% fly ash is obtained as optimum amount as stabilizer for a slope of certain height.

Keywords Fly ash · Slope · Stabilized soil

T. K. Rajak (✉) · L. Yadu
Department of Civil Engineering, National Institute of Technology Raipur,
Raipur 492001, India
e-mail: rajak.tarun18@gmail.com

L. Yadu
e-mail: lkyadu.ce@nitrr.ac.in

S. K. Pal
Department of Civil Engineering, National Institute of Technology Agartala,
Agartala, Tripura 799046, India
e-mail: skpal1963@gmail.com

A. I. V. and V. B. Maji (eds.), *Geotechnical Applications*, Lecture Notes in Civil Engineering 13, https://doi.org/10.1007/978-981-13-0368-5_13

1 Introduction

Most of the thermal power plants in India are facing difficulty for disposal and utilization of fly ash. Total generation of fly ash is about 184.14 million tons from 145 Nos. of thermal power plants of India (Report of central electricity authority, 2014–15). Only 55.69% of fly ash is utilized for the construction work, i.e., brick, cement, concrete production, etc. Remaining fly ash has to be suitably disposed off. Several researchers have worked for its effective utilization as embankment materials, stabilization of soft soils, road sub-base constructions, and other geotechnical fields. Still, there is requirement to explore different fields for the utilization of fly ash and its safe disposal. Notification issued by Ministry of Environment and Forest (MoEF) made compulsory to utilize fly ash for construction of road and embankments, reclamation and compaction of low lying areas within the radius of 100 km from thermal power plants. Fly ash is a non-plastic material, having very negligible cohesion in dry condition, while it shows some cohesion under wet conditions due to its self-cementing properties (Porbaha et al. 2000). Fly ash has friction angle in the range of 29 to 37° (Pandian 2004). Fly ash consists of hollow spheres of silicon, aluminum, iron oxides, and unoxidized carbon. Fly ash is pozzolanic in nature and contains some lime. Brooks et al. (2011) studied the stabilization of expansive soil with fly ash and suggested that mixing of fly ash improves the engineering properties of soil. Few researchers have worked on the stabilization of soil slope using fly ash. Construction of highway embankments with incorporation of fly ash has also been investigated by Martin et al. (1990). Pradhan et al. (2014) studied the optimization of dump slope geometry and dump slope stability with randomly mixed fly ash and suggested that mixing of fly ash in dump materials increases the slope stability.

In the present study, fly ash is used as soil stabilizer for embankment slope of northeast region soil (Agartala, Tripura, India) which is highly prone to landslide. The shear strength parameter of native soil and stabilized soil with different percentages of fly ash (10, 20, 30, and 40%) has been analyzed. An assessment of stability of fly ash stabilized soil slope under summer and rainy season were carried out using FLAC/Slope version 7.0. The parametric study has been done to calculate the factor of safety (FoS) by considering different slope height at constant slope angle under summer and rainy season. Further optimum percentage of fly ash has been suggested for stabilizing the soil slope of certain height.

2 Materials and Experimental Program

Bulk quantities of soil sample were collected from an embankment slope of northeast region, i.e., Agartala, Tripura, India. Fly ash was collected from Kolaghat Thermal Power Station, Kolaghat, West Bengal, India. Particle size analysis of soil and fly ash were carried out in the laboratory as per IS 2720 (Part 4)—1985 and

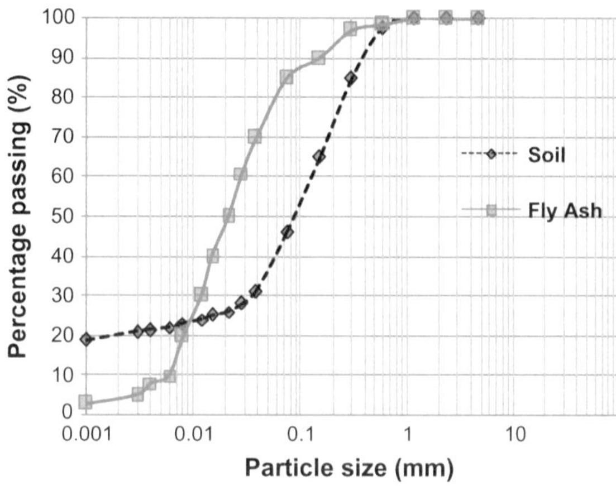

Fig. 1 Particle size distribution curve of soil and fly ash

Table 1 Geotechnical properties of soil and fly ash

Properties	Experimental results	
	Soil	Fly ash
Specific gravity	2.56	2.13
Plasticity	Low plastic	Non-plastic
Maximum dry density (kN/m^3)	19.30	11.75
Optimum moisture content (%)	12.08	29.8

Table 2 List of sample prepared for analysis

Name of sample	Proportion
100S	100% soil
90S + 10F	90% soil + 10% fly ash
80S + 20F	80% soil + 20% fly ash
70S + 30F	70% soil + 30% fly ash
60S + 40F	60% soil + 40% fly ash

particle size distribution curve is shown in Fig. 1. Geotechnical properties of the soil and fly ash were determined which are listed in Table 1. The sample has been prepared by mixing fly ash at 10, 20, 30, and 40% by dry weight of soil for stabilization as shown in Table 2. The modified Proctor compaction test and direct shear test were conducted as per IS: 2720 (Part 7)—1980 and IS: 2720 (Part 13)—1986 respectively for all the samples.

3 Analysis of Slope Stability Using Numerical Modeling

The analysis of slope stability of embankment soil and stabilized soil has been done by Fast Lagrangian Analysis of Continua (FLAC) slope software version 7.0. FLAC/Slope. It is a mini version of FLAC which is specially designed to analyze slope stability using FoS. It is an alternative approach to limit equilibrium analysis for FoS calculation. FLAC/Slope provides full solution of coupled stress/ displacement, equilibrium, and constitutive equations. FoS were calculated for all the samples under different geometric conditions.

4 Geometry of Soil Slope

The slope stability analysis of embankment with native soil and fly ash stabilized soil were carried out with different slope height and constant slope angle (37°). The slope heights were varied from 8.0 to 14.0 m at every 2.0 m interval under both summer and rainy season. Under rainy season the water table were simulated at toe level, i.e., 1.0 and 2.0 m from the toe of the slope. The geometry of embankment slope with native soil of 8.0 m slope height and 37° slope angle at summer season is shown in Fig. 2.

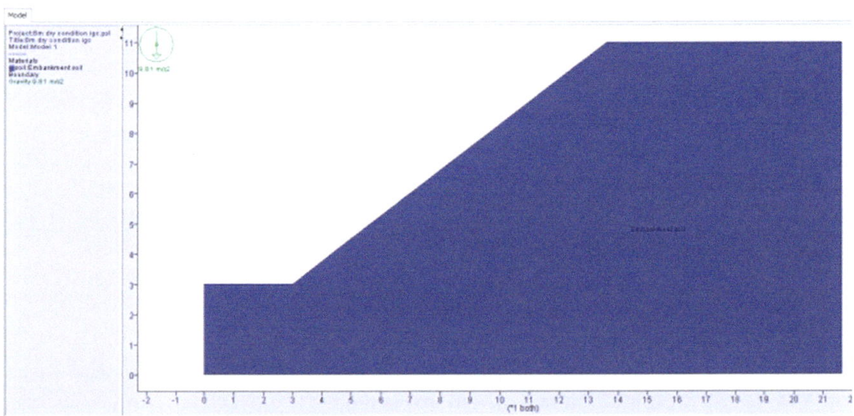

Fig. 2 Geometric model of embankment slope with native soil during summer season

5 Results and Discussions

5.1 Soil Stabilization

5.1.1 Compaction Characteristics

Modified Proctor test were conducted on all five samples as presented in Table 2. The values of Maximum Dry Density (MDD) and Optimum Moisture Content (OMC) were evaluated in the laboratory and compaction curve of all five proportion are shown in Fig. 3. MDD and OMC value of these proportions are tabulated in Table 3. As the percentage of fly ash increases from 0 to 40%, the MDD decreases and OMC increases. This could be due to the lower unit weight and lesser content of sand size particle in the fly ash as compared to the soil sample which causes reduction in dry density (Deb and Pal 2014). Similar trend was also reported by Brooks et al. (2011).

5.1.2 Shear Strength Characteristics

Direct shear tests were conducted on all the samples. Based on normal stress versus shear stress curve the cohesion (c) and angle of internal friction (ϕ) were obtained as presented in Table 3. From Table 3, it can be revealed that, with increase in percentage of fly ash the angle of internal friction increases and cohesion decreases. The increase in angle of internal friction may be due to the increase in silt size particle of the stabilized sample. Similar trend was reported by Sezer et al. (2006).

5.2 Analysis of Slope Stability

The analysis of slope stability of native soil and stabilized soil has been studied by FLAC/Slope software. Parametric study has been done to evaluate the FoS by considering different slope height at constant slope angle (37°) under summer and

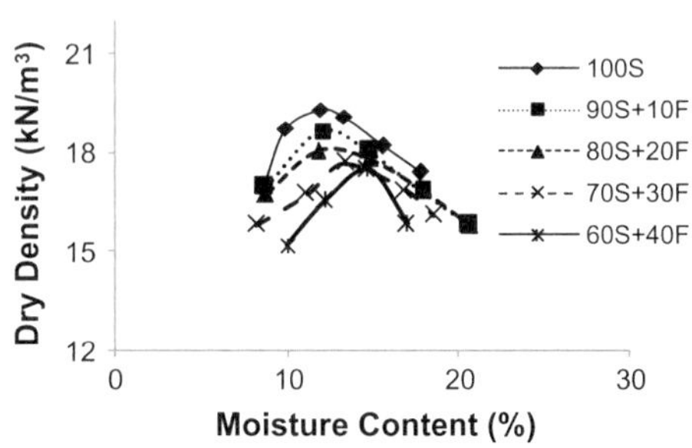

Fig. 3 Compaction curve for all samples

Table 3 Compaction and shear strength characteristics

Sample	MDD (kN/m^3)	OMC (%)	c (kN/m^2)	Φ (°)
100S	19.30	12.08	38.25	29
90S + 10F	18.70	12.50	32.28	32
80S + 20F	18.10	13.05	29.08	35
70S + 30F	17.68	14.25	20.10	37
60S + 40F	17.50	14.70	10.18	39

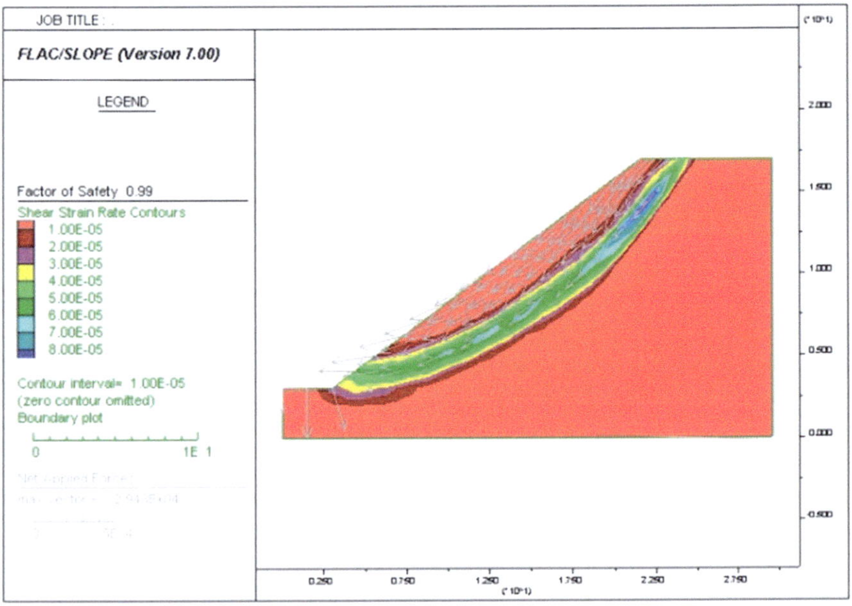

Fig. 4 Shear strain rate contour for native soil slope of 14.0 m height and 2.0 m water level (FoS = 0.99)

rainy season. The slope heights were varied from 8.0 to 14.0 ms at every 2.0 ms interval. The shear strain rate contour for slope of 14.0 m height at 2.0 m water level with native soil and stabilized soil with 30% fly ash are shown in Figs. 4 and 5 respectively.

Similarly FoS has been evaluated for other conditions. The variations in FoS for all the conditions are shows in Fig. 6. From Fig. 6, it can be revealed that the slope with native soil was stable up to 12.0 m height under both summer and rainy season as its FoS is greater than 1. With further increase in height, the slope becomes critically stable and failed under rainy season at 14.0 m height (2.0 m water level). The addition of fly ash enhances the strength and provides resistance to slope instability under both summer and rainy season up to 14.0 m height.

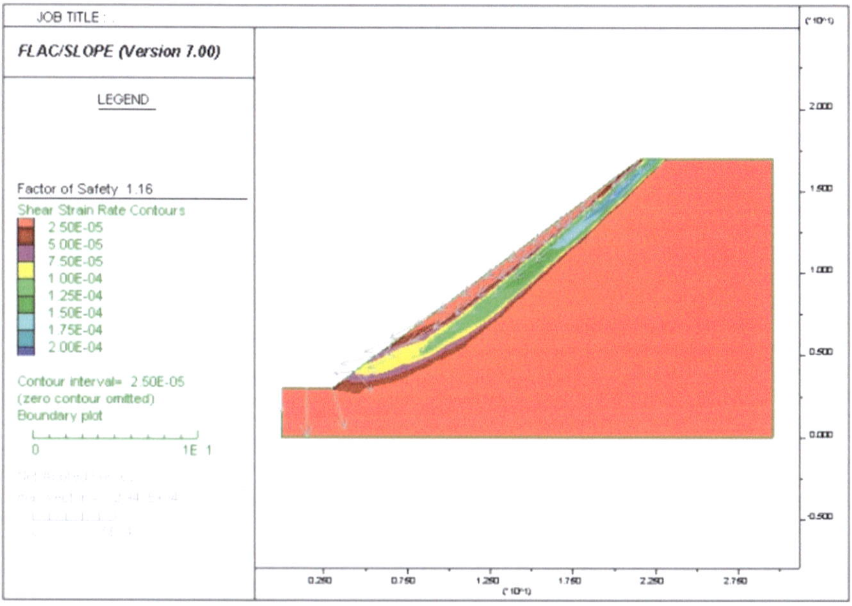

Fig. 5 Shear strain rate contour for stabilized soil with 30% fly ash slope of 14.0 m height and 2.0 m water level (FoS = 1.162)

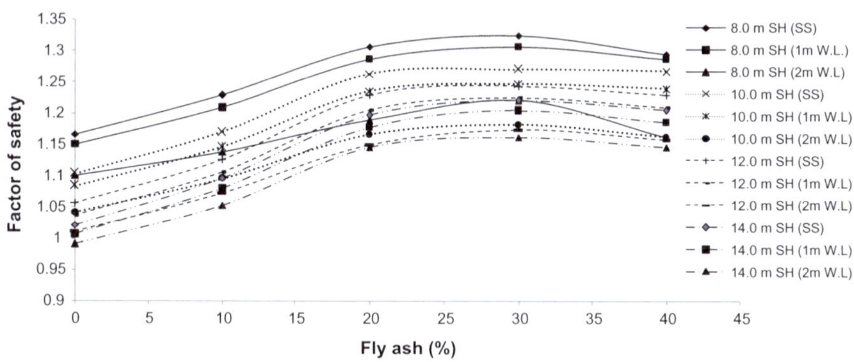

Fig. 6 Variation in FoS with different slope height and fly ash content at summer and rainy season. *SH*—slope height; *SS*—summer season: *W.L.*—water level at rainy season

It has been observed from the analysis that the FoS increases with increase in percentage of fly ash up to 30%, after that slight reduction in FoS is observed. From the present study, 30% fly ash is considered as optimum amount as stabilizer for a slope of certain height.

6 Conclusion

The soil has been stabilized with different proportion of fly ash. Further study has been conducted to analyze the slope stability of fly ash stabilized soil slope. Based on the test results and numerical modeling following conclusions has been made.

- The MDD of the soil decreases and OMC increases with increase in the percentage of fly ash in soil.
- The increase in the percentage of fly ash increases the angle of internal friction and decreases the cohesion of the soil.
- The increase in the percentage of fly ash increases the FoS up to addition of 30% fly ash, after that there is a slight reduction in FoS.
- Numerical modeling infers that by addition of certain amount of fly ash, slope height can be raised even with the presence of water level.
- Based on the value of FoS optimum value of fly ash is 30% for stabilizing the soil.

References

Brooks, R., Udoeyo, F. F., & Takkalapelli, K. V. (2011). Geotechnical properties of problem soils stabilized with fly ash and limestone dust in Philadelphia. *Journal of Materials in Civil Engineering, 23*(5), 711–716.

Deb, T., & Pal, S. K. (2014). Effect of fly ash on geotechnical properties of local soil-fly ash mixed samples. *International Journal of Research in Engineering and Technology (IJERT), 03*(05), 507–516.

Martin, Joseph P., Collins, Robert A., Browning, John S., & Bieh, Francis J. (1990). Properties and use of fly ashes for embankments. *Journal of Energy Engineering., 116*(2), 71–86.

Pandian, N. S. (2004). Fly ash characterization with reference to geotechnical applications. *Journal of the Indian Institute of Science, 84*(6), 189–216.

Porbaha, B. A., Pradhan, T. B. S., & Yamane, N. (2000). Time effect on shear strength and permeability of fly ash. *Journal of Energy Engineering, 126*(1), 15–31.

Pradhan, S. P., Vishal, V., Singh, T. N., & Singh, V. K. (2014). Optimisation of dump slope geometry vis-à-vis flyash utilisation using numerical simulation. *American Journal of Mining and Metallurgy, 2*(1), 1–7.

Sezer, A., Inan, G., Yilmaz, H. R., & Ramyar, K. (2006). Utilization of a very high lime fly ash for improvement of Izmir clay. *Building and Environment, 41*(2), 150–155.

Slope Stability Studies of Excavated Slopes in Lateritic Formations

R. Shivashankar, Biji Chinnamma Thomas, K. T. Krishnanunni and D. Venkat Reddy

Abstract The study area for this paper is coastal Karnataka in India, which has laterites and lateritic soils. The soil stratification in this area mainly consists of lithomargic clay, which is a product of laterization, sandwiched between the hard and porous weathered laterite crust at the top and the hard granite or granitic gneiss underneath. This lithomargic clay, locally called as 'shedi soil' behaves as dispersive soil and is also highly erosive. In the first stage of this study, laboratory erosion studies are conducted by using the hole erosion test apparatus on controlled shedi soil samples. Erosion observed in the HET is accelerated due to slaking irrespective of dispersive nature of the soil. Erosion problems were also dealt with using a stabilizer, calcium lignosulfonate and resulted in high increase in its erosion resistance. In the second stage of this study, slope stability studies of excavated slopes in lateritic formations are conducted considering intensity of rainfall, ponding and seepage, apart from the usual geotechnical parameters. The slopes steeper than 60° are not stable in the case of shedi soil considered here.

Keywords Lithomargic clay · Erosion · Hole erosion test · Slope stability

R. Shivashankar (✉) · B. C. Thomas · K. T. Krishnanunni · D. V. Reddy
Department of Civil Engineering, National Institute of Technology Surathkal, Karnataka 575025, India
e-mail: shivashankar.surathkal@gmail.com

B. C. Thomas
e-mail: bijithomas4@gmail.com

K. T. Krishnanunni
e-mail: nunni1991@gmail.com

D. V. Reddy
e-mail: dvr1952@gmail.com

A. I. V. and V. B. Maji (eds.), *Geotechnical Applications*, Lecture Notes in Civil Engineering 13, https://doi.org/10.1007/978-981-13-0368-5_14

1 Introduction

The present study area considered is the coastal Karnataka in India. The soil stratification in this area, consists of hard (vesicular layer) and highly porous laterite at the surface followed by lithomargic clay. This lithomargic clay is locally called as 'shedi soil'. There are excavated slopes being devoid of natural cover and protection becomes more susceptible to erosion during heavy rainfall.

In the present study hole erosion tests Wan and Fell (2002, 2004), Benahmed and Bonelli (2012) have been conducted to study the erodibility characteristics of the shedi soil. The erosion resistance of controlled shedi soil samples is also studied after treating it with calcium lignosulfonate (Indraratna et al. 2013) that could score over the traditional stabilizers like cement and lime as it does not alter the pH of soil or affect the quality of groundwater below. Slope stability analysis is conducted using the software Plaxis 2D considering the effect of precipitation and ponding at the top of excavated slopes for varying cut slope angles.

2 Experimental Investigation

2.1 Preliminary Investigation

Controlled shedi soil samples were prepared by mixing varying percentage of fines, i.e. 90% fines, 70% fines and 50% fines (passing through 150μ sieve) with river sand (passing through 1.18 mm) and were designated as F1, F2 and F3 respectively. The basic geotechnical properties of these soil samples are listed in the Table 1.

The soil samples treated with optimum amount of 0.6% calcium lignosufonate by dry weight of soil are referred as F1T, F2T and F3T respectively. The optimum amount of lignosulfonate to be added was found from the UCC tests conducted in the laboratory. A series of hole erosion tests were then performed on both treated and untreated samples to measure their erodibility.

Table 1 Properties of controlled shedi soil samples

Parameter	F1	F2	F3
Specific gravity	2.55	2.60	2.63
Plastic limit (%)	34.1	29.4	25.0
Liquid limit (%)	50.2	41.2	33.5
Plasticity index (%)	15.9	11.8	8.5
Optimum moisture content (OMC)	27.1	22.7	18.2
Maximum dry density (g/cc)	1.45	1.60	1.73
Clay size (%)	36.2	28.7	21.2
Silt size (%)	53.8	41.3	28.8
Sand size (%)	10	30	50

2.2 *Hole Erosion Test*

2.2.1 Experimental Setup

The HET assembly comprises of three parts, i.e. the upstream water tank, eroding unit and the downstream water tank. The eroding unit comprises of three chambers: (a) inlet chamber (b) erosion chamber (c) outlet chamber.

The inlet and the outlet chambers are made up of an acrylic tube of 8 cm diameter and 8 cm length and are connected to the upstream and downstream tanks respectively. They are also connected to standpipes to measure the hydraulic gradient. The erosion chamber consists of a MS split mould of 8.3 cm diameter and 16 cm length. The entire eroding unit were kept inside an acrylic tank of size 60 cm × 40 cm × 40 cm which acts as the downstream tank. All the three chambers of the eroding unit are connected together by using steel plates (having 6 mm hole drilled at the centre) which in turn are connected by steel rods.

2.2.2 Specimen Preparation

Test specimens were prepared at maximum dry density and optimum moisture content of the untreated soil. The treated samples were prepared by the addition of 0.6% of lignosulfonate (LS). The LS is first mixed with water and then added to the dry soil. Both the treated and untreated samples were then kept for curing in a desiccator for 7 days.

2.2.3 Procedure

A hole of 6 mm diameter was predrilled along the central longitudinal axis of the soil sample. The sample is then placed into the test apparatus in which water flows through the hole under a constant hydraulic head for up to 45 min. The flow rate is measured at the downstream side of the apparatus and at different time intervals during the test until 45 min from the initiation of the erosion. This may have to be continued more depending on the level of flow rate and the progression of piping. Simultaneously, the pressure drop is also monitored to obtain the hydraulic gradient with time as the downstream head is continuously rising because of accelerating flow rate.

The specimen of eroded soil is then retrieved out of the apparatus and melted paraffin is poured into the eroded hole. After the paraffin solidifies, the specimen is cut out and the paraffin is carefully extracted. This paraffin represents the shape of the final eroded hole. The volume of the paraffin is used for the calculation of the final average radius of the eroded hole and the total eroded mass of dry soil during the test.

2.2.4 Results

The hydraulic shear stress in case of sample F1 has shown to reach a value more than 200 kPa. It can be noted that as the percentage of fines increased from about 50 to 90%, the eroded mass has increased about 10 times under the corresponding test head conditions and compaction level (relative compaction = 100%) (Table 2). This could be due to the presence of high amounts of silt in the fines of shedi soil which is easily eroded even under low hydraulic shear stresses.

In the case of treated specimens, it has been found that all the samples have shown resistance to erosion after treating it with optimum amount of 0.6% of LS. There has been considerable improvement especially in case of F1 sample as F1T sample showed very little erosion (3 g). It is also noted that the lignosulfonate treatment is effective when there are large quantity of fines, and especially clay. In the case of F3 sample there was good binding that occurred between the coarse sands and fine fraction as the fines have filled the spaces between the coarse particles and good interlocking is obtained when compacted at maximum dry density and optimum moisture content. The cumulative masses eroded and final hole diameters of both treated and untreated samples are shown in Table 2.

For both treated and untreated samples the decrease in erosion rate with increasing hydraulic shear stress (Fig. 1) suggests that critical shear stress has not been exceeded or the applied hydraulic shear stresses are still below the critical

Table 2 Cumulative mass eroded and final hole diameter of both treated and untreated samples

Sample	Cumulative mass eroded (g)	Final diameter (mm)
F1	106	25
F2	18	11.3
F3	9.5	8.9
F1T	3	7.2
F2T	8.26	7.5
F3T	6	8

Fig. 1 Erosion rate versus hydraulic shear stress for F1 sample

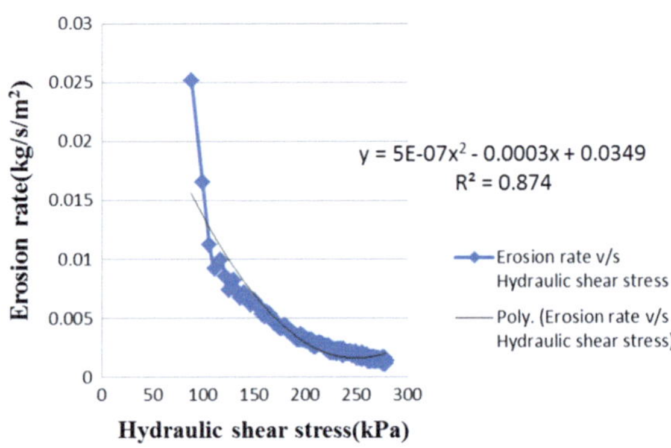

stress necessary to initiate progressive erosion. The cumulative mass eroded with time indicated that erosion has decreased with the decrease in the percentage of fines. However the mechanism of erosion has been complicated and further accelerated by slaking. Slaking is the breakdown of unconfined soil after exposure to air and subsequent immersion in water and no external confining pressure is assumed to act over the soil prior to immersion. One of the important characteristics associated with slaking is that it can occur under low hydraulic shear stresses. The F1 sample showed considerable slaking along with dispersion.

3 Numerical Analysis on Excavated Slopes Considering Precipitation and Ponding

The analysis is conducted to simulate a situation in excavated slope that occurs during the months of heavy rainfall during which the slope becomes fully saturated. The slope comprises of laterite for the top 3 m, underlain by lithomargic clay for a depth of 8 m and a hard rock (granitic gneiss) below it. Ponding for a considerable width and for varying depths (i.e. 1, 2 and 3 m respectively), each for a period of 3 days are considered. The cut slope angles for the excavated slope were varied as 50°, 60° and 70° with the ponding situation described above was separately analysed. Figure 2 shows the geometric profile of the slope.

The kind of calculation used for this study is fully coupled flow deformation analysis (Hamdhan and Schweiger 2011) in which the full interaction between deformations, consolidation and groundwater flow are solved simultaneously in the same phase. The safety analysis is carried out by phi-c reduction method. The slope was analysed as a plane strain model and Mohr–Coulomb model is selected as the material model. Table 3 shows the properties assigned for laterite and shedi soil respectively.

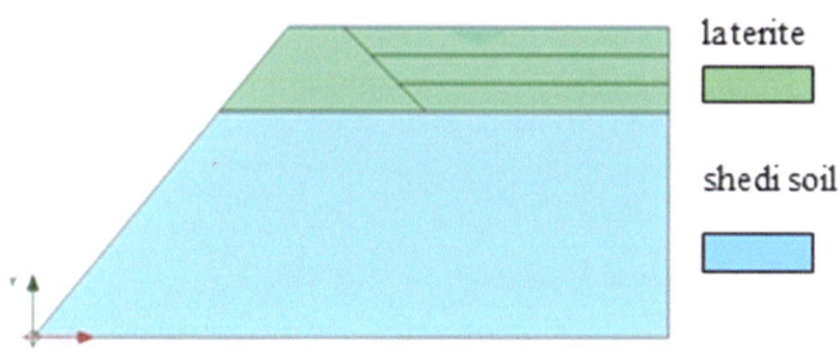

Fig. 2 Geometric profile of the slope

Table 3 Properties of laterite and shedi soil

Properties	Laterite	Shedi soil
Young's modulus	40,000 kPa	10,000 kPa
Poisson's ratio (μ)	0.33	0.33
Cohesion (c)	35 kPa	27 kPa
Angle of internal friction (ϕ)	30°	20°
Material type	Drained	Undrained
Hydraulic conductivity (k)	0.1 m/day	0.00321 m/day

3.1 Defining the Calculation Phases

The calculation phase has been divided into four phases to simulate the situation for a particular geometry with fixed slope angle.

3.1.1 Initial Phase

The initial phase was defined with the excavated slope geometry and assuming the slopes is almost fully saturated during the months of June and July in the district. Initial stresses are generated in this phase using the gravity loading.

3.1.2 First Phase/Precipitation Phase

In this phase, precipitation is brought into picture on a fully saturated slope. To simulate this, infiltration rate was applied at each of the surface boundaries. The effect of rainfall for a period of 3 days was given. On the sloping surface, precipitation is modelled perpendicular to the surface and a value $q\cos\theta$ is applied where θ is the angle of slope and q the infiltration rate on a horizontal surface (i.e. $\theta = 0$). At first, surface ponding effects on the horizontal ground surface is neglected considering the porous and free draining nature of the laterite soils, the water might easily move out through the sloping part rather than accumulating over the surface.

3.1.3 Second Phase (1 m Ponding)

In the second phase, the effect of ponding is introduced along with the precipitation. Ponding was provided for a depth of 1 m at a distance of 2 m from the crest for a period of 3 days.

3.1.4 Third Phase (2 m Ponding)

The procedure was repeated with ponding depth increased to 2 m. The analysis was carried out for the next 3 days.

3.1.5 Fourth Phase (3 m Ponding)

This is the last and final phase in which the ponding depth almost covers whole of the top lateritic layer, i.e. 3 m and the effect on the slope stability was carried out as done in other phases. It is to be noted that all these phases including this final phase has got the precipitation effect.

3.2 Results

The study done in PLAXIS 2D 2016 has given an insight to the effect of precipitation, ponding and also to the slope angle beyond which stability is a concern especially in excavated slopes during the peak rainfall periods. Three excavated slopes with different cut angles of 50°, 60° and 70° were analysed.

In the precipitation phase, from the incremental displacements obtained for the cut slope angles of 50° and 60°, it can be seen that there is much tendency of the failure surface to extend deeply into the weak lithomargic clay layer leading to a progressive failure of the entire excavated slope. Reduction by around 40% in the safety factor are seen in the both cases due to the rainfall infiltration affecting the saturated slope. However when the precipitation was provided on a 70° slope whose initial safety factor came out to be 1.75 in the saturated state failed (i.e. safety factor less than 1) after the precipitation phase (Table 4).

After the introduction of ponding, the safety factors (Table 5) have not altered much from the precipitation phase in both the 50 and 60° slopes. The steeper slope has more chances of failing in the presence of precipitation and higher ponding depths which is expected and observed from the analysis. Also in the case of 50° slope, there was more flow into the inner lithomargic clay layer during the initial ponding phase (ponding depth = 1 m) making it weaker and the failure surface had a tendency to widen creating more displacement resulting in a lesser factor of safety. It is always seen that the critical failure surface moved with advancement in ponding depth.

Table 4 Factor of safety in the precipitation phase

Cut slope angle	Initial FOS when slope is fully soaked	FOS after the precipitation phase
50	2.35	1.43
60	2.02	1.18
70	1.75	<1 (slope failed)

Table 5 Factor of safety for 50° and 60° slope including ponding effect and precipitation

Phase	FOS for 50° slope	FOS for 60° slope
2nd phase	1.340	1.185
3rd phase	1.345	1.155
4th phase	1.370	1.150

4 Conclusions

From the HET conducted, critical shear stress could not be calculated since progressive erosion was not observed in any of the soil samples for the given compaction and head conditions. Erosion rate decreased with increased hydraulic shear stress confirming there is no progressive erosion under the given compaction and head conditions. By this we can conclude that in case of a well compacted shedi soil there will be no progressive erosion. Under higher hydraulic gradient or poorer compaction, progressive erosion may occur. Erosion observed in the HET is accelerated due to slaking irrespective of dispersive nature of soil. Among the untreated samples, Soil F1 showed more erosion than F2 and F3 which occurred mainly due to slaking process. This is mainly due to the presence of higher percentage of clay in F1 sample when compared to other samples. Addition of LS to the untreated samples has resulted in high increase in its erosion resistance and this increase is specifically seen in F1 soil sample.

From slope stability analysis, it is observed that slopes with slopc angle steeper than 60° could pose considerable instability problems and result is failure. The ponding had little effect on the slopes in the analysis considered where factor of safety values did not vary much.

References

Benahmed, N., & Bonelli, S. (2012). Investigating concentrated leak erosion behaviour of cohesive soils by performing hole erosion tests. *European Journal of Environmental and Civil Engineering, 16*(1), 43–58.

Hamdhan, I. N., & Schweiger, F. H. (2011). *Slope stability analysis of unsaturated soil with fully coupled flow-deformation analysis* (pp. 1133–1150). Salzburg, Austria: IAMG.

Indraratna, B., Athukorala, R., & Vinod, J. (2013). Estimating the rate of erosion of a silty sand treated with lignosulfonate. *Journal of Geotechnical and Geoenvironmental Engineering, ASCE, 5*(139), 701–714.

Wan, C. F., Fell, R. (2002). Investigation of internal erosion and piping of soils in embankment dams by the slot erosion test and the hole erosion test. UNICIV Report, no. R-412.

Wan, C. F., & Fell, R. (2004). Laboratory tests on the rate of piping erosion of soils in embankment dams. *Geotechnical Testing Journal, 3*(27), 295–303.

Effect of Varying Geometrical Configuration of Sheet Piles on Exit Gradient and Uplift Pressure

Priyanka Talukdar and Arindam Dey

Abstract Seepage becomes an inevitable concern when dealt with earthen dams, as the impounded water seek paths of least resistance through the dam and its foundation. Seepage, if uncontrolled, can erode soil from the embankment or foundation causing progressive or rapid erosion and piping of the embankment or foundation soils. Out of the several approaches, increasing the length of the flow path by using sheet piles has proved to be effective in seepage reduction. This study examines the effect of using multiple sheet piles of varying length positioned at different distances from the upstream and downstream of a homogeneous earthen dam on the variation of exit gradient, uplift pressure, porewater pressure, and flux along the toe of the dam. The analysis is done using a Finite Element modeling with the aid of SEEP/W module in Geo-Studio 2007 software. The results of the numerical analysis gives an insight about the effect of varying geometrical arrangement of the sheet piles on the reduction of exit gradient, uplift pressure, porewater pressure, and flux along the downstream portion of the dam during the reservoir filling condition. The study highlights the arrangement of the sheet piles to obtain the maximum reduction of exit gradients and porewater pressures during the rise-up condition.

Keywords Earthen dam · Transient analysis · Finite element modeling
Exit gradient · Porewater pressure

P. Talukdar (✉) · A. Dey
Department of Civil Engineering, Indian Institute of Technology Guwahati, Guwahati 781039, India
e-mail: priyanka.talukdar@iitg.ernet.in

A. Dey
e-mail: arindam.dey@iitg.ernet.in

A. I. V. and V. B. Maji (eds.), *Geotechnical Applications*, Lecture Notes in Civil Engineering 13, https://doi.org/10.1007/978-981-13-0368-5_15

1 Introduction

Hydraulic structures such as dams, reservoirs, barrages, used for controlling the flow of water are subjected to seepage and the water seeping underneath the hydraulic structures endanger the safety of the structure resulting in its failure. Therefore, the seepage under hydraulic structures can be considered as one of the most important problems in the hydraulic structures safety. After the construction of dams, the water level rises from their initial position during the first reservoir filling, this condition is termed as "Rise Up"—a state in which the rate of rising water level and the permeability of the associated earthen dam is vital, as during the initial rise up, the porewater pressure changes and volume changes may cause slope failure as well as piping failure. During the rise-up condition the embankment dam is affected by the differential head formed between the upstream and downstream level and the seepage occurring may have detrimental effects at the downstream side of the dam. To prevent such hazards it is necessary to lengthen the seepage path and one of the methods of lengthening the seepage path is to introduce sheet piles as cutoff below the dam to reduce the exit gradient. The reduction of exit gradient plays a vital role because if the exit gradient is greater than the critical gradient for the foundation, piping may occur due to the progressive washing and removal of the fines of the subsoil. The exit gradient being the main design criterion in determining the safety of hydraulic structures against piping, this study is focused on a numerical method to investigate the performance of sheet piles against exit gradient.

Solutions for various problems of seepage under dams with embedded vertical cutoff wall were given using different approaches by (Terzaghi and Peck 1967). Failure mechanism during the reservoir filling condition have been studied by different research workers (Lawton et al. 1992) aiming to provide a better understanding of the phenomenon. Based on the previous studies, an attempt has been to conduct a parametric study to obtain the influence of the length of the sheet piles, their spacing and number of sheet piles under hydraulic structure on the exit gradient, uplift pressure and porewater pressure distribution.

2 Model for the Study

For the present study the model dam is presented with its dimension and relevant soil parameters in Fig. 1.

The governing differential equations (Moharrami et al. 2014) for the plane strain seepage can be expressed as

$$\frac{\partial}{\partial x}\left[K_x \frac{\partial h}{\partial x}\right] + \frac{\partial}{\partial y}\left[K_y \frac{\partial h}{\partial y}\right] + Q = \frac{\partial \theta}{\partial t}, \tag{1}$$

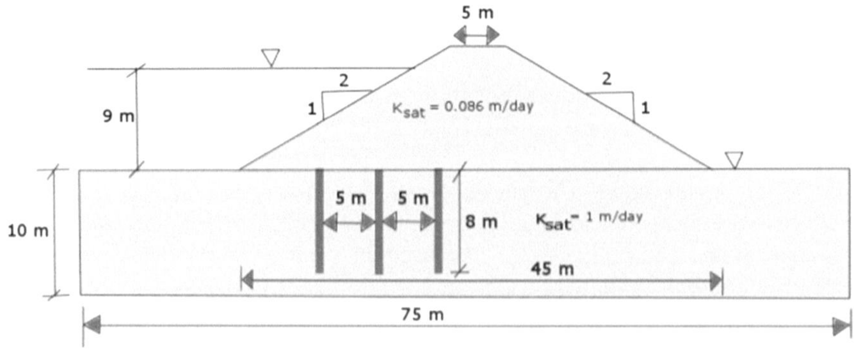

Fig. 1 Model dam for the present study

where, K_x and K_y are the hydraulic conductivity in the horizontal and vertical directions, respectively; h is the total head; θ is the volumetric water content; Q is the applied boundary flux; t is the elapsed time.

Under steady-state conditions, the flux entering and leaving an elemental volume is the same at all times. Therefore the equation reduces to the following form:

$$\frac{\partial}{\partial x}\left[K_x \frac{\partial h}{\partial x}\right] + \frac{\partial}{\partial y}\left[K_y \frac{\partial h}{\partial y}\right] + Q = 0 \quad (2)$$

3 Numerical Analysis

In this study, the analysis was done using the SEEP/W module of a finite element modeling software Geo-Studio. SEEP/W is used for analyzing water seepage and water pressure problems within porous materials for both steady-state and transient conditions in a plane strain configuration. For this study all the three material models, i.e., saturated/unsaturated, saturated only and interface have been used for the embankment, foundation and the sheet piles respectively. The saturated permeability of the foundation and the sheet piles in foundation were taken to be 1 m/day and impermeable, respectively. The permeability field was assumed to be isotropic ($K_x = K_y = K$) and a steady-state seepage analysis to set up the initial conditions followed by a transient analysis during the rise up condition was performed for all the different sheet pile combinations. The rate of rise up in the transient analysis was taken to be 1.9 m/day. Two boundary conditions in terms of total head were specified in the upstream and downstream level in the problem.

4 Results and Discussions

The safety along the downstream end of the hydraulic structures can be ascertained with the decrease of the exit gradient and uplift pressures. The study carried out includes the variation of the length, position and number of sheet piles along the base of the embankment. The embankment width was considered as B, and the distance of the sheet piles from the upstream end was taken to be x thus, the various positions of the sheet piles were explained in $\frac{x}{B}$ ratio. The total width of the embankment in the model was 45 m and the position of the sheet piles i.e., x from the upstream end was varied starting from the toe followed by a spacing of 5 m. The length of the sheet piles considered for this analysis was 8, 4, 2, and 1 m. Moreover, the effect of the variation of the number of sheet piles has been studied.

The embankment when analyzed without any sheet pile the value of the exit gradient, porewater pressure and flux were obtained to be 0.45, 16.23 kPa, −0.73 m^3/days respectively.

4.1 Effect of Length and Position for a Single Sheet Pile

The variation of exit gradient, uplift pressure, porewater pressure and flux with respect to length and position for a single sheet pile has been plotted as shown from Figs. 2, 3 and 4. From the figures, the maximum reduction of all the four parameters near the toe region was obtained for a length of 8 m at a $\frac{x}{B}$ ratio of 0.78. The variation of the uplift pressure and porewater pressure are basically the same (as in Fig. 3) since the uplift pressure below the hydraulic structure is due to the change in the porewater pressure which results in the reduction of effective stress of the soils.

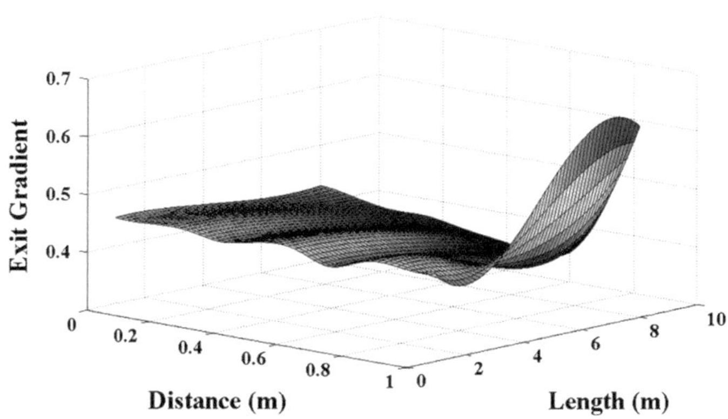

Fig. 2 Variation of exit gradient

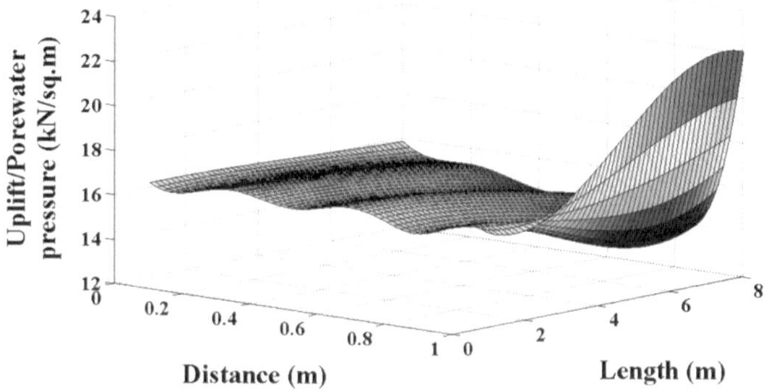

Fig. 3 Variation of uplift and toe porewater pressure

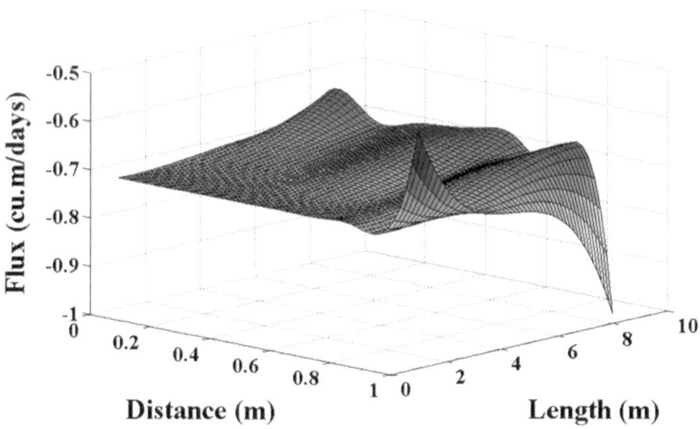

Fig. 4 Variation of flux

4.2 *Effect of Length and Position for Multiple Sheet Piles*

The same analysis was carried out by increasing the number of sheet piles to two and three with a spacing (S) of 5 m between the sheet piles. Increasing the number of sheet piles decreases the value of the parameters, however, the length for which the maximum reduction of the parameters was obtained is the same as that obtained for one sheet pile, i.e., 8 m and the $\frac{x}{B}$ ratio for a combination of two and three sheet piles which results in the maximum reduction of the parameters also starts at 0.78 for the first sheet pile in the combination followed by a spacing of 5 m between the remaining ones. The variation of the parameters for two and three number of sheet piles for varying lengths and positions is shown in Table 1 and it is evident that

Table 1 Variation of parameters for different lengths

Number of sheet piles = 2, $\frac{x}{B}$ ratio values = 0.78, 0.67			
Length (m)	Parameters		
	Exit gradient	Porewater pressure (kPa)	Flux (m^3/days)
8	0.35	12.63	−0.59
4	0.42	14.89	−0.69
2	0.44	15.74	−0.73
1	0.45	15.74	−0.73
$\frac{x}{B}$ ratio values = 0.11, 0			
8	0.41	14.91	−0.67
4	0.44	15.81	−0.72
2	0.45	16.11	−0.73
1	0.45	16.11	−0.73
Number of sheet piles = 3, $\frac{x}{B}$ ratio values = 0.78, 0.67, 0.56			
8	0.32	11.66	−0.54
4	0.41	14.63	−0.68
2	0.40	14.91	−0.70
1	0.40	15.10	−0.70
$\frac{x}{B}$ ratio values = 0.22,0.11,0			
8	0.38	14.32	−0.64
4	0.43	15.89	−0.71
2	0.45	16.41	−0.73
1	0.45	16.39	−0.73

irrespective of the position of the sheet piles, the increase in length of the sheet piles results in the decrease of the parameters.

Similarly the effect of varying the position of the sheet piles has been presented in Table 2.

The result of varying the position of the sheet piles in this analysis shows that the maximum reduction of the parameters takes place when the sheet piles are placed near the downstream end at some distance from the toe, thereafter reducing the $\frac{x}{B}$ values results in an increase of the parameters.

4.3 *Effect of Increasing the Number of Sheet Piles*

The effect of increasing the number of sheet piles on the rise up condition for different parameters has been obtained by plotting its variation with time. The variation of one such parameter near the toe region has been shown below in Fig. 5. The figure clearly indicates that the increasing number of sheet piles results in reduction of exit gradient. The rise up of water results in the increase of the exit gradient till 200 days after which it attains a steady state.

Table 2 Variation of parameters for different positions

	Number of sheet piles = 2, length of the sheet piles = 8 m		
$\frac{x}{B}$ ratio	Parameters		
Position	Exit gradient	Porewater pressure (kPa)	Flux (m^3/days)
1, 0.89	0.62	25.53	−0.85
0.89, 0.78	0.41	18.72	−0.58
0.78, 0.67	0.35	12.63	−0.59
0.67, 0.56	0.36	13.23	−0.59
0.56, 0.44	0.36	13.61	−0.60
0.44, 0.33	0.37	13.48	−0.61
0.33, 0.22	0.38	13.98	−0.63
0.22, 0.11	0.39	14.54	−0.65
0.11, 0	0.41	14.91	−0.67
	Number of Sheet Piles = 3		
1,0.89, 0.78	0.56	22.79	−0.80
0.89, 0.78, 0.67	0.37	17.27	−0.54
0.78, 0.67, 0.56	0.32	11.66	−0.53
0.67, 0.56, 0.44	0.33	12.29	−0.55
0.56, 0.44, 0.33	0.34	12.76	−0.57
0.44, 0.33, 0.22	0.34	12.78	−0.58
0.33, 0.22, 0.11	0.36	13.45	−0.61
0.22, 0.11,0	0.38	14.32	−0.64

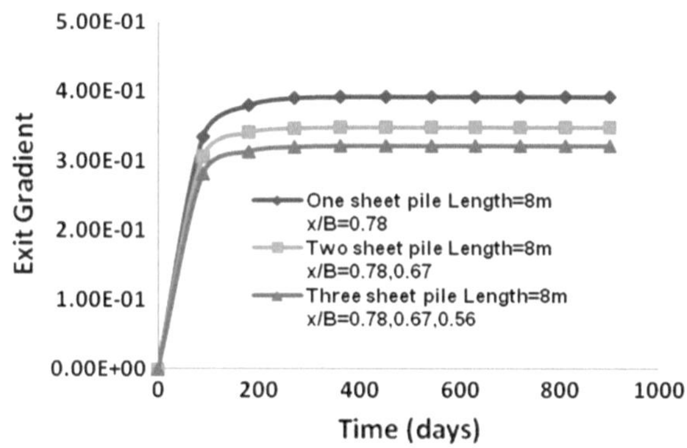

Fig. 5 Variation of exit gradient with time

From the output of the analyses, though it can be stated that the length and position of the sheet piles plays an important role in reducing the exit gradient yet a proper sensitivity analysis will be required to conclude the most influential parameter resulting in the safety of the structure.

5 Conclusions

The following conclusions are drawn from the present study based on the results discussed above:

- Use of multiple sheet piles results in the reduction of exit gradient and uplift pressure as compared to the structure without sheet piles. The most beneficial number for this study is three.
- Increasing the length of the sheet piles results in the reduction of exit gradient and uplift pressure. For this study the most effective length has been 8 m.
- The increase in distance of the sheet piles from the toe results in the reduction of the exit gradient and uplift pressure till a particular point but thereafter further increase in the distance results in the increase of the parameters. The best position of the sheet piles for this study is at $\frac{x}{B}$ of 0.78 for the first sheet pile in the combination of two and three sheet piles.
- The increase in spacing between the sheet piles results in the increase of the exit gradient and uplift pressure. The spacing which resulted in maximum reduction of the parameters was $\frac{B}{9}$ for this study.
- The lengths of the sheet piles were varied equally for this study, however further research can be made if unequal variation of sheet pile lengths can be made to guide the flow resulting in more safety of the hydraulic structures.
- Optimization of the length and position of the sheet piles will be very useful as merely increasing the number of sheet piles does not give the best results.

References

Lawton, E., Fragaszy, R., & Hetherington, M. (1992). Review of wetting-induced collapse in compacted soil. *Journal of Geotechnical Engineering, 118,* 1376–1394.

Moharrami, A., Moradi, G., Bonab, M. H., et al. (2014). Performance of cutoff walls under hydraulic structures against uplift pressure and piping. *Journal of Geotechnical and Geo Engineering, 33,* 95–103.

Terzaghi, K., & Peck, R. B. (1967). *Soil mechanics in engineering practice*. New York: Wiley.

Stability Analysis of a Tailings Dam—A Comparative Study

Rungta Vipul, Raj Dhiraj and S. Mukerjee

Abstract Tailings dams are used to impound waste tailings generated by the mining industry. The tailings material is fine-grinded particles of sizes similar to clay particles however they behave as cohesionless material. The tailings dam under consideration is a zoned dam raised in six stages using the centre line method and having three different types of fill materials. The zones, from upstream to downstream, are consisted of (i) impervious material (ii) compacted tailings and (iii) pervious random fill. An inclined chimney drain and a connecting horizontal filter are provided to keep most of the dam section on the downstream side dry. The main objective of this study is to perform static and pseudo-static analyses by the Finite Element Method, using the Strength Reduction Technique, and comparing the results with those from various Limit Equilibrium Methods, such as Bishop Simplified, Janbu Simplified, Janbu Modified, Spencer, Corps of Engineers-1, Corps of Engineers-2, Lowe Karafiath and Morgenstern-Price for both circular and non-circular failure surfaces. The results are presented in the form of normalised plots considering the following cases: (i) Stagewise (Without Tailings) (ii) Stagewise (With Tailings) (iii) Considering circular failure surface, and (iv) Considering non-circular failure surface.

Keywords Slope stability · Tailings dam · Limit equilibrium method
Finite element analysis · Strength reduction method

R. Vipul · R. Dhiraj (✉) · S. Mukerjee
Department of Earthquake Engineering, IIT Roorkee, Roorkee, India
e-mail: dhirajraj.iitr@gmail.com

R. Vipul
e-mail: vipulrungta91@gmail.com

S. Mukerjee
e-mail: shyamal.mukerjee@gmail.com

A. I. V. and V. B. Maji (eds.), *Geotechnical Applications*, Lecture Notes in Civil Engineering 13, https://doi.org/10.1007/978-981-13-0368-5_16

1 Introduction

To recover metals and minerals, large quantities of rocks are mined, crushed, pulverised and processed. Hence, the mining industries produce enormous quantities of fine rock particles, in sizes ranging from sand to as low as a few microns. A tailings dam is typically an earth-fill embankment dam used to store byproducts of mining operations after separating the valuable fraction from the uneconomic fraction of an ore. This fine-grained waste (byproducts of mining operations) is known as Tailing, which is frequently released during metals and minerals processing (Mukerjee and Bharathi 2016).

With the advent of industrialization, large quantities of tailings are being produced and their disposal has become a significant part of the overall mining operation at most hard rock mining projects. The most common method of disposal of the tailings material is by impounding it in slurry form behind embankment dams, generally known as Tailings Dam. Failure of tailing dam can cause widespread damage to the surrounding environment especially if the tailings are from corrosive or radioactive ores, hence its stability is of prime importance for both static and seismic loading.

Duncan (1996) has compared various methods for slope stability analyses and explored their advantages and disadvantages. In this study, the main objective was to perform static and pseudo-static stability analyses of a typical tailings dam by the Finite Element Method (FEM), using the Strength Reduction Technique, and comparing these results with those obtained from various Limit Equilibrium Methods (LEM). The Limit Equilibrium methods used were Bishop Simplified, Janbu Simplified, Janbu Modified, Spencers, Corps of Engineers-1, Corps of Engineers-2, Lowe Karafiath and Morgenstern-Price. In LEM, both circular and non-circular failure surfaces have been considered, while in FEM failure surfaces were calculated automatically by shear strain contours. Finite element analyses have been carried out using the geotechnical software Phase^2 (2011), while Limit Equilibrium analyses have been carried out using the geotechnical software Slide (2011).

2 The Tailings Dam

In this study, a tailings dam has been considered, as a zoned dam raised in six stages using the centre line method of construction (EQ 2000-9). It had three different types of fill materials. The zone, from upstream to downstream, consisted of (i) Impervious material (ii) Compacted tailings and (iii) Pervious random fill. An inclined chimney drain and a connecting horizontal filter were provided to keep most of the dam section dry on the downstream side. Figure 1 shows a typical cross-section of the tailings dam showing the different constituent materials. The material properties are given in Table 1.

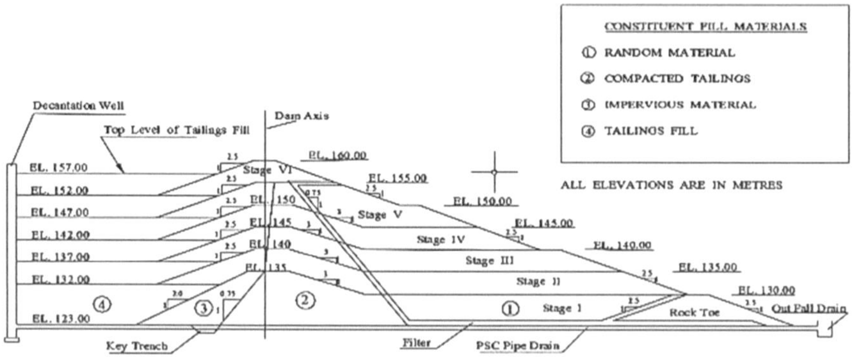

Fig. 1 Cross-section of the tailings dam

Table 1 Material properties for the constituent fill materials

Soil type	c (kPa)	ϕ (°)	G (MPa)	υ	γ_b (kPa)
Pond tailing	0	15	45.64	0.35	19.2
Compacted tailing	0	26	95.39	0.35	19.2
Impervious material	14.72	18	53.56	0.40	19.6
Random material	0	28	190.25	0.30	21.1
Rock toe	0	28	190.25	0.30	21.1
Foundation rock	0	28	217.35	0.20	22.2

Tailings dams are constructed in stages. Initially the embankment is constructed up to the first stage height, 12 m in this case. Then the tailings are deposited, on the upstream side, in the form of a slurry up to the first stage height, 9 m in this case. Now the dam embankment is raised up to the second stage height, 17 m in this case. Then the tailings are deposited up to the second stage height, 14 m in this case. In this manner, the tailings dam is raised in six stages to its full height of 37 m. The tailings are deposited to a height of 34 m at the end of the sixth stage. Details of the construction stages can be seen in Fig. 1.

3 Analyses

The different stages of construction and filling should be reflected in the stability analyses (Rungta 2016). Therefore the analyses were also carried out for the different stages, with and without the tailings fill for that particular stage of construction. This resulted in $6 \times 2 = 12$ sets of analyses for each condition considered. The objective of this study was to perform static and pseudo-static stability analyses of the tailings dam described above by the Finite Element Method (FEM) and comparing these results with those obtained from various Limit Equilibrium Methods (LEM). Both circular and non-circular failure surfaces have been considered in LEM. For the pseudo-static stability analysis horizontal seismic coefficient, $\alpha_h = 0.06$ has been considered.

In FEM analyses the Strength Reduction Technique, where the strengths of the constituent materials are gradually reduced till collapse occurs, has been used, Griffith and Lane (1999) has defined Factor of Safety (FOS) as the factor by which the soil shear strength must be reduced to bring a slope to the verge of failure. Numerically this is represented as,

$$\text{FOS} = \frac{\tau}{\tau_f}, \tag{1}$$

where, τ is the shear strength of the slope material. From Mohr–Coulomb criterion,

$$\tau = c + \sigma_n \tan\phi \tag{2}$$

and τ_f is the shear stress developed on the sliding surface. It is given by,

$$\tau_f = c_f + \sigma_n \tan\phi_f \tag{3}$$

The shear strength parameters c_f and ϕ_f are given by:

$$c_f = \frac{c}{\text{SRF}} \tag{4}$$

$$\phi_f = \tan^{-1}\left(\frac{\tan\phi}{\text{SRF}}\right), \tag{5}$$

where, SRF is a strength reduction factor. The Limit Equilibrium methods considered were Bishop's Simplified, Janbu Simplified, Janbu Modified, Spencers, Corps of Engineers-1, Corps of Engineers-2, Lowe Karafiath and Morgenstern-Price.

4 Results and Discussion

The stage wise variation of FOS obtained from FEM for all considered cases is shown in Fig. 2. In order to compare static and pseudo-static stability analyses results from the FEM, using the strength reduction technique, with those obtained from various LEM, the results were normalised with respect to those obtained from the FEM. These normalised results were then plotted in two groups. In the first group, tailings deposit for the relevant stage of construction was not considered. In the second group, tailings deposit for the relevant stage of construction was considered. Each group was then subdivided into four cases, (i) Static with circular failure surfaces (ii) Static with non-circular failure surfaces (iii) Pseudo-static with circular failure surfaces (iv) Pseudo-static with non-circular failure surfaces. The results and comparisons are shown in Figs. 3, 4, 5 and 6, for the cases where the tailings deposit for the relevant stage is not considered and in Figs. 7, 8, 9 and 10, for the cases where the tailings deposit for the relevant stage is considered.

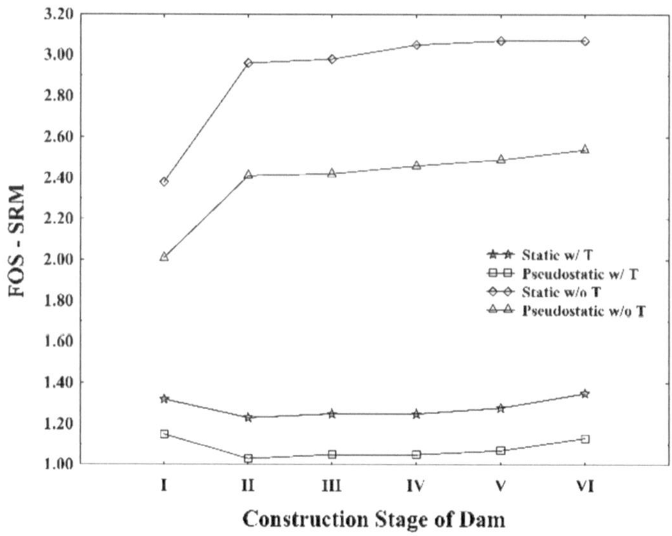

Fig. 2 Stage wise variation of FOS obtained from FEM

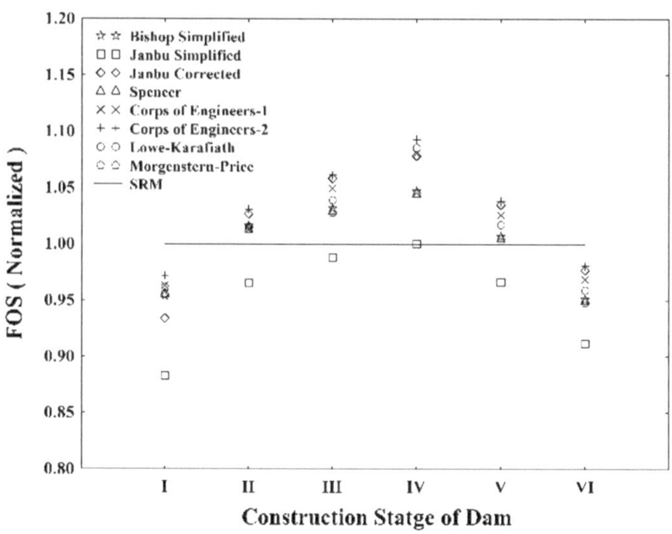

Fig. 3 Static analysis without tailings considering circular failure surface

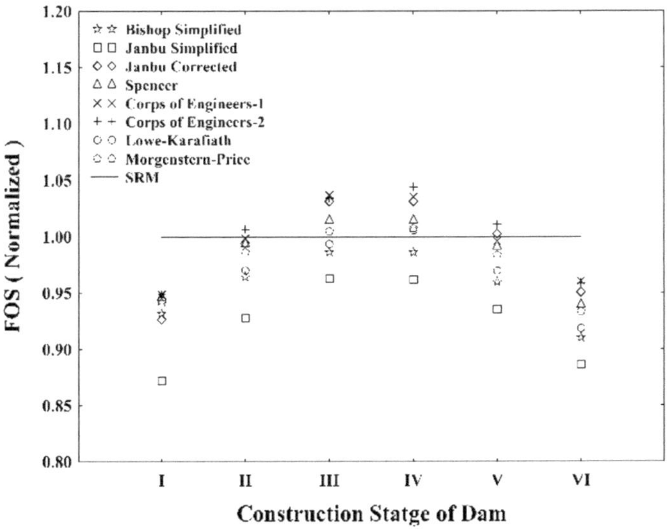

Fig. 4 Static analysis without tailings considering non-circular failure surface

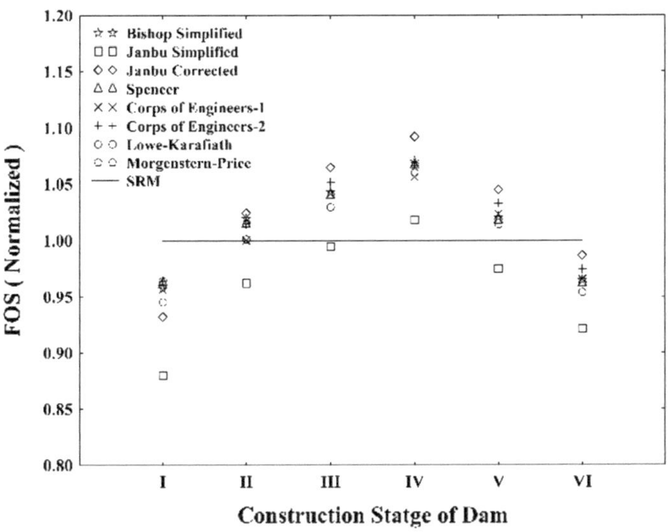

Fig. 5 Pseudo-static analysis without tailings considering circular failure surface

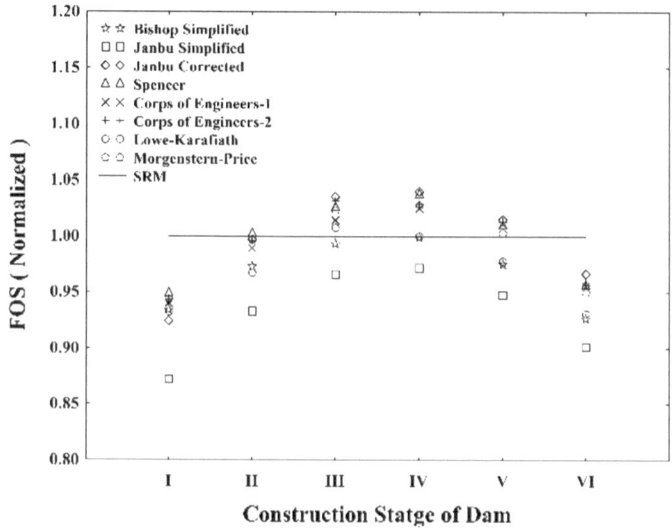

Fig. 6 Pseudo-static analysis without tailings considering non-circular failure surface

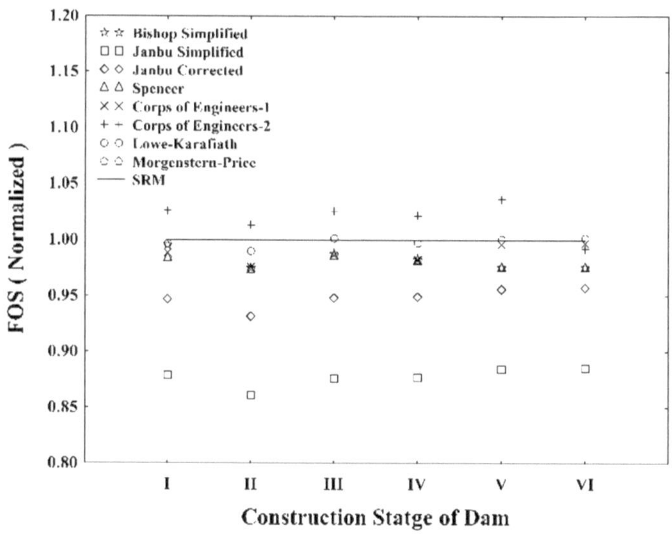

Fig. 7 Static analysis with tailings considering circular failure surface

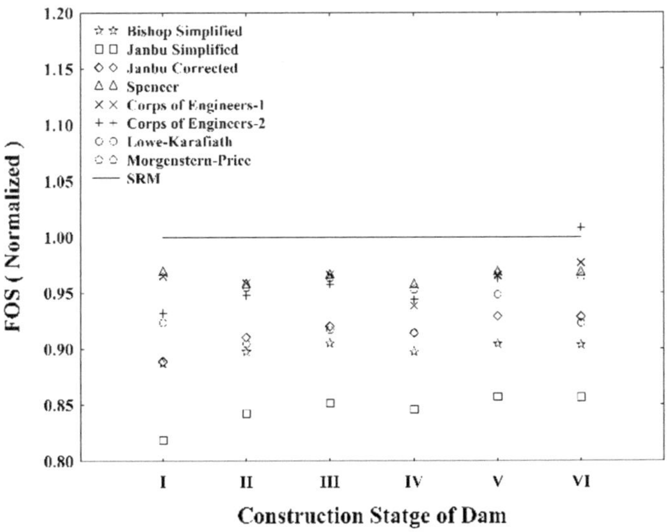

Fig. 8 Static analysis with tailings considering non-circular failure surface

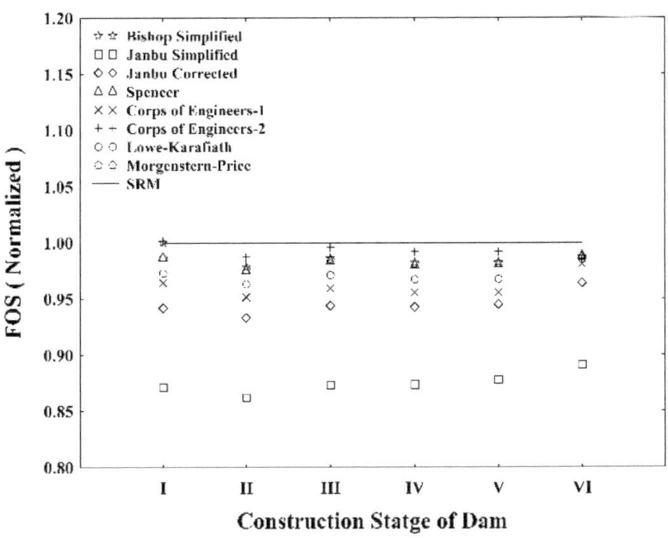

Fig. 9 Pseudo-static analysis with tailings considering circular failure surface

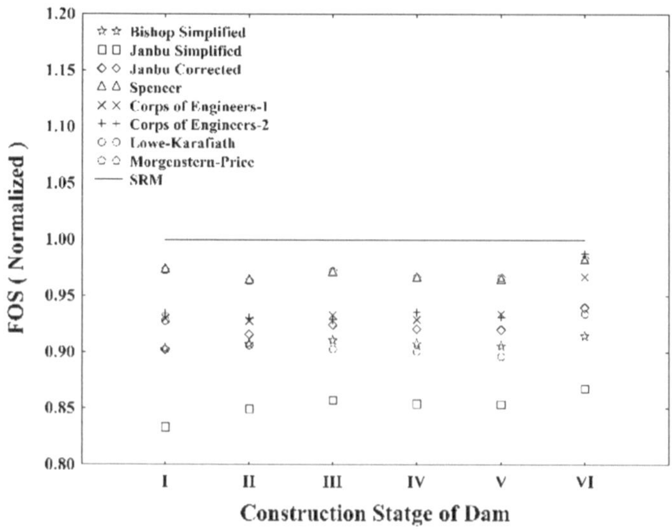

Fig. 10 Pseudo-static analysis with tailings considering non-circular failure surface

5 Conclusion

Among the Limit Equilibrium Methods considered in this study, the simplified method suggested by Janbu is the most conservative, as it gives the lowest Factor of Safety. In the Limit Equilibrium Method, considering a non-circular failure surface results in a lower Factor of Safety value, when compared to that obtained considering a circular failure surface. FOS obtained from the Pseudo-static analysis of the dam with Tailings considering both circular and non-circular failure surface gives a lesser FOS value for all considered LEM than FEM. Factor of Safety values, obtained from the various limit equilibrium methods considered in this study, lie within +10 to −20% of the Factor of Safety value obtained from the Strength Reduction Method.

Acknowledgements The authors are grateful to the Indian Institute of Technology Roorkee, Roorkee, for providing the facilities for conducting this investigation.

References

Duncan, J. M. (1996). State of the art: Limit equilibrium and finite-element analysis of slopes. *Journal of Geotechnical Engineering, 122*(7), 577–596.

EQ. (2000-9). *Seismic safety analysis of tailings dam.* Report No. EQ: 2000-9, Department of Earthquake Engineering, IIT Roorkee.

Griffiths, D. V., & Lane, P. A. (1999). Slope stability analysis by finite elements. *Géotechnique, 49* (3), 387–403.

Mukerjee, S., & Bharathi, M. (2016). *Evaluation of Insitu Shear Modulus for a Tailings Dam, 6th International Conference on Recent Advances in Geotechnical Earthquake Engineering and Soil Dynamics*, Greater Noida, India.

Phase^2. (2011). Version 8. *Finite Element Analysis for Excavations and Slopes*. Rocscience Inc., Toronto, Ontario, Canada.

Rungta, V. (2016). Seismic analysis of tailings dam. *M. Tech. Dissertation*, Department of Earthquake Engineering, IIT Roorkee, Roorkee, India.

Slide. (2011). Version 6. *2D Limit Equilibrium Slope Stability Analysis*, Rocscience Inc., Toronto, Ontario, Canada.

Stability Analyses for Reclamation Bund on Marine Clay

M. Jeevan Reddy, N. Kumar Pitchumani and Aminul Islam

Abstract A container yard is being developed near Mumbai by creating around 90 ha of land through reclamation in sea over soft, compressible marine clay. The thickness of the marine clay varies from 4 to 18 m over the reclamation area underlain by Basaltic rock. The strength of the soft clay varies from very soft to stiff clay, and susceptible to large settlements and shear failure. This stratum cannot support the reclamation fill by its own, hence ground improvement is essential to improve the strength and stiffness properties so that the finished reclamation is able to satisfy the serviceability criteria. Stability analyses of perimeter bund confining the marine reclamation were examined for all probable surfaces including circular, non-circular, and sliding failures as per BS 6031. The analyses show that overfill sections beyond the permanent reclamation cope line are required along with a layer of tension geotextile in order to achieve stability during filling to support the full design load until the cope line. Once the surcharge is removed, the overfill is cut back to the final geometry and revetment is provided along the perimeter bund.

Keywords Marine clay · Ground improvement · PVD · Surcharge
Slope stability

M. J. Reddy (✉) · N. K. Pitchumani
AECOM India Pvt. Ltd., VBC Solitaire, 2nd Floor, Bazullah Road, T Nagar, Chennai 600017, India
e-mail: M.Reddy@aecom.com; Jeevan.iitg@gmail.com

N. K. Pitchumani
e-mail: Nkumar.pitchumani@aecom.com; kpitchumani@gmail.com

A. Islam
ITD Cementation India Limited, National Plastic Building, A-Subhash Road, Vile Parle (East), Mumbai 400057, India
e-mail: Aminul.Islam@itdcem.co.in

A. I. V. and V. B. Maji (eds.), *Geotechnical Applications*, Lecture Notes in Civil Engineering 13, https://doi.org/10.1007/978-981-13-0368-5_17

1 Introduction

In any marine reclamation project, construction of a confinement bund around the area of reclamation is the first activity. The filling is carried out within this confined area either hydraulically, if marine sand is available, or by end on dumping if filling is material from nearby borrow sites. The stability of the slopes of this reclamation bund is crucial as the slightest failure can be destructive in terms of material loss into the sea, loss of time thus resulting in financial losses. Time becomes all the more critical in private projects because the Liquidity damages levied for delay in completion is huge. In this context the reclamation bund slopes need to be carefully analyzed for stability, prior to filling, during filling and post construction.

The project considered in this paper comprises of a proposed container yard with rail and road corridor over marine clay near Mumbai with a finished level of +7 mCD to cater to permanent surcharge loads of 50 and 30 kPa. CD implies Chart Datum and is a local co-ordinate system specific to a port. In this case 0 mCD indicates a level 2.51 m below MSL, i.e. 0 mCD = −2.51 MSL. The ground had to be improved using PVD and preloading until 95% consolidation is achieved.

The section described in this paper pertains to the rail/road portion. The requirement is that the permanent surcharge load of 30 kPa is expected to act within the rail/road portion and traffic load of 10 kPa to act all the way to the cope line, i.e., to the edge of the finished geometry. Hence additional filling beyond the cope line was warranted with ground improvement. This guaranteed that the Factors of Safety as described in Sect. 3.2 are achieved.

The current paper briefly describes the stability analyses for edge slopes of the reclamation during the construction at different stages of loading with fill materials. The edge structure is also analyzed for the permanent edge slopes for long-term serviceability.

2 Geotechnical Data

2.1 Geotechnical Profile, Parameters-Virgin State

A detailed review of the geotechnical investigation data was carried out to identify the variations in the geology across the footprint of the reclamation areas. The sub soil profile over the entire site comprises marine clay with thickness varying from 4 to 18 m followed by basaltic rock.

The shear strength profile, parameters and corresponding design line for the selected sections are presented in Fig. 1 and the corresponding design shear strength parameters for marine clay are presented in Table 1. It is seen that the clay is extremely soft close to the sea bed (+1 mCD) and increases gradually with depth.

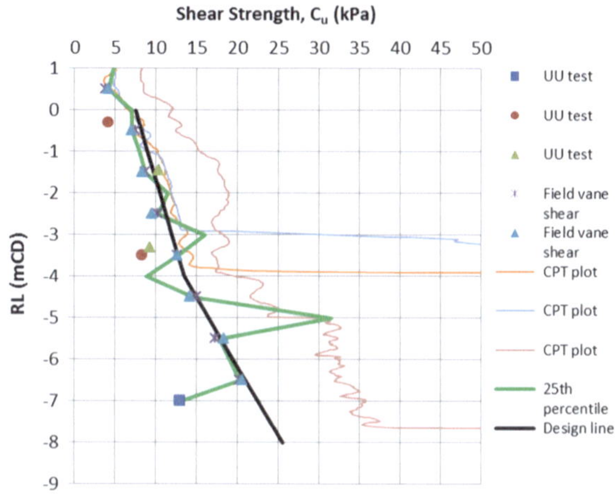

Fig. 1 Plot of shear strength versus RL

Table 1 Geotechnical design parameters (virgin state)

Unit weight (kN/m^3)	Shear strength parameters
	c_u (kPa)
15	7.5 + 1.5 kPa/m from 0.0 to −4.0 mCD 13.5 + 3.0 kPa/m below −4.0 mCD

2.2 Geotechnical Profile, Parameters-After Stage 1 Filling

The filling proceeded with land based filling until the fill reached a level of +5.5 mCD above the maximum water level from which level PVD were installed for ground improvement. This was termed as stage 1 filling. The filling indicated that the top 2–3 m of the soft clay was getting displaced thus resulting in a mud-wave. This displaced mud was getting replaced by the fill material. Confirmatory boreholes were conducted to witness the actual thickness of replacement and the shear strength of the soft clay.

Figure 2 presents the top of marine clay before and after stage 1 filling, as evidenced from the confirmatory boreholes.

From Fig. 2 it is clear that 2.5–3.5 m of the soft clay had displaced and was replaced by the murrum fill.

Figure 3 presents the design shear strength after the replacement.

It is seen that the shear strength of soft clay, 2–3 m below the fill are higher than the virgin shear strength. Beyond this depth the shear strength is close to the virgin shear strength. In the analysis a replacement of the top 2.5 m during the filling process is considered and shear strength to a level of −4 mCD is considered a constant 13.5 kPa. Below −4 mCD the shear strength follows the same trend as given in Fig. 1 or Table 1.

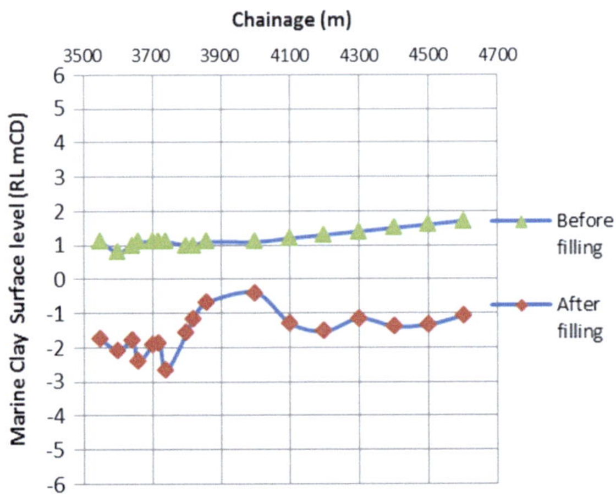

Fig. 2 Top of marine clay: before and after filling

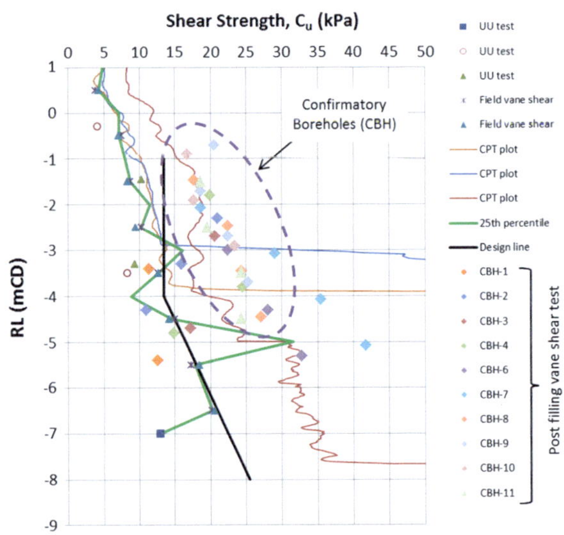

Fig. 3 Design shear strength profile considered in analysis of edge slope

Table 2 Geotechnical design parameters for reclamation fill

Description	Unit weight (kN/m^3)	Shear strength parameters	
		c_u (kPa)	Φ (°)
Stage 1 (below +5.5 mCD)	17	5	30
Stage 2 (from +5.5 mCD to +7.0 mCD)	19	5	32
Stage 3 (above +7.0 mCD)	17	5	30

2.3 Geotechnical Parameters for Reclamation Fill

Table 2 presents the design parameters for the reclamation fill that are used for analyses.

3 Ground Improvement and Reclamation

3.1 Construction Methodology and Sequence

The ground improvement scheme for the reclamation comprises of the installation of pre-fabricated vertical drains (PVD) from stage 1 spaced at 1.0 m c/c in a triangular grid pattern and the application of a surcharge preloading to expedite consolidation and generate strength gain.

The filling is carried out in stages so as to achieve the shear strength required to raise to the next level. The stages and the required gain in shear strength are presented in Table 3. In order to confirm that the required shear strength is achieved, confirmatory vane shear tests were carried out at every stage after the design waiting period.

Table 3 Shear strength required at top of marine clay

S. No.	Fill levels	c_u (kPa)
1	Seabed to +5.5 mCD	13.5
2	From +5.5 mCD to +7.0 mCD	26
3	+7.0 mCD to surcharge top	34
Permanent condition (after 95% consolidation)		
4	At +10.5 mCD	56

Table 4 Factors of safety requirement

Condition	FOS
Temporary	1.3
Permanent static and seismic	1.5 and 1.1

3.2 Design Requirements

The required factors of safety for edge stability are tabulated below. It may be noted that the FOS during temporary stage is as high as 1.3 mainly to avoid excessive deformation resulting in bending of PVD (Table 4).

3.3 Tension Geotextile

It was proposed to use 1 layer of tension geotextile at +5.5 mCD, so as to satisfy the FOS requirements in the temporary and permanent stages. In the stability analysis the following parameters for the geotextile reinforcement are adopted based on the data sheet given by the suppliers.

$$\text{Partial Factor (PF) for Tensile Capacity} = 1.6$$

This is derived as = PF for creep × PF for installation damage × PF for Environmental effect × PF for material = 1.45 × 1.05 × 1.05 × 1.00 and Partial Factor for Pull out = 1.5.

The Geotextile reinforcement at +5.5 mCD level has a length of 30 m and an ultimate tensile capacity of 600 kN/m.

4 Slope Stability Analysis

Static slope stability analyses were carried out using the industry recognized software Slope/W licensed by Geoslope Inc. (version 2012). The Morgenstern-Price Method was used to calculate factors of safety reported in this section. Both circular and non-circular slips were considered.

4.1 Temporary Condition (Construction Stage)

A traffic load of 10 kPa was assumed on the top of each stage loading as per CIRIA C580.

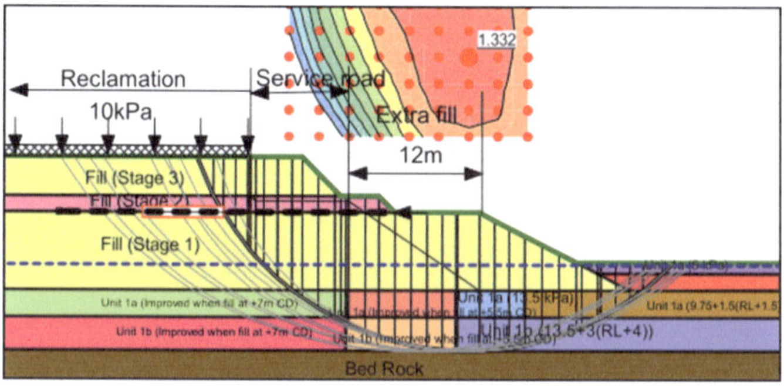

Fig. 4 Slope stability model for +10.5 mCD level fill (circular failure)

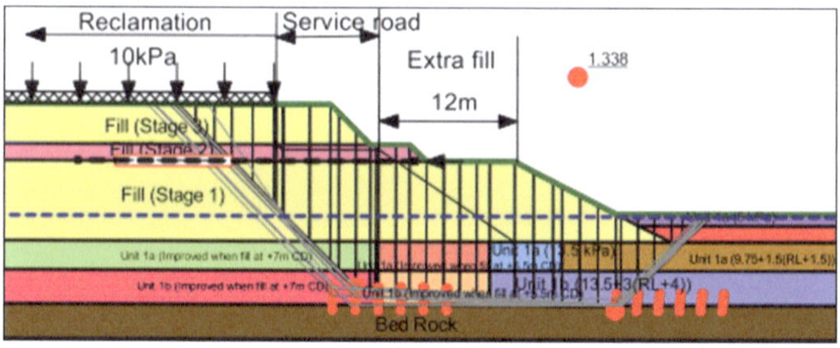

Fig. 5 Slope stability model for +10.5 mCD level fill (sliding failure)

Critical Circular and Sliding block slip surfaces are checked and presented in Figs. 4 and 5 for the final stage, i.e., stage 3. After every stage the increased shear strength of the marine clay, due to the partial consolidation resulting from previous stage was considered in the analyses. The FOS values for the corresponding stages of fill levels are shown in Table 5. In selecting the sliding block surface, care is taken to ensure that the middle segment line is longer than the two end projection segment lines. This is to avoid unrealistic failure mechanisms.

From Table 5 it is clear that the minimum FOS of 1.3 is achieved in stage 3 filling. Hence this stage dictated the geometry of the temporary filling which necessitated 12 m overfilling beyond the cope line.

4.2 Permanent Condition

After the achievement of 95% consolidation, the surcharge was removed to achieve a final level of +7 mCD and the over fill was cut back to the required geometry. The perimeter bund is checked for the long-term static and seismic cases using Slope/W.

Table 5 FOS values for different stages of fill levels

Stages	Fill levels	Critical FOS circular/ non-circular	
1	Bed level to +5.5 mCD	Circular	1.43
		Sliding shallow	1.96
		Sliding deep	1.64
2	+5.5 mCD to +7.0 mCD	Circular	1.53
		Sliding shallow	2.26
		Sliding deep	1.77
3	+7.0mCD to +10.5mCD	Circular	1.33
		Sliding shallow	1.59
		Sliding deep	1.33

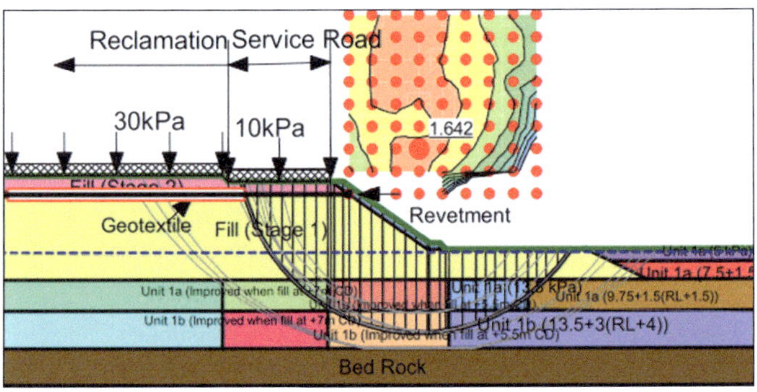

Fig. 6 Slope stability model for permanent static case

For this case a surcharge loading of 30 kPa for main reclamation and 10 kPa for Service Road are considered for static case.

For the seismic case, the horizontal (k_h) and vertical (k_v) seismic coefficients considered in the analysis are derived based on IITK-GSDMA (2007) guidelines for seismic zone 4. These are derived as under

$k_h = 1/3 \times Z \times I \times S$ and $k_v = 0$; where $Z = 0.24$, $I = 1.5$ and $S = 1.2$ and surcharge loading is reduced by 50%.

Figures 6 and 7 present the critical slip surfaces for the static and seismic cases respectively. Table 6 presents the FOS values for both cases which are greater than 1.5 and 1.1 respectively.

Figure 8 shows the geometry that is arrived at from the slope stability analyses, for the three stages of loading with the overfill and one layer of geotextile.

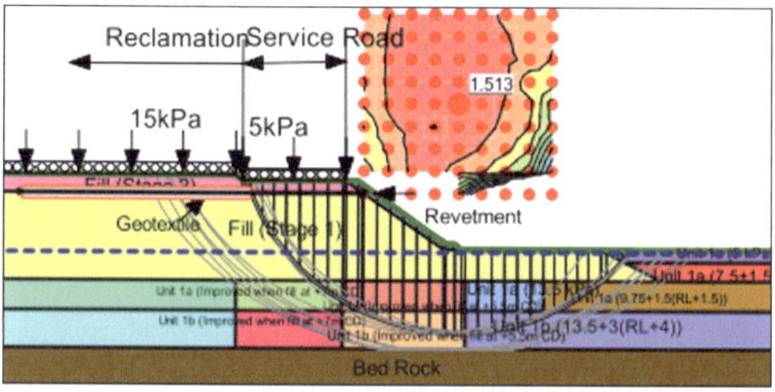

Fig. 7 Slope stability model for permanent seismic case

Table 6 FOS values for permanent static and seismic case

S. No.	Fill levels	Critical FOS circular/ non-circular	
1	+7.0 mCD static	Circular	1.64
		Sliding shallow	2.20
		Sliding deep	1.86
2	+7.0 mCD seismic	Circular	1.51
		Sliding shallow	1.97
		Sliding deep	1.68

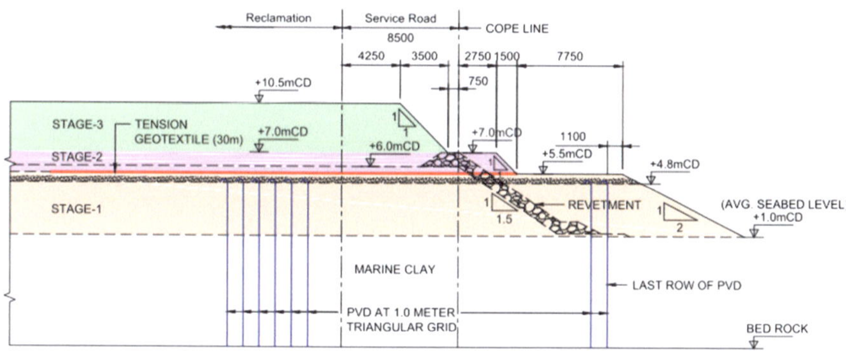

Fig. 8 Geometry arrived from slope stability analyses

5 Summary and Conclusions

This paper presents the slope stability analyses of reclamation bund on soft clay improved with PVD and surcharge loading. The analyses show that the filling has to be carried out in stages so as to achieve the required FOS of 1.3 in the temporary stage. An over fill of 12 m beyond the cope line and a layer of geotextile was necessitated in order to support a surcharge load to +10.5 mCD corresponding to a design load of 30 kPa. At every stage of construction the theoretical gain in shear strength was confirmed with confirmatory vane shear tests before raising to the next stage. This paper demonstrates the importance of carrying out detailed slope stability analyses for marine reclamation.

References

BS: 6031. (2009). *Code of practice for earth works*. British standards.
IITK-GSDMA. (2007). *Guidelines for seismic design of earth dams and embankments*.
Stability modelling with Slope/W Manual. *Geo-Slope International Ltd.*, December 2012.

Slope Stability Analysis of Steep-Reinforced Soil Slopes Using Finite Element Method

Animesh Sharma, P. T. Raju, V. Sreedhar and Hemant Mahiyar

Abstract Recently, use of geosynthetics has gained widespread popularity and has been increasingly used as reinforcing element in several engineering practices of earth retention structures like reinforced earth walls, reinforced slopes, etc. Often the required soil type of requisite characteristic soil properties considered for design are not available locally, hence the available soil type has to be used. With that under consideration, this study deals with the analysis of a steep soil slope embankment reinforced with geogrids under different soil backfill properties. The steep slope is basically designed to widen and elevate the existing natural ground in order to provide road pavement over it. Initial design consideration included conventional limit equilibrium method along with classical earth pressure theories. The design is further modeled and analyzed meticulously according to the staged construction sequence undertaken at site, in a geotechnical finite element analysis software, i.e., PLAXIS. This study mainly investigates the effects of use of different types of soil backfill on stability of slope. Effect of variation of soil parameters, i.e., cohesion and angle of internal friction on steep slope embankments of different heights (6, 12, 18, 24, and 30 m) are studied through this analysis. Since very steep and very high slope embankments were considered for the study, tiered structure, i.e., provision of berm was considered. The effect of width of berm was also considered as a parameter and corresponding effect on the stability of slope was

A. Sharma (✉)
Afcons Infrastructure Ltd., Kolkata East West Metro, Howrah, West Bengal, India
e-mail: animesharma009@yahoo.com

P. T. Raju · V. Sreedhar
Coastland Geotechnics LLP, Hyderabad, Telangana, India
e-mail: raju@coastlandgeo.com

V. Sreedhar
e-mail: vs@coastlandgeo.com

H. Mahiyar
S.G.S.I.T.S., Indore, Madhya Pradesh, India
e-mail: hemantmahiyar01@gmail.com

A. I. V. and V. B. Maji (eds.), *Geotechnical Applications*, Lecture Notes in Civil Engineering 13, https://doi.org/10.1007/978-981-13-0368-5_18

studied. The observed trends of results were further compared with available literatures and previous studies. The results suggested that with increase in soil parameters like cohesion (c') and angle of internal friction (Φ) the factor of safety for the global stability further increased. From the study it was observed that the most optimum soil type was found to be the soil with low cohesion and high angle of internal friction, because upon increasing the cohesion of soil, although increase in global stability was observed, it also led to increase in horizontal displacements and axial forces acting over the geogrids. From the results it was also observed that use of tiered structure or provision of berm increased the factor of safety for global stability marginally and reduced the forces and horizontal displacements significantly on the lower tier. It was concluded that in case of non-availability of recommended granular soil, the available soil fill be utilized only for low height embankments. For high embankments, only recommended granular soils should to be used.

Keywords Plaxis · Slope stabilization · Reinforced soil slope · Geosynthetics Geogrids

1 Introduction

Use of geosynthetics has become a widespread engineering practice and is increasingly used for different purposes such as reinforcement, drainage, separation etc. Present study deals with analysis of steep soil slope reinforced with geogrids using 2-D finite element analysis, i.e., Plaxis 2D. Often the required and characterized design soil backfill, i.e., recommended granular soil is not locally available. Hence through this study, effects of use of different soil parameters as soil backfill on the slope stability were analyzed. Slope of different heights, i.e., 6, 12, 18, 24, and 30 m heights were considered in the study. Because of consideration of steep slopes and very high structures, provision of berm or tiered construction was considered, thereby making the structure a tiered structure. Provision of berm was selected at every 6 m interval. The width of the berm was also considered as a parameter and its effect was studied.

Although the spacing of reinforcement can be varied, for simplicity it has been considered fixed as 0.6 m. The preliminary design analysis for the slope is done using Limit Equilibrium method. The designs are then modeled using Plaxis. The working site conditions were modeled very closely for analysis. The designs have been analyzed according to the under construction site of Central University of Jammu, Samba, Jammu. The ground water table as per site condition was well below the ground level, hence its effect of water table was not considered.

2 Earlier Studies

Since earlier times, limit equilibrium method has been used for analysis of slope stability. Initially slope stability analysis was mainly done using Swedish slip circle method. This method was further upgraded by Bishop which is used even today. Stability of slope is usually expressed as Factor of safety which is basically ratio of resisting forces and driving forces. Later Morgenstern-Price and Janbu further incorporated methods which included non-circular slip analysis. Calculations for slope stability analysis involve complex iterative procedures which are more suitable computationally. Recently many advanced numerical methods are used for slope stability analysis. Amongst the computational methods, finite element analysis is commonly followed. Few earlier studies have been made for slope stability analysis using finite element methods. Naeini et al. (2015) studied and compared the effects of using steel and geogrid as reinforcements using finite element analysis. Ozcelik et al. (2014) compared the results of limit equilibrium analysis and finite element analysis on a 70° slope (Ozcelik et al. 2014). Mirmoradi and Ehrlich (2013) studied the behavior of soil slopes with different facing type's, i.e., concrete blocks and wrap-around facing (Mirmoradi and Ehrlich 2013).

3 Design Analysis and Considerations

The procedure for design of reinforced slopes involves use of conventional limit equilibrium analysis. A circular- or wedge-shaped potential failure pattern is assumed and the factor of safety is determined. In this study, the stiffness of reinforcements was considered using classical theories of earth pressure. Maximum axial force acting over the reinforcement ($T_{\max}$) were calculated using Eq. (1) and the stiffness of reinforcements were provided accordingly.

$$T_{\max} = \mathrm{Ka} \cdot \gamma \cdot H + \mathrm{Ka} \cdot Q, \quad (1)$$

where,

Ka Earth pressure co-efficient
γ Density of soil (kN/m^3)
H Height at the given level (m)
Q Surcharge (kN/m^2)

Partial material reduction factors according to BS: 8006-2010 were considered for the long-term design strength (LTDS) of the reinforcements (Naeini et al. 2015). The embedment lengths of the reinforcements must be sufficient to offer adequate resistance and hence considered according to Jewels Chart recommended by BS-8006-2010 (2010). For ease in construction methodology, uniform allowed spacing between the reinforcements was considered as 0.6 m. The design considerations were further analyzed using finite element technique.

4 Finite Element Modeling and Analysis

Plaxis allows users to define geotechnical problems in a realistic way. With provision of staged construction analysis, stability of structure during and after construction can be analyzed. Present study deals with modeling and analysis of high steep embankments of 70° slope. The heights of the embankments considered are 6, 12, 18, 24, and 30 m. The basic purpose of the embankments is to widen and elevate the existing slopes.

In this study, the geometrical prototype is modeled closely in accordance with the construction methodology and construction sequences. The typical geometry considered for analysis is shown in Fig. 1. In this study, the soil parameters such as cohesion (cohesion) and angle of internal friction (Φ) of soil have been varied and their corresponding effects on the stability of slope embankments of various heights have been studied.

Construction methodology adopted for the construction was bottom-to-up technique which involved use of a facing (welded wire mesh) on the slopes. The primary reinforcements adopted were geogrids (geosynthetics), which are flexible tensile reinforcing elements. The stiffness of the geogrids has been calculated using classic earth pressure theories. With due consideration to the partial material factors, the axial stiffness of the geogrids were modeled and corresponding reduction factor was applied accordingly (BS 2010; Dechasakulsom et al. 2012; Siavoshnia et al. 2010).

The study dealt with the effect of variation in soil backfill parameters, i.e., values of cohesion (c) and angle of internal friction (Φ) on the deformation and displacement of the slope. For stability analysis in Plaxis, phi-c reduction calculation type was used, which reduces the values of cohesion (c) and angle of internal friction (Φ) in steps until the soil body fails. The following equation is used:

$$\sum Msf = \frac{\tan(\tan \Phi\text{input})}{\tan(\tan \Phi\text{reduced})} = \frac{c\text{input}}{c\text{reduced}} \tag{2}$$

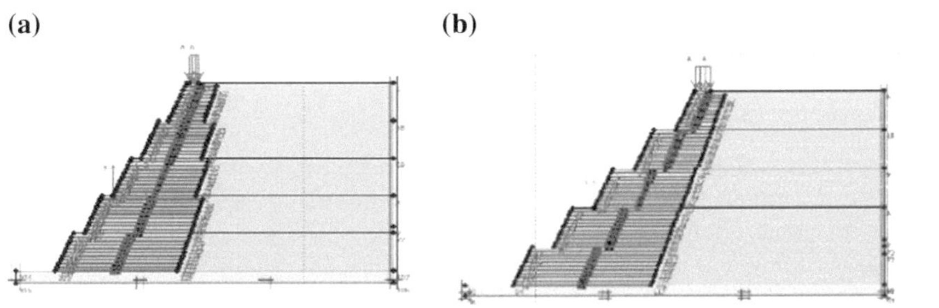

Fig. 1 Typical geometry model, **a** 30 m height with 1.5 m berm width, **b** 30 m height with 3.0 m berm width

The parameters considered for the study are mentioned in Table 1. Although soil properties were varied, default properties considered are mentioned.

Since the slopes considered for analyses were very high and the slope angle very steep, provision of berm or tiered structures was considered. For simplicity, tiered structure was considered with provision berm at every 6 m was considered. For example, 18 m height was modeled as three-tiered structures with every tier being 6 m in height. Similarly 12, 24, and 30 m structures were modeled as four- and five-tiered structures with height of every tier being 6 m. The considered berm width was a parameter and was varied as 1.5 and 3.0 m. Corresponding effect of the variation of berm width was studied.

The Plane strain model of 15-node elements was used in the analysis. The coarseness of mesh was considered as fine. Since water table was not considered for the analysis, the soil fill was modeled using Mohr–Coulomb criterion and plastic staged construction was adopted for analysis. The facing type was modeled using plate. The facing and geogrids were modeled with five-node line. Axial and bending stiffness for the facing were input accordingly. The stiffness and lengths of geogrids were modeled according to the results of analytical methods of limit equilibrium (BS 2010; Dechasakulsom et al. 2012; Siavoshnia et al. 2010). In order to consider long-term assessment of geogrids, reduction factor was considered for soil-geogrid interaction and long-term design strength of geogrids. Traffic loads according to traffic conditions were modeled.

Table 1 Input parameters for finite element modeling

Parameter	Value	Unit
Soil backfill parameters		
Cohesion	0.5	kN/m^2
Angle of internal friction (Φ)	30	Degrees
Unit weight of soil backfill (γ)	20	kN/m^3
Modulus of elasticity (Esoil)	2.00E+04	kN/m^2
Reduction factor	0.7	
Cohesion (c-foundation)	10	kN/m^2
Angle of internal friction, $\Phi_{\text{foundation}}$	30	Degrees
Reinforcements		
Tensile strength (kN/m)	Elastic stiffness	Unit
250	1250	kN/m
200	1000	kN/m
150	750	kN/m
120	600	kN/m
100	500	kN/m
Vertical spacing	0.6	M
Facing		
Axial stiffness	2.00E+06	kN/m
Bending stiffness	2.00E+04	kNm^2/m

5 Results and Discussions

5.1 Axial Forces

The typical trend of the axial forces acting over geogrids obtained through finite element analyses in context with variation in cohesion, friction angle, and berm widths with respect to increase in height are graphically represented in Fig. 2. It was observed that acting horizontal thrust or axial forces increased with respect to the height. It was also observed that the axial forces acting over the geogrids increased with increase in cohesion of soil. The observed trend is in good agreement with the theoretical trend of horizontal thrust or axial forces or tensile forces over geogrids obtained using classical Rankine Equation or the equation suggested by FHWA Design Manual.

The observed trend can be attributed to the fact that the acting stress due to overburden and surcharge are function of height and they increase with increase in height, hence acting axial forces increase accordingly.

5.2 Horizontal Displacements

Observed trend for maximum horizontal displacements at failure (*c*-phi reduction phase) are shown in Fig. 3. It was observed that with increase in cohesion, the horizontal displacements increased. Optimum results were observed for soils with low cohesion and high friction angle. The observed trends are in agreement with earlier studies and previous literatures (Mirmoradi and Ehrlich 2013). In order to reduce the acting axial forces and to minimize the horizontal displacements, the code recommendations suggest use of granular soil (non-cohesive) soils. Existence

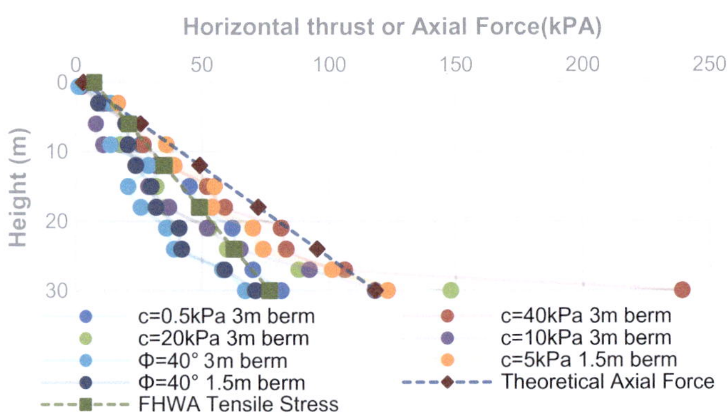

Fig. 2 Graphical representation of observe trend of acting axial forces over geogrids with respect to height (30 m)

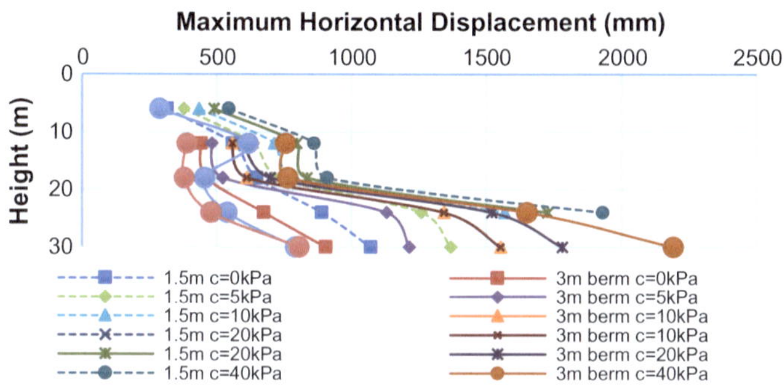

Fig. 3 Graphical representation of maximum horizontal displacement on facing at failure (*c*-phi reduction phase) with respect to height

of ground water table further develops excess pore pressure in cohesive soils, which would increase in acting axial stresses and horizontal displacements.

Figure 4 represents typical trend of horizontal displacement observed after plastic analysis. Observed trends are similar with the theoretical correlations provided by FHWA-NHI-10-024 and results presented in earlier studies (Dechasakulsom et al. 2012; Mirmoradi and Ehrlich 2013). Figure 3 depicts observed maximum horizontal displacements at failure (*c*-phi reduction phase), hence relatively higher range of values. This portrays the flexible nature of reinforcements which allow structural deformation at relatively higher strains.

Figure 5 represents the trend of variation of Global Factor of safety with different heights and different berm widths upon variation in corresponding soil parameter, i.e., cohesion and angle of internal friction. The observed trend with variation in geotechnical parameters was comparable with the results of limit equilibrium analyses. It was observed that on increasing the cohesion and friction angle of soil, the factor of safety increases. This can be attributed to the fact that shear strength of soil is a function of cohesion and friction angle, which can be

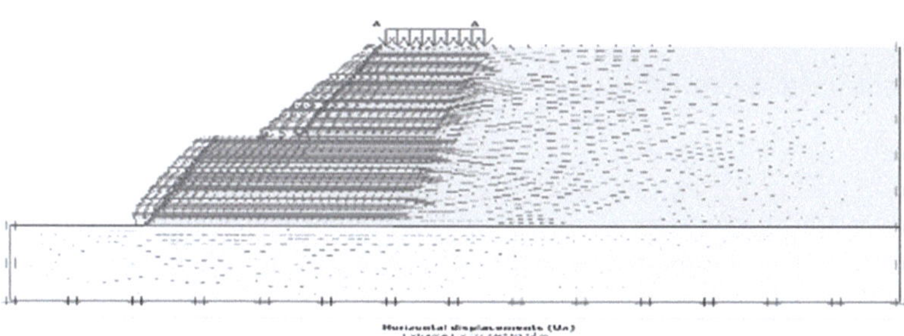

Fig. 4 Screenshot: typical observed trend for maximum horizontal displacement (after staged construction analysis)

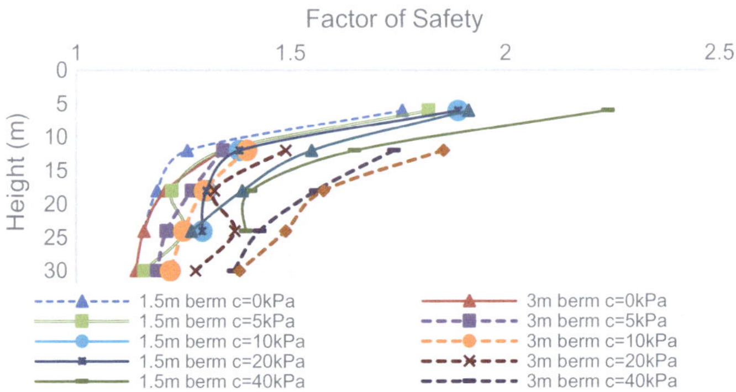

Fig. 5 Graph depicting trend of factor of safety w.r.t height of slope embankment

represented by following equation. Where τ denotes shear strength of soil and σ denotes normal acting stress on soil.

$$\tau = c + \sigma(\tan(\Phi)) \tag{3}$$

For present study with finite element analysis of the slope, maximum displacement, and incremental shear strains gave a better understanding for the results and possible failure shear surfaces (slip circles) of the slope. A typical plot for maximum displacements and possible shear surfaces are shown in Fig. 6.

It can be inferred from the above figure that with increase in berm width, the displacements and axial forces on the lower tiers reduced marginally. This can be attributed to the simple load distribution theory. With increased berm width, structure offered relatively supplementary area for the load to be distributed, hence the acting stresses reduced, hence the subsequent reduction in displacements.

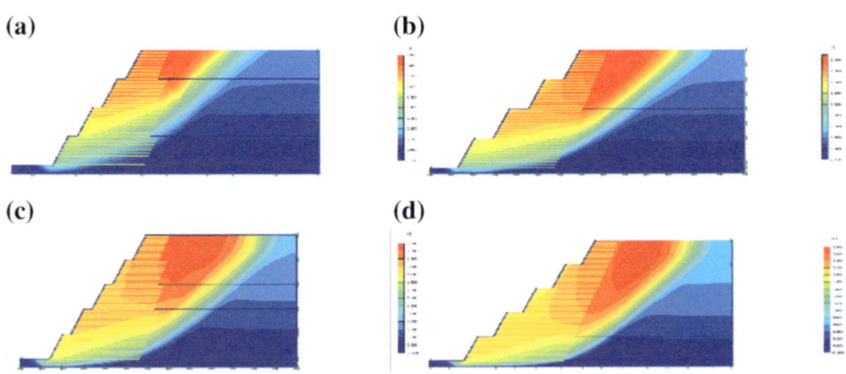

Fig. 6 Typical graphical plot of total maximum displacements (shear surface or slip circles), **a** 24 m 1.5 m berm, **b** 24 m 3.0 m berm, **c** 30 m 1.5 m berm, **d** 30 m with 3.0 m berm

6 Conclusions

The following conclusions were drawn from the analyses:

- Factor of safety of slope increased with increase in cohesion and friction angle of soil. Soil body collapsed for friction angle less than 20°, hence, soil having such properties must be avoided.
- Axial forces and horizontal displacements on geogrids and facing are directly correlated with cohesion of the soil, as they further increased with increase in cohesion of soil.
- Optimum results were attained for soil with low cohesion values (<10 kPa) and high friction angle ($\geq 30°$), as the observed axial forces and horizontal displacements were minimum.
- A general trend observed was that horizontal displacement and axial forces increased with increase in height and hence reduced the factor of safety with increase in height.
- Provision of increased berm width significantly reduced the acting axial forces and corresponding displacements on the bottom tier. This led to further increase in Factor of safety.
- From the study it can be suggested that in case of non-availability of recommended granular soil, the available soil fill be utilized only for low height embankments. For high embankments, only recommended granular soils should to be used.

References

BS 8006-1. (2010). *Code of practice for strengthened/reinforced soils and other fills.*

Dechasakulsom, M., Sukolrat, J., & Pensomboon, G. (2012). *Parametric study of a reinforced highway slope in Thailand.* Bureau of Road Research and Development, Department of Highways, Thailand.

Lakshmi, M., & Praseeja, V. (2015). Numerical simulation of geogrid reinforced soil wall using Plaxis-2d. *International Journal of Advanced Research Trends in Engineering and Technology (IJARTET), II*(X).

Mirmoradi, S. H., & Ehrlich, M. (2013). Numerical evaluation of the behaviour of reinforced soil retaining walls. In *Proceedings of the 18th International Conference on Soil Mechanics* and Geotechnical Engineering, Paris, 2013.

Naeini, S. A., Gholampoor, N., & Rezaei Kashki, R. (2015). The effect of reinforcement on stability of slopes. *International Journal of Chemical, Environmental & Biological Sciences (IJCEBS), 3*(1). ISSN 2320–4087.

Ozcelik, G., Pasaoglu, O., & Huvaj, N. (2014). Analyses of reinforced soil slopes with limit equilibrium and finite element methods.

Siavoshnia, M., Kalantari, F., & Shakiba, A. (2010). Assessment of geotextile reinforced embankment on soft clay soil. In *The 1st International Applied Geological Congress, Department of Geology*, Islamic Azad University—Mashad Branch, Iran, April 26–28, 2010.

Reinforcement Tensile Forces in Back-to-Back Retaining Walls

Sasanka Mouli Sravanam, Umashankar Balunaini and Madhira R. Madhav

Abstract Back-to-back reinforced retaining walls are mostly used in approach embankments of bridges and flyovers. In the internal stability check for mechanically stabilized earth (MSE) walls, earth pressure theory is used to obtain the tensile forces in the reinforcement. However, tensile forces calculated from this method are found to be very conservative (by a factor of two) in comparison to the actual field measurements. The objective of this study is to examine the mobilization of reinforcement tensile forces at different levels within back-to-back MSE wall at working stress condition. Tensile forces in the reinforcements with reinforcement connected in the middle (i.e., reinforcement extending from one wall to the other) are also obtained. Parametric study is carried out with the stiffness of reinforcement varying from 500 to 50,000 kN/m and the ratio of spacing between two walls to wall height (*W*/*H*) varying from 1.4 to 2.0 to investigate their effect on tensile forces at every level of the reinforcement. Charts are also proposed showing the variation of the maximum tensile forces along the height of the wall.

1 Introduction

Back-to-back retaining walls are retaining walls that are relatively close to one another. Back-to-back walls are used in construction of railroad bridge embankments or two- to four-lane highway bridge approach embankments. Figure 1 shows the schematic cross-section of back-to-back walls. The behaviour of single reinforced walls is well understood, however studies on back-to-back walls are very

---

S. M. Sravanam (✉) · U. Balunaini · M. R. Madhav
Department of Civil Engineering, IIT Hyderabad, Hyderabad, Kandi 502285, India
e-mail: sasankamouli7@gmail.com

U. Balunaini
e-mail: buma@iith.ac.in

M. R. Madhav
e-mail: madhavmr@gmail.com

A. I. V. and V. B. Maji (eds.), *Geotechnical Applications*, Lecture Notes in Civil Engineering 13, https://doi.org/10.1007/978-981-13-0368-5_19

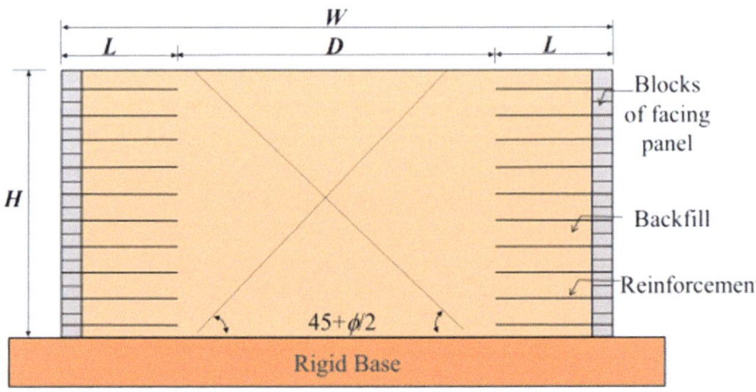

Fig. 1 Schematic of back-to-back MSE walls

limited. Anubhav and Basudhar (2012) studied the behaviour of back-to-back walls using experimental setup. Numerical analysis was carried by Han and Leshchinsky (2010); Hardianto and Turong (2010); Mouli and Umashankar (2014); El-Sherbiny et al. (2013); Mouli et al. (2015). Katkar and Viswanadham (2012) performed centrifugal studies on back-to-back walls.

Tensions mobilized in single MSE wall was analysed experimentally by Yang et al. (2010), Bathurst et al. (2009); etc. Numerical modelling of MSE wall for studying tensile loads was also done (e.g. Hatami and Bathurst 2005; Guler et al. 2007; etc.). The stiffness of reinforcement plays an important role in the calculation of reinforcement loads under working stress condition. In the present study, the effect of stiffness of reinforcement in back-to-back walls is analysed.

2 Problem Definition

The objective of the present work is to study the effects of reinforcement stiffness (J) and ratio of the wall spacing to height of the wall (W/H) on the tensile forces mobilized in the reinforcement. Effect of connecting the reinforcement at the middle of the wall is also analysed. A finite-difference based numerical analysis was conducted to perform the analysis of mechanically stabilized back-to-back walls.

3 Numerical Modelling

The Finite difference programme, FLAC (Fast Lagrangian Analysis of Continua), was used for the analysis (Itasca 2011). Back-to-back walls of height 6 m was considered, and the length of reinforcements for both the walls was fixed as 4.2 m (i.e., equal to 0.7 times the height of wall). The distance between the ends of the reinforcements extending from the two walls was varied from 0 to 0.6 H so that

W/H ratio ranges from 1.4 to 2.0. The vertical spacing between the reinforcement layers was taken as 0.6 m.

Bottom of the foundation soil was fixed in both horizontal and vertical directions. Mesh convergence was done and the size of the grid was taken as approximately 0.1 m. Large-strain mode was activated so that the coordinates of the grid points are updated at every step. This ensures accuracy in the numerical model, especially when high strains are developed in the material.

The foundation soil was assumed to be rigid. Reinforced soil was simulated as homogenous, isotropic, elastic-perfectly plastic using Mohr–Coulomb failure criterion. Table 1 provides the properties of reinforced soil, foundation soil, and facing panel.

Elastic modulus of the soil was dependent on the confining stress (Duncan et al. 1980). It was updated at every stage using the procedure given in Hatami and Bathurst (2005). Equation given by Duncan et al. (1980) was used Eq. (1).

$$E_t = \left[1 - \frac{R_f(1 - \sin\phi)(\sigma_1 - \sigma_3)}{2c \cdot \cos\phi + 2\sigma_3 \cdot \sin\phi}\right]^2 \cdot K_e \cdot P_{atm} \cdot \left(\frac{\sigma_3}{P_{atm}}\right)^n, \quad (1)$$

where, E_t is the tangent elastic modulus, R_f is the failure ratio, K_e is the elastic modulus number, n is the elastic modulus exponent, P_{atm} is the atmospheric pressure, ϕ is the angle of shearing resistance of soil, c is the cohesion intercept of soil, σ_1 is the effective vertical pressure (overburden), and σ_3 is the effective lateral confining pressure.

The wall facing was modelled as modular blocks of size 0.3 × 0.2 m. Material properties of modular blocks were assumed to be equal to that of concrete material (Table 1).

Reinforcement was simulated as cable element. Cable element in FLAC is a two-noded, one-dimensional element with high tensile stiffness and negligible compressive stiffness. Reinforcement was assumed to be rigidly fixed at the left end of the cable element to nodes of the wall facing to simulate the rigid connection that exists in the field.

Table 1 Properties of the foundation soil, reinforced and retained backfills

Properties	Reinforced soil	Foundation soil	Modular blocks
Material type	Mohr–Coulomb	Mohr–Coulomb	Elastic
Cohesion (kPa)	0	1000	–
Angle of shearing resistance (ϕ) in deg.	34	35	–
Dilation angle in deg.	10	0	–
Shear modulus (kPa)	3.846e4	3.846e4	8.70e6
Bulk modulus (kPa)	8.333e4	8.332e4	9.52e6
Density (kg/m^3)	1800	1800	2400

Table 2 Constants used in the equation for stress dependent modulus of backfill soil

Properties	Reinforced soil
Elastic modulus number (K_e)	1150
Bulk modulus number (K_b)	575
Elastic modulus exponent (n)	0.5
Bulk modulus exponent (m)	0.5
Failure ratio (R_f)	0.86

Table 3 Reinforcement properties

Properties	Cable element
Stiffness (J) (kN/m)	500, 50,000
Poisson's ratio (υ)	0.3

Table 3 provides the reinforcement properties.

The model wall was constructed in stages. Each lift of 0.3 m height was first placed, and the model solved for equilibrium. The elastic modulus was then updated using Eq. (1) (using the constants mentioned in Table 2), and again solved for equilibrium. The next layer of soil was now placed on the deformed grid of the previous layer. At every stage, the maximum unbalanced force ratio was maintained to be less than 1e−3.

4 Results

4.1 Connected and Unconnected Case with W/H *= 1.40*

Tensile forces mobilized in the reinforcement are a function of the stiffness of the reinforcement. Hence, the variations of tensile forces are provided for different stiffness values of the reinforcement. Tensile forces in reinforcements were given in back-to-back walls with W/H = 1.40 and 2.00 (Fig. 2). Tensile forces in reinforcements were calculated in back-to-back walls with W/H = 1.40 and 2.00 (Fig. 2). The maximum tensile load profiles were also drawn. Comparison was made between connected and unconnected reinforcement cases. Figure 2 shows the tensile force profile for reinforcements at different heights of wall (reinforcements near the top of the wall (5.7 m), near to the middle of the wall (3.3 m) and near the bottom of the wall 0.9 m) for W/H = 1.40. It can be observed that for J = 500 kN/m, the location of maximum tensile force in the reinforcement near the top of the wall and for the reinforcement near the bottom of the wall were at a distance of about 3.5 m (A') and 0.9 m (B′) from the facing respectively. There was no clear peak tensile force observed in the reinforcement near the middle height. In the case of reinforcement near the bottom of the wall, position of peak tensile force had shifted towards the facing of the wall. The location of maximum tensile force can be justified from Fig. 3. Figure 3 plots the maximum shear strain increment contours

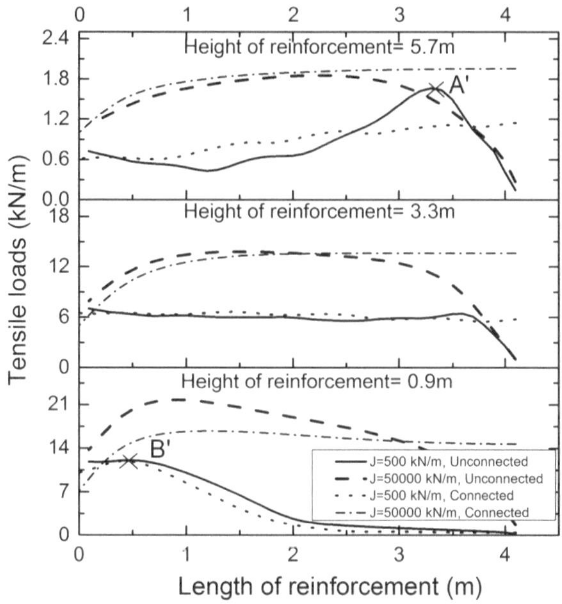

Fig. 2 Tensile force profile along the reinforcement length at different heights for connected and unconnected case (W/H = 1.40)

for the case of J = 500 kN/m and reinforcement unconnected in the middle. Lines were drawn along the maximum of maximum shear strain increments. This line intersected the top and bottom reinforcements of the wall at point A and B, respectively. The horizontal distances from the facing to point A and B were around 3.5 and 0.9 m which coincide with that of A' and B' (Fig. 2). It can be concluded that the maximum tensile force occurred at along the plane of maximum shear strain.

For the case of J = 50,000 kN/m, unconnected reinforcement, there was no clear peak observed along the reinforcement length. As the wall was analysed in working stress condition, the displacements due to gravity loads (which are much lesser in case of stiffer reinforcement) were not sufficient enough to mobilize the critical slip surface. Near the bottom of the wall, the tension values in reinforcement with J = 50,000 kN/m case were much higher when compared to that of J = 500 kN/m case, which is expected. The tensile forces for both connected and unconnected cases had almost equal values in J = 500 kN/m case. In the case of J = 50,000 kN/m, reinforcement near the bottom of the wall, the tensile forces were higher in the unconnected case than that of connected case.

In most of the cases, the tensile forces in the unconnected cases were tending towards zero value at the end of the reinforcement, but there was significant tension value in the cases of connected reinforcement at that location as the reinforcement was extending up to the adjacent wall. The distribution of maximum tension in each layer along the height of the wall is plotted in Fig. 4. It can be observed that for the case of reinforcement with low stiffness (i.e. J = 500 kN/m), difference between the connected case and unconnected case was insignificant. But when stiffness of

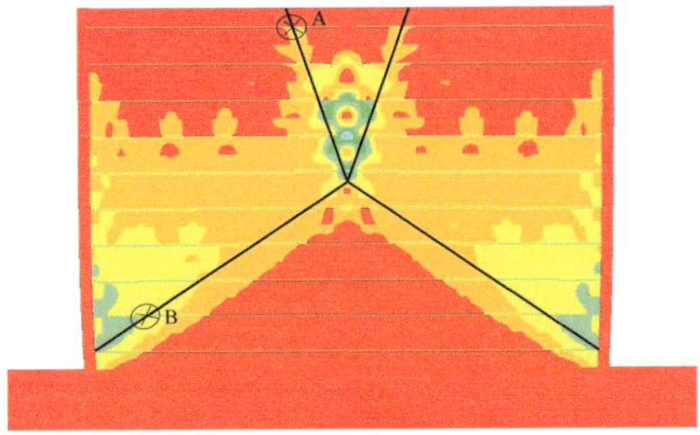

Fig. 3 Maximum shear strain increment contour for the case $J = 500$ kN/m and $W/H = 1.40$ unconnected reinforcement

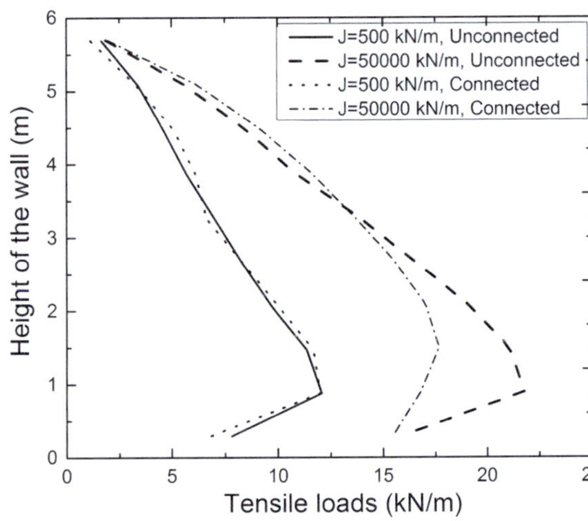

Fig. 4 Maximum tensile forces profile for connected and unconnected reinforcement cases ($W/H = 1.40$)

reinforcement was increased to 50,000 kN/m, unconnected reinforcement case showed higher maximum tension than that of the connected case. In the case of connected and stiffer reinforcement case, the interaction of other wall was much higher than the unconnected case.

4.2 Unconnected Case with W/H *= 2.0*

Figure 5 represents the tensile profiles of the reinforcement at different heights of the walls for $W/H = 2.0$. As the distance between the two walls was considerable large, the walls were behaving as independent walls (Berg et al. 2009). Higher tension values were mobilized in the stiffer reinforcement. It can be observed that, except for the reinforcement near the top of the wall, the reinforcements had the

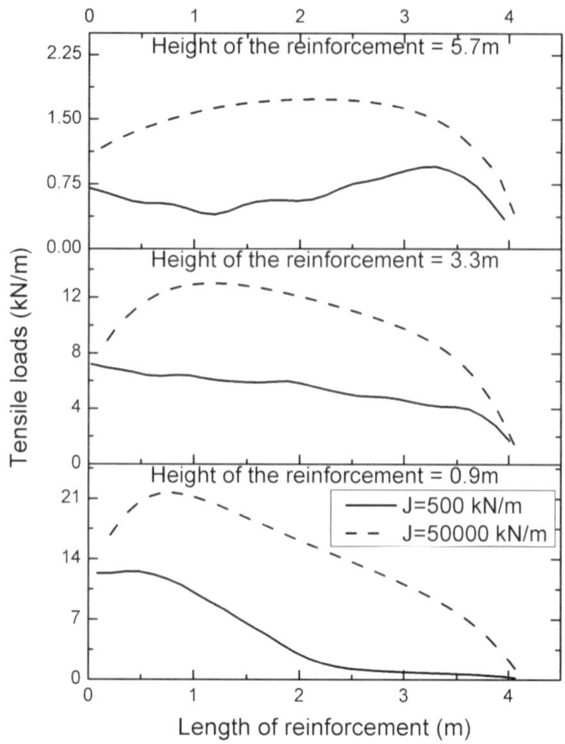

Fig. 5 Tensile force profile along the reinforcement length at different heights for unconnected case ($W/H = 2.0$)

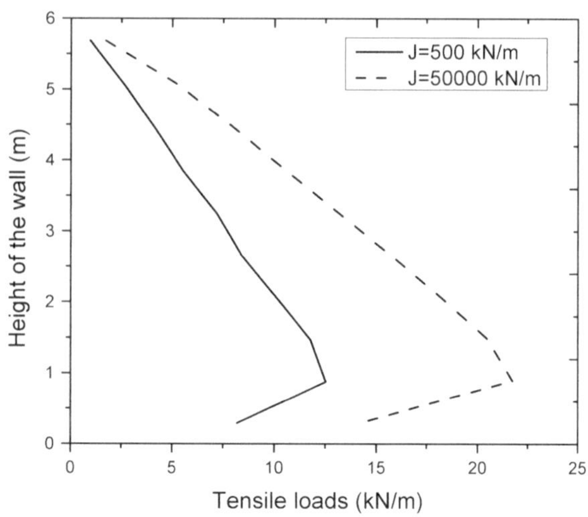

Fig. 6 Maximum tensile forces profile for unconnected case ($W/H = 2.0$)

maximum tension near the facing. Figure 6 shows the maximum tension distribution along the height of the wall. It was observed that the magnitude of tensions were almost the same for both W/H = 1.4 and W/H = 2.0 unconnected cases. But the tension profiles along the reinforcement length were different. In the latter case, the maximum tensile forces along the reinforcement occurred at the connection in all reinforcement except in the reinforcement at the top of the wall.

5 Conclusions

Tension profiles along the length of the reinforcement were plotted for both W/H ratios 1.4 and 2.0. Maximum tension distribution along the height of the wall was also plotted. Following conclusions can be drawn from the present study:

- Maximum of maximum tension occurred at 0.9 m height from the bottom of the wall.
- Maximum of maximum tension in connected case is lesser than that of the unconnected case, when the reinforcement stiffness is large (J = 50,000 kN/m). But for the case of reinforcement with low stiffness value (J = 500 kN/m), both connected and unconnected cases showed insignificant difference.
- W/H ratio has no significant effect with respect to the magnitude of maximum tension profile. But the location of maximum tension along the reinforcement length changes when the distance between the two walls increases.

References

Anubhav, A., & Basudhar, P. K. (2012). Footing on double-faced wrap-around reinforced soil walls. *Ground Improvement, 167*(G12), 73–87.

Bathurst, R. J., Nernheim, A., Walters, D. L., Allen, T. M., Burgess, P., & Saunders, D. D. (2009). Influence of reinforcement stiffness and compaction on the performance of four geosynthetic reinforced soil walls. *Geosynthetics International, 16*(1), 43–59.

Berg, R. R., Christopher, B. R. & Samtani, N. C. (2009). Design and construction of mechanically stabilized earth walls and reinforced soil slopes–Volume I. *U.S. Department of Transportation Federal Highway Administration*, Publication No. FHWA-NHI-10–024.

Duncan, J. M., Byrne, P., Wong, K. S., & Mabry, P. (1980). Strength, stress–strain and bulk modulus parameters for finite element analysis of stresses and movements in soil masses. Dept. of Civil Engg., University of California, Berkeley, California. *Report No.UCB/GT/80-01*.

El-Sherbiny, R., Ibrahim, E., & Salem, A. (2013). Stability of back-to-back mechanically stabilized earth walls. *Proceedings of Geo-congress* (pp. 555–565). March 3–7, 2013, San Diego, California, USA.

Guler, E., Hamderi, M., & Demirkan, M. M. (2007). Numerical analysis of reinforced soil-retaining wall structures with cohesive and granular backfills. *Geosynthetics International, 14*(6), 330–345.

Hardianto, F. S., & Truong, K. M. (2010). Seismic deformation of back-to-back mechanically stabilized earth (MSE) walls. *Earth Retention Conference 3* (pp. 704–711), August 1–4, 2010. ASCE: Bellevue, Washington.

Han, J., & Leshchinsky, D. (2010). Analysis of back to back mechanically stabilized earth walls. *Journal of Geotextiles and Geomembranes, 28,* 262–267.

Hatami, K., & Bathurst, R. J. (2005). Development and verification of a numerical model for the analysis of geosynthetic reinforced soil segmental walls under working stress conditions. *Canadian Geotechnical Journal, 42,* 1066–1085.

Itasca. (2011). *FLAC-Fast Langragian Analysis of Continua Version 7.00.* Itasca Consulting Group Inc., Minneapolis, Minnesota.

Katkar, B. H., & Viswanadham, B. V. S. (2012). Centrifuge Studies on the behavior of back-to-back geogrid reinforced soil walls. In *Proceedings of the 1st Asian Workshop on Physical Modelling in Geotechnics*, November 14–16, 2012, Bombay, India.

Mouli, S. S., & Umashankar, B. (2014). Numerical analysis of MSE walls considering wall friction and reinforcement stiffness. In *Proceedings of 14th International Association Computer Methods and Advances in Geomechanics (IACMAG)*, September 22–25, 2014, Kyoto, Japan.

Mouli, S. S., Umashankar, B., & Madhav, M. R. (2015). Analysis of back-to-back reinforced retaining walls with modular block facing. In *Proceedings of 50th Indian Geotechnical Congress (IGC)*, December 17–19, 2015, Pune.

Yang, K., Zornberg, J., & Bathurst, R. (2010). Mobilization of reinforcement tension within Geosynthetic-Reinforced soil structures. In *Earth Retention Conference 3*, Bellevue, Washington, August 1–4, 2010.

Limit Analysis of Full-Scale MSE Wall—A Comparative Study

Vikrant Patel, Shantanu Patra and P. V. Pavan Kumar

Abstract In this paper, a case study is considered to study the behavior of a 3.6 m high MSE wall reinforced with geogrid reinforcement. Reinforcement spacing is 0.6 m. A limit analysis has been carried out using LimitState:GEO (which used discontinuity layout optimization DLO) to study the performance of a full-scale wall. The maximum reinforcement tension at each level was computed and the results are compared with the measured values. Computation of load coming on reinforcement under different surcharge load is done by analytical approach using MATLAB. In this paper analytical model (current AASHTO Simplified Method) are also prepared with different surcharge loading. The computed values of maximum load on reinforcement are finally compared with the measured values. The results show the LimitStateGEO analysis gives a better estimation of the maximum reinforcement tension compared to the measured values.

Keywords Reinforced soil wall · LimitState:GEO · Analytical model

1 Introduction

The present work deals with the Mechanically Stabilized Wall (MSW) reinforced with geogrid. For the internal stability of the reinforced soil wall which is built with frictional backfill, the design methods are based on the limit equilibrium method of

V. Patel (✉) · S. Patra (✉) · P. V. Pavan Kumar (✉)
Department of Civil Engineering, National Institute of Technology, Rourkela, Rourkela, India
e-mail: vikrant.patel89@gmail.com

S. Patra
e-mail: patrashantanu@nitrkl.ac.in

P. V. Pavan Kumar
e-mail: pvpavn416@gmail.com

A. I. V. and V. B. Maji (eds.), *Geotechnical Applications*, Lecture Notes in Civil Engineering 13, https://doi.org/10.1007/978-981-13-0368-5_20

analysis. This approach is based on the assumption that the loads in reinforcement are developed because of active earth pressure state in the reinforced soil which is calculated using the peak friction angle of soil. The active earth pressure is then distributed to the reinforcement based on their spacing.

Allen et al. (2003) reviewed reinforcement strain measurement techniques those are reported case studies. From the corrected reinforcement strains combined with the reinforcement stiffness values, the loads on reinforcement can be estimated.

Bathurst et al. (2008) studied case studies of reinforced soil wall. They observed that the geosynthetic reinforcement loads for walls with cohesionless backfill soils were three times lesser than the predicted values using AASHTO (2002) simplified method. Also, the reinforcement loads are more uniform with the depth than the predicted

2 Model Dimension and Material Properties

The measurements are taken from full-scale modular block wall in the Royal Military College of Canada (RMCC) research program (Bathurst et al. 2009). The height of the wall is 3.6 m, reinforcement length is 2.52 m with a vertical spacing of $S_v = 0.6$ m and the batter angle $\omega = 8°$ (inclination of the wall with vertical).

Properties of backfill soil is shown in Table 1, however dilation angle can't be mentioned in LimitState GEO. The modular facing units have a size of LXHXW = $300 \times 150 \times 200$ mm and mass of 20 kg.

3 Limitstate:GEO

This is a general purpose software program which utilizes Discontinuity Layout Optimization (DLO) to directly identify the critical collapse mechanism for a wide variety of geotechnical problems. This software can be used to model 2D models of any geometry as specified by the user. The solution is presented as an adequacy factor which can be applied on load and material strength. The solution is displayed as a failure mechanism which involves a number of blocks sliding relative to one another.

The prepared model in LimitState:Geo is shown in Fig. 1.

Table 1 Properties of backfill soil

Property	Value
Model	Mohr–Coulomb
Angle of internal friction, ϕ (degree)	44
Cohesion, c (kN/m^2)	0
Unit weight, γ (kN/m^3)	16.8
Drainage condition	Always drained

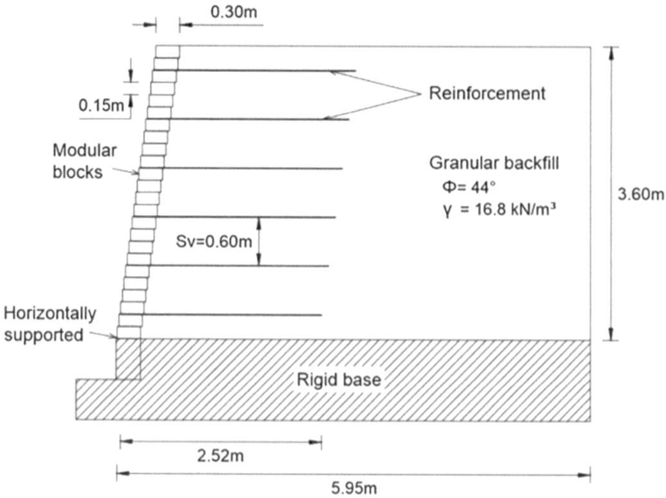

Fig. 1 Geometry of the investigated model

3.1 *Theory*

In the Limitstate:GEO the geogrid is defined as engineered material, and soil as Mohr–Coulomb (strength is defined in terms of cohesion intercept and angle of shearing resistance) material. The adequacy factor obtained from the software is applied only for soil strength parameters.

The Engineered Material is used to represent the one-dimensional object that has a pullout resistance T_1 per unit length, N_1 per unit length to lateral displacement and a plastic moment of M_{p1}. If there are m objects present per unit width then $T = mT_1$ is the pullout resistance per unit width, and $N = mN_1$ is the resistance per unit length per unit width to lateral displacement. Interaction between soil and geogrid in Limitstate:GEO is defined by Eqs. (1) and (2) and is a linear function of vertical effective stress.

The pullout resistance per unit length per unit width

$$T = T_c + T_q \sigma_v^{'} \tag{1}$$

The resistance to lateral displacement per unit length per unit width

$$N = N_c + N_q \sigma_v^{'}, \tag{2}$$

where,

T_c Pullout factor, the contribution of the material cohesion to the pullout resistance of the element.

T_q Pullout factor, the contribution of the overburden to the pullout resistance of the element.
N_c Lateral factor, the contribution of the material cohesion to the lateral displacement of the element.
N_q Lateral factor, the contribution of the overburden to the lateral displacement of the element.
$\sigma_v^{'}$ The vertical effective stress.

In the present case T_c, N_c, N_q are zero.

4 Analytical Solution

AASHTO (2002) simplified method (tie-back wedge method) is used to evaluate the reinforcement tension analytically. The AASHTO calculation for maximum reinforcement load $T_{\max}$ is expressed as

$$T_{\max} = K(\gamma z + q)S_v,$$

where,

z the depth of reinforcement layer below the crest of the wall and
K the active earth pressure coefficient, calculated as

$$K = \frac{\cos^2(\phi + \omega)}{\cos^2 \omega[1 + (\sin \phi / \cos \omega)]^2}$$

5 Results

The tension in the reinforcement at different levels by AASHTO and Limitstate: GEO and measured experimentally by Bathurst et al. (2009) are shown in Figs. 2, 3, 4 and 5. The factor of safety at the end of construction and for different surcharge is presented in Table 2. Different failure surfaces at the end of construction, for 50 and 100 kPa surcharge are shown in Fig. 6.

The LimitState:GEO data shows that the reinforcement loads are varying with depth, and average loads are lower than the values computed using the AASHTO (2002) simplified method for geosynthetic reinforced soil walls. The difference in LimitState:GEO and predicted load values using the simplified method increases with depth of layer below the wall crest.

Fig. 2 Comparison of LimitState GEO values with AASHTO and measured values (Bathurst et al. 2009) at the end of construction

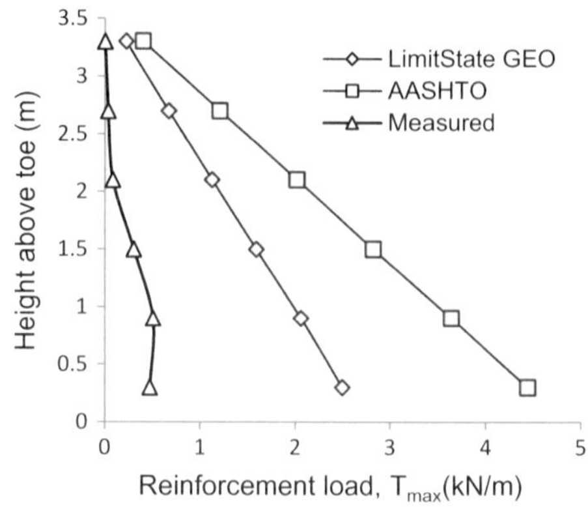

Fig. 3 Comparison of LimitState GEO values with AASHTO values for 25 kPa surcharge

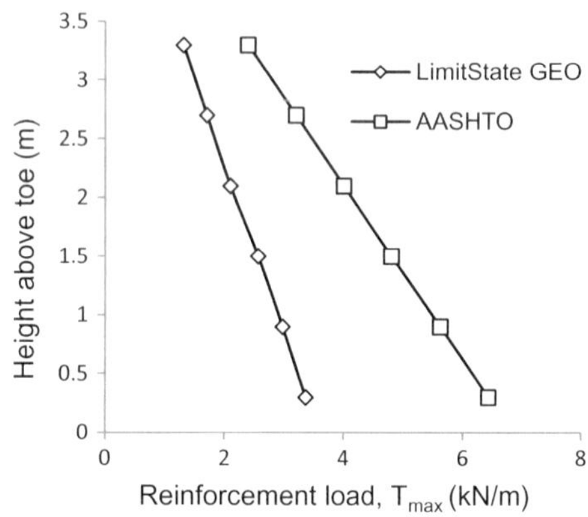

Fig. 4 Comparison of LimitState GEO values with AASHTO and measured values (Bathurst et al. 2009) for 50 kPa surcharge

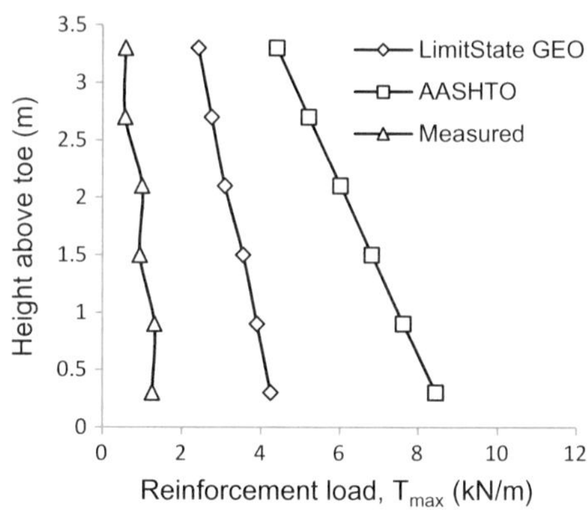

Fig. 5 Comparison of LimitState GEO values with AASHTO values for 100 kPa surcharge

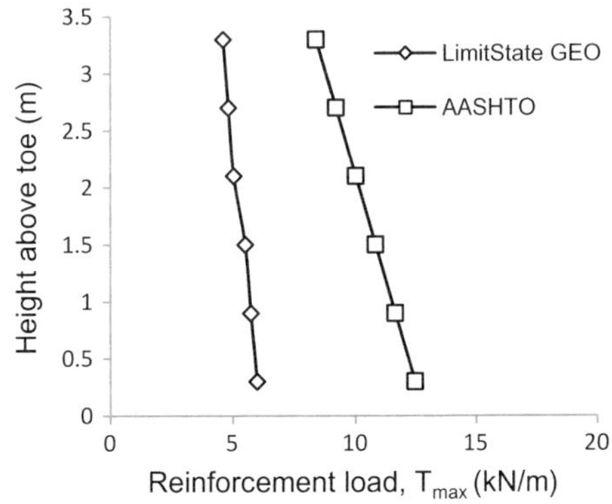

Table 2 Factor of safety

Surcharge	Factor of safety
At the end of construction	1.444
25 kPa	1.153
50 kPa	1.075
100 kPa	1.033

Fig. 6 Failure surface

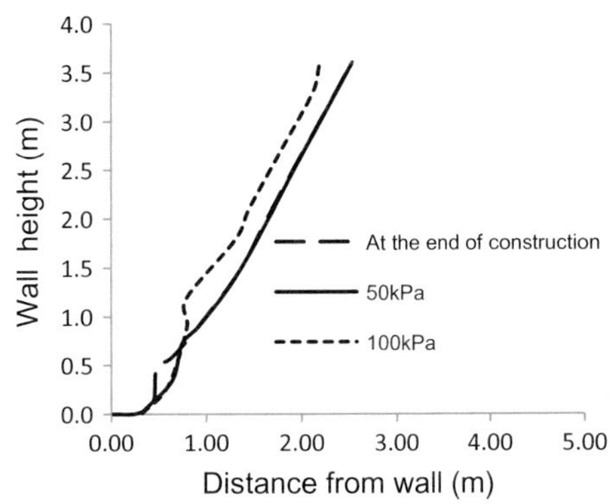

6 Conclusion

- The axial forces in reinforcement obtained by LimitStateGEO lies between the forces calculated by AASHTO (2002) method and experimentally measured by Bathurst et al. (2009).
- Factor of safety against pullout failure by LimitStateGEO is much lesser than that is calculated analytically.

- Failure surface shown in output environment of LimitStateGEO is close to wall as shown in Fig. 3a, b.
- Variation of axial forces in reinforcement with height of wall is almost linear with higher value near toe of wall and lower value near crest of wall.

References

AASHTO (2002). *Standard specifications for highway bridges* (17th ed.). Washington, D.C.: American Association of State Highway and Transportation Officials (AASHTO).

Allen, T. M., Bathurst, R. J., Holtz, R. D., Walters, D. L., & Lee, W. F. (2003). A new working stress method for prediction of reinforcement loads in geosynthetic walls. *Canadian Geotechnical Journal, 40*(5), 976–994.

Bathurst, R. J., Miyata, Y., Nernheim, A. & Allen, T. M. (2008). Refinement of K-stiffness Method for geosynthetic reinforced soil walls. *Geosynthetics International, 15*(4), 269–295.

Bathurst, R. J., Nernheim, A., Walters, D. L., Allen, T. M., Burgess, P., Saunders, D. D. (2009). Influence of reinforcement stiffness and compaction on the performance of four geosynthetic-reinforced soil walls. *Geosynthetics International, 16*(1).

Effect of Toe Cutting on Hillslope Stability

Rubi Chakraborty and Arindam Dey

Abstract This paper reports about the numerical analyses conducted to investigate the instability induced due to toe cutting of simple homogeneous slopes during possible construction and widening of roads. A simple homogeneous hill slope made up of *c*-*φ* soil and overlying a rigid bedrock foundation material has been analyzed using Morgenstren-Price limit equilibrium method (Morgenstern and Price 1965) with the aid of Geostudio SLOPE/W 2007. Toe cutting analyses have been conducted for a wide variation of geotechnical, hydraulic and seismic parameters. The stability of the slope is assessed by the evaluated factor of safety which defines the mobilization of the shear strength along a particular slip line. Analyses are done in three stages. In the first stage, simplified hill slopes are analyzed solely under self-weight of the slope. In the second stage, pseudo-static analysis has been conducted. In third stage, water table variation is incorporated in the toe cutting analysis. Results are presented in graphical form and based on present study discussions are made regarding the critical horizontal extent of vertical toe cutting of hill slopes depending upon the slope type and different geotechnical parameters.

Keywords Toe cutting · Hillslope stability · Morgenstern-price method
Pseudo-static analysis · SLOPE/W 2007

R. Chakraborty (✉)
Department of Civil Engineering, National Institute of Technology Meghalaya,
Shillong 793003, Meghalaya, India
e-mail: rubi.rc.nit@gmail.com

A. Dey
Department of Civil Engineering, Indian Institute of Technology Guwahati,
Guwahati 781039, India
e-mail: arindamdeyiitk@gmail.com

A. I. V. and V. B. Maji (eds.), *Geotechnical Applications*, Lecture Notes in Civil
Engineering 13, https://doi.org/10.1007/978-981-13-0368-5_21

1 Introduction

Major slope excavations are routinely carried out to facilitate construction and widening of roads. The objectives of routine road cuts are to create space for the road template and driving surface, to remain stable over time, to not be a source of sediment and minimize long-term costs. A major cause of cut slope failure is related to the reduced confining stress within the soil upon excavation. Undermining the toe of the slope or increasing the slope angle results in slope failure. The major cut slope design parameters are slope geometry, soil shear strength, and predicted or measured ground water levels. For cohesionless soil, stability of a cut slope is independent of height and therefore slope angle becomes the key parameter of concern. For cohesive soils, the height of the cut becomes the critical design parameter. For c-φ and saturated soils, slope stability is dependent on both slope angle and height of cut.

Preliminary slope stability analysis can be performed using simple stability charts. These charts can be used to determine if a proposed slope might be subjected to failure. If slope instability appears possible, or if complex condition exits beyond the scope of the charts, more rigorous methods can be employed. Ling et al. (1997) had given a simple closed form solution of slope stability by limit equilibrium (LE) method considering both the horizontal as well as vertical seismic accelerations using pseudo-static approach. Choudhury et al. (2007) used a simplified method of vertical slices by limit equilibrium approach for the determination of dynamic factor of safety of any generalized slope under seismic condition considering pseudo-static seismic forces both in horizontal and vertical directions. Recently, nonlinear programming based on minimization of factor of safety approach has been used frequently for locating the critical slip surface. Effective use of these programs requires accurate determination of site geometry including surface profiles, soil unit boundaries and location of the water table, as well as unit weight and strength parameters. Presently, stability of slope can be analyzed using several available geotechnical software which uses the LE or FE methodologies. GEOSTUDIO is developed specifically for the analysis of deformation and stability of geotechnical structures based on limit equilibrium principles and finite element modules. Saikia et al. (2014) investigated numerically the failure of virgin slopes characterized by different strength parameters, slope heights and slope angles using LE method with the aid of SLOPE/W module of Geostudio and also suggested the different types of stabilization measures. Chakraborty and Dey (2016) has investigated numerically the stability of slopes for a wide range of geotechnical, seismic, and hydraulic parameters under different circumstances using the SLOPE/W. This paper mainly reports the SLOPE/W analyses conducted to investigate the slope instability induced due to toe cutting of simple homogeneous hill slopes during possible construction or widening of roads.

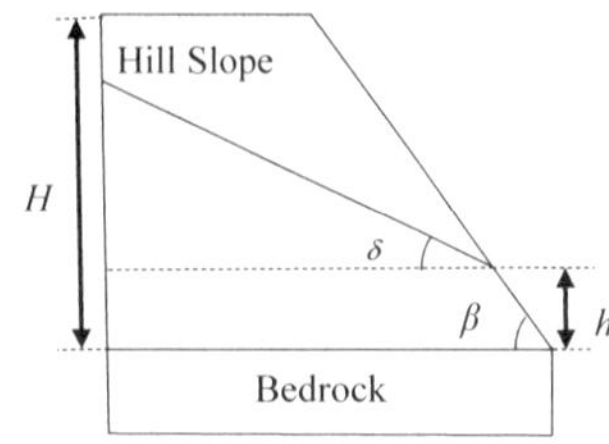

Fig. 1 Schematic diagram of the problem statement

2 Numerical Model

A 40 m high slope made up of c-φ soil and overlying a rigid bedrock foundation material has been analyzed for instability due to toe cutting for a wide range of cohesion (c) and angle of internal friction (φ) using Geostudio *SLOPE/W 2007* utilizing Morgenstren-Price limit equilibrium method (Fig. 1).

Determining the position of the critical slip surface with the lowest FoS remains one of the key issues in a stability analysis. There are many different ways for defining the positions of trial slip surfaces in *SLOPE/W 2007*, namely Grid and Radius method, Entry and Exit method, Fully specified slip surfaces. In the present analysis, *Entry and Exit method* has been used.

The most realistic position of the critical slip surface is obtained when effective strength parameters are used in the analysis. Effective strength parameters, however, are only meaningful when they are used in conjunction with porewater pressures. There are different ways to specify the pore pressure conditions in *SLOPE/W 2007*, namely Single piezometric line, Multiple piezometric line, Porewater pressure head with spatial function. In the present analysis, *single piezometric line* method has been used. The variation of water table has been considered by specifying height and inclination of the piezometric line. Pseudo-static analysis has been conducted by adopting horizontal and vertical pseudo-static coefficients as 0.18 and 0.09 respectively, corresponding to the Seismic Zone V in which the regions of the Northeast India are located.

3 Results and Discussions

3.1 Effect of Toe Cutting on Dry Slopes

Results obtained are shown in Figs. 2, 3, and 4. Analyses conducted shows that flatter slopes (slope inclination varying from 30° to 40°) having low c and φ values are failing under self-weight of the slopes only [e.g., for (a) $\beta = 30°$, $c = 10$ kPa, $\varphi = 20°$ the FoS value is 0.893 and (b) $\beta = 40°$, $c = 10$ kPa, $\varphi = 20°$ FoS value is 0.681 respectively under self-weight of the slope]. Therefore, vertical toe cutting for road construction or widening is not possible for such slopes without adequate safety measures. Higher c and φ values may lead to safe toe cutting even up to a horizontal extent (b_t) of more than 15 m [e.g., for (a) $\beta = 30°$, $c = 50$ kPa, $\varphi = 20°$ toe cutting of more than 25 m and (b) $\beta = 40°$, $c = 30$ kPa, $\varphi = 35°$ more than 15 m]. On the other hand, for comparatively steeper slopes (slope inclination

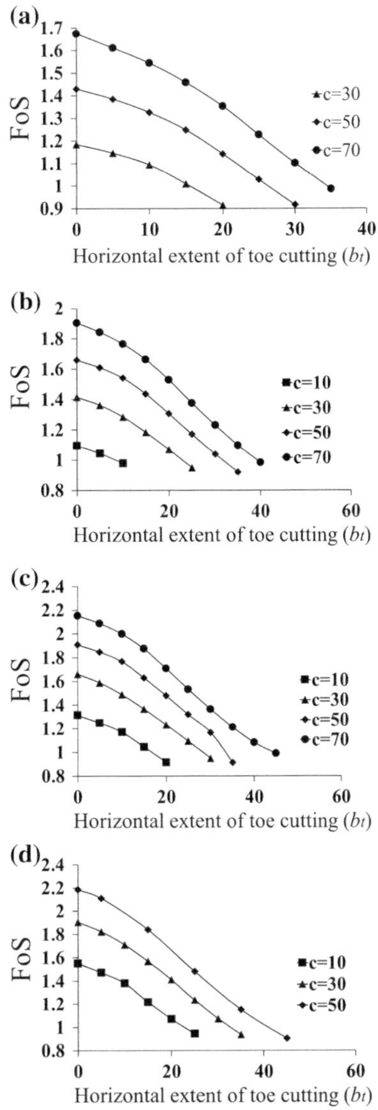

Fig. 2 Effect of toe cutting on dry hill slopes for $\beta = 30°$, **a** $\varphi = 20°$, **b** $\varphi = 25°$, **c** $\varphi = 30°$, **d** $\varphi = 35°$

Fig. 3 Effect of toe cutting on dry hill slopes for $\beta = 40°$, **a** $\varphi = 20°$, **b** $\varphi = 25°$, **c** $\varphi = 30°$, **d** $\varphi = 35°$

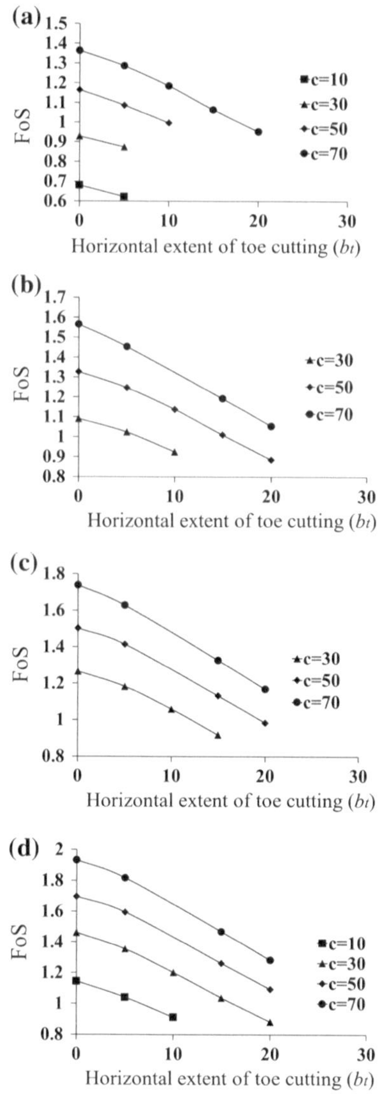

varying from 50° to 60°) having higher c and φ values also fail under self-weight [e.g., for $\beta = 60°$, $c = 30$ kPa, $\varphi = 30°$ the FoS value is 0.842]. A reasonably high value of c and φ (e.g. for $\beta = 60°$, $c = 70$ kPa, $\varphi = 35°$) allows a horizontal extent (b_t) of toe cutting of around 10 m, but such combination of c and φ values may not be practically feasible.

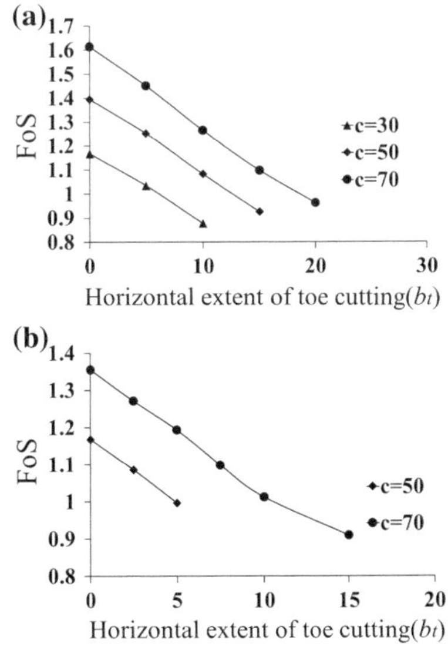

Fig. 4 Effect of toe cutting on dry hill slopes for **a** $\beta = 50°$, $\varphi = 35°$, **b** $\beta = 60°$, $\varphi = 35°$

3.2 *Effect of Toe Cutting on Dry Slopes with Pseudo-static Earthquake*

Results obtained are shown in Fig. 5. Analyses conducted shows that flatter slopes (slope inclination varying from 30° to 40°) having low c and φ values are failing under the action of pseudo-static earthquake [e.g., for (a) $\beta = 30°$, $c = 10$ kPa, $\varphi = 25°$ the FoS value is 0.775 and (b) $\beta = 40°$, $c = 30$ kPa, $\varphi = 25°$ FoS value is 0.819 respectively]. Therefore, vertical toe cutting for road construction or widening is not possible for such slopes without adequate safety measures. Higher c and φ values lead to safe toe cutting even up to a horizontal extent (b_t) of more than 15 m [e.g., for (a) $\beta = 30°$, $c = 30$ kPa, $\varphi = 35°$ and (b) $\beta = 40°$, $c = 50$ kPa, $\varphi = 30°$]. On the other hand, for comparatively steeper slopes (slope inclination varying from 50° to 60°), having higher c and φ values, also fail under the action of pseudo-static earthquake only [e.g., for (a) $\beta = 50°$, $c = 30$ kPa, $\varphi = 35°$ the FoS value is 0.896 and (b) $\beta = 60°$, $c = 50$ kPa, $\varphi = 35°$ the FoS value is 0.912]. A reasonably high value of c and φ [e.g., for (a) $\beta = 50°$, $c = 70$ kPa, $\varphi = 30°$ and (b) $\beta = 60°$, $c = 70$ kPa, $\varphi = 35°$] allows a horizontal extent of toe cutting of around 2 m, but such combination of c and φ values may not be practically feasible.

3.3 *Effect of Toe Cutting on Partially Saturated Slopes*

The height of piezometric line from toe (h) is considered to be $2H/3$ and at an inclination of $\delta = \beta/3$. Results obtained are shown in Fig. 6. Analyses conducted

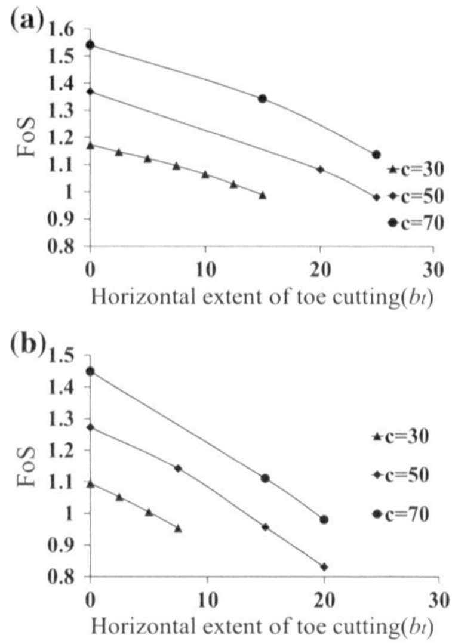

Fig. 5 Effect of toe cutting on dry hill slopes with pseudo-static earthquake for **a** $\beta = 30°$, $\varphi = 30°$, **b** $\beta = 40°$, $\varphi = 35°$

shows that flatter slopes (slope inclination 30°) having low c and φ values are failing due to presence of water table [e.g., for (a) $\beta = 30°$, $c = 10$ kPa, $\varphi = 20°$ the FoS value is 0.776 and (b) $\beta = 40°$, $c = 10$ kPa, $\varphi = 20°$ FoS value is 0.562 respectively]. Toe cutting for such slopes without adequate safety measures may cause failure of the slope. Higher c and φ values lead to safe toe cutting even up to a horizontal extent (b_t) of more than 15 m [e.g., for (a) $\beta = 30°$, $c = 30$ kPa, $\varphi = 30°$ and (b) $\beta = 40°$, $c = 70$ kPa, $\varphi = 30°$] On the other hand, FoS values of comparatively steeper slopes (slope inclination varying from 50° to 60°) having

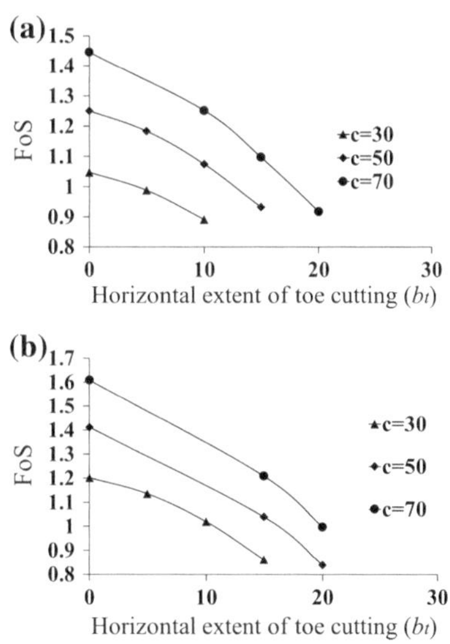

Fig. 6 Effect of toe cutting on partially saturated slopes for **a** $\beta = 40°$, $\varphi = 30°$, **b** $\beta = 40°$, $\varphi = 35°$

higher c and φ values also fall below unity [e.g., for $\beta = 50°$, $c = 30$ kPa, $\varphi = 35°$ the FoS value is 0.917]. A reasonably high value of c and φ [e.g., for $\beta = 50°$, $c = 70$ kPa, $\varphi = 35°$] allows a horizontal extent of toe cutting of around 5 m, but such combination of c and φ values may not be practically feasible.

4 Conclusion

In this paper, simple homogeneous hill slopes are analyzed for vertical toe cutting for construction or widening of roads. Study shows that slopes with very low c and φ values fail under self-weight of the slope only, therefore vertical toe cutting without adequate stabilization measures are not recommended in such cases. Higher c and φ values may lead to a safe toe cutting up to a certain extent depending on the geometrical, seismic and hydraulic conditions. Moreover, steeper slopes having comparatively high c and φ values also fail under self-weight. A reasonably high c and φ values may allow safe vertical toe cutting, but such combination of c and φ values may not be practically feasible.

References

Chakraborty, R., & Dey, A. (2016). Numerical investigation of slope instability induced by hydraulic and seismic actions. In *North-east students geo-congress: NESGC 2016*, Agartala, India, pp. 237–244.

Choudhury, D., Basu, S., & Bray, J. D. (2007). Behavior of slopes under static and seismic conditions by limit equilibrium method. In *Geo-Denver 2007*, Denver, CO, USA.

Ling, H. I., Leshchinsky, D., & Mohri, Y. (1997). Soil slopes under combined horizontal and vertical seismic accelerations. *Earthquake Engineering and Structural Dynamics, 26*, 1231–1241.

Morgenstern, N. R., & Price, V. E. (1965). The analysis of the stability of general slip surfaces. *Geotechnique, 15*(1), 79–93.

Saikia, R., Deka, P., Kalita, S., & Dey, A. (2014) Analysis and behavior of hill slopes and their stabilization measures. In *Indian geotechnical conference (IGC-2014)*, Kakinada, India, pp. 2183–2190.

Stability of Earthen Embankment with Clay Core Under Tidal Fluctuation

Smita Tung, Sibapriya Mukherjee and Gupinath Bhandari

Abstract Seepage through and below earth dam plays an important role in determining the stability of earthen embankment. Stability analysis during rapid drawdown is an important aspect in earthen embankment. Reducing porewater pressure through improved drainage could possibly ensure longer life of embankments. Introduction of filters, drains, clay blankets, clay core, and flatter side slopes are the potential remedial measures to avoid the failure of embankment, due to seepage. Out of these provision of an embankment core is an economic solution. The embankment core is usually installed at the center of the embankment, has significance in controlling the seepage through embankment body. In the present investigation stability analysis of an embankment using clay core under tidal fluctuation has been studied. The current study has been done by seepage and stability analysis of a typical section of an embankment with base width of 17 m and side slope 2(H):1(V) made of homogeneous soil, under rapid drawdown condition. The study has been done by finite element method using SEEP/W software varying thickness of thin clay core of embankment. Pore pressure variations, Flownet, and Phreatic surface have been obtained for the numerical models by SEEP/W software. In this study the pore pressure development of the embankment, by a mixed clay core of different thicknesses, is investigated and evaluated. The stability of the slope during water level drawdown conditions has also been analyzed, considering the variation of porewater pressures calculated from the transient seepage analysis. It has been observed that, factor of safety against slope failure, decreases when river side slope of the embankment is subjected to drawdown condition. The paper brings out the effect of variation on thickness of thin clay core on improvement of stability of earthen embankments through numerical modeling.

S. Tung (✉) · S. Mukherjee · G. Bhandari
Department of Civil Engineering, Jadavpur University, Kolkata 700032, India
e-mail: smitatung@yahoo.in

S. Mukherjee
e-mail: sibapmukh@yahoo.co.in

G. Bhandari
e-mail: g.bhandari.ju@gmail.com

A. I. V. and V. B. Maji (eds.), *Geotechnical Applications*, Lecture Notes in Civil Engineering 13, https://doi.org/10.1007/978-981-13-0368-5_22

Keywords Stability · Pore pressure variations · Flow net · Phreatic surface
Finite element method · SEEP/W

1 Introduction

When a totally or partially submerged slope experiences a decrease of the external water level, the slope may collapse by the increased seepage forces. Failures in slopes or earth dams induced by the water drawdown have been reported and investigated by Morgenstern (1963), Pauls et al. (1999) and Dai et al. (2004). Gao et al. (2014) invented a slope stability chart for 3D slope subjected to drawdown condition. Reducing porewater pressure through improved drainage could possibly ensure longer life of embankments using clay core. In thin core embankment, core or impervious soil zone is flanked on either side by a zone of rock fill. Tidal induced water-level fluctuations have been specifically studied for a long time. Drawdown of an external water level is important in the stability analysis. During a transient process if the total stresses remain constant, the differential equation governing three dimensional transient case through a porous medium when the controlling parameters change with respect to has been can be written as,

$$\frac{\partial}{\partial x}\left(k_x\frac{\partial h}{\partial x}\right)+\frac{\partial}{\partial y}\left(k_x\frac{\partial h}{\partial y}\right)+\frac{\partial}{\partial z}\left(k_x\frac{\partial h}{\partial z}\right)=m_\gamma\frac{\partial h}{\partial t}, \tag{1}$$

where h = total available head under which unsteady seepage occurs; x, y = two mutually perpendicular directions, i.e., horizontal and vertical direction respectively; k_x, k_y = permeability in horizontal and vertical directions respectively; m_γ = storage coefficient. With this in view an attempt has been made in the present study to analyze stability of an embankment using with and with out clay core. With this in view, an attempt to analyze the stability of an embankment has been made in the present study. The present analysis has been done for an embankment of 5 m height with side slope of 2(H):1(V) on seaside and 1.5(H):1(V) on countryside. A 1 m-wide earth filled berm has been considered at 3 m from the bottom with varying clay core thickness. Pore pressure, flow net, and phreatic surface have been obtained by numerical models in numerical analysis. The stability of the slope during water level drawdown conditions is also analyzed incorporating the porewater pressures calculated from the seepage analysis.

2 Analytical Model

The model dam studied is presented with its dimension and the relevant soil parameters are also indicated in Fig. 1. Tables 1 and 2 represent the typical clay core geometry and properties.

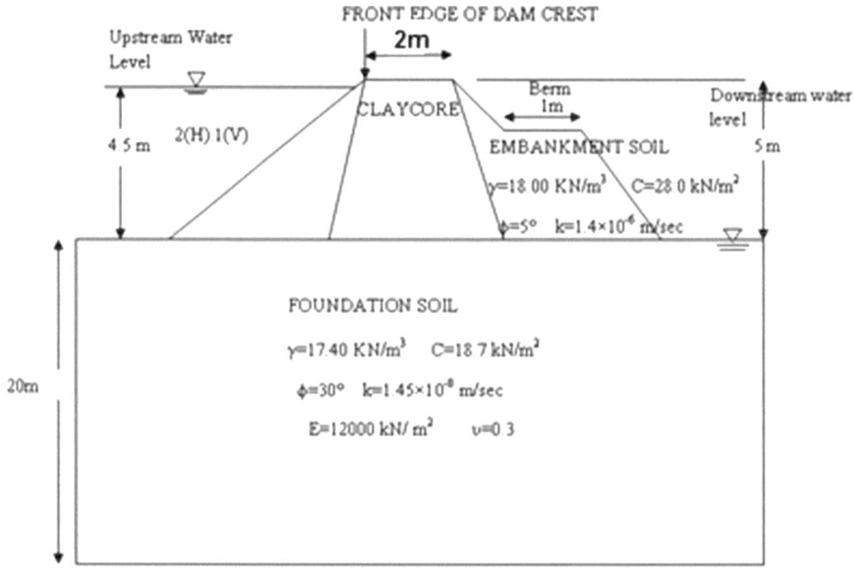

Fig. 1 Schematic diagram of dam with details

Table 1 Geometry of clay core

Core crest width	2 m
Slope of core (x_1)	0.4(H):1(V)
Slope of core (x_2)	0.6(H):1(V)

Table 2 Properties of clay core

Bulk density	19.0 kN/m^3
Cohesion	43.0 kN/m^2
Friction	15°
Coefficient of permeability	1×10^{-10} m/s

3 Numerical Modeling Using Geo-Studio

Seepage and Slope stability analyses have been carried out using Geo-Studio (GEO-SLOPE International 2012) software packages considering uncoupled transient seepage. The soil is modeled by the linear elastic perfectly plastic Mohr–Coulomb (M–C). In this numerical model, at the upstream face of embankment Spline data point function have been applied. At the downstream end of the embankment in this numerical model, potential seepage face boundary condition has been applied. Limit equilibrium method is adopted for the present methods using uncoupled transient seepage analyses to evaluate the stability of slopes affected by change of hydraulic boundary conditions. In finite element calculation, the model has been discretized into smaller finite numbers of 1084 no of quadratic elements.

4 Result and Discussion

The seepage analysis in the present study has been carried out for the earthen dam with clay core. Results obtained from SEEP/W with and with out clay core, at the rate of 0.85 m/h in draw down condition for time (t) = 0, 216 s, 535 s, 1004 s, 1696 s, 2715 s, 4218 s, 6433 s, 9697 s, 14,509 s, 18,109 s and 21,600 s respectively. The results of the parametric studies are discussed below.

4.1 Flownet and Phreatic Surface

Figure 2 shows the flow net diagram obtained from SEEP/W without clay core, at the rate of 0.85 m/h for time (t) = 9697 s in drawdown condition. Figures 3 and 4 show the flow net diagram obtained from SEEP/W with clay core, at the rate of 0.85 m/h for time (t) = 9697 s, 14509 s respectively. Figure 5 show the iso-line of phreatic surface for drawdown rate of 0.85 m/h at time (t) = 0.0 s to 21,600 s respectively obtained from SEEP/W. Phreatic surface shifts in downward direction slowly with drawdown of water level. In all the cases the nature of phreatic surface is in tangent direction to the water level and its proceeds towards downstream side.

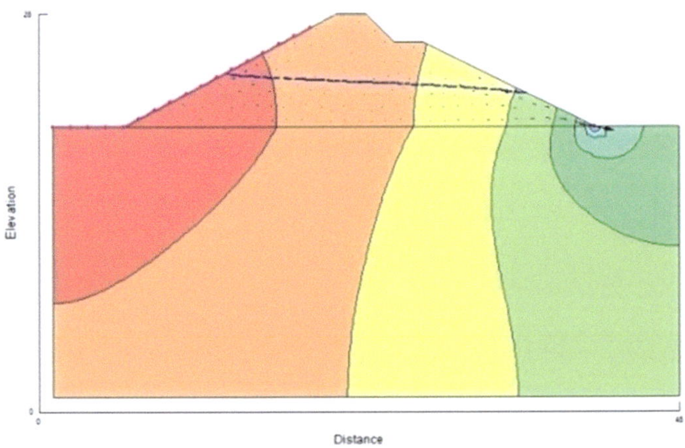

Fig. 2 Earthen dam flow net for drawdown rate = 0.85 m/h and t = 9697 s

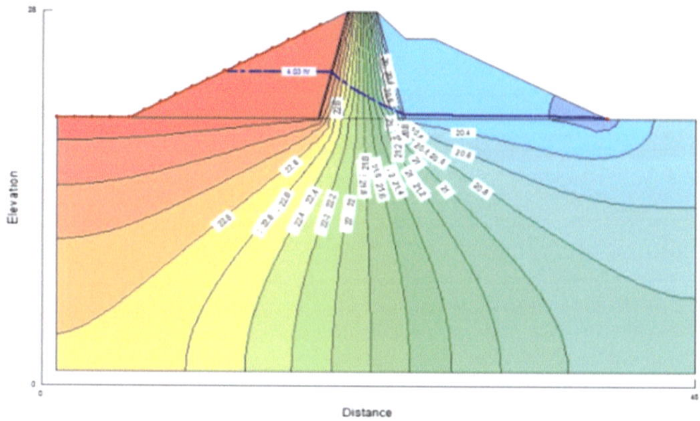

Fig. 3 Earthen dam flow net with clay core for draw down rate = 0.85 m/h and t = 9697 s

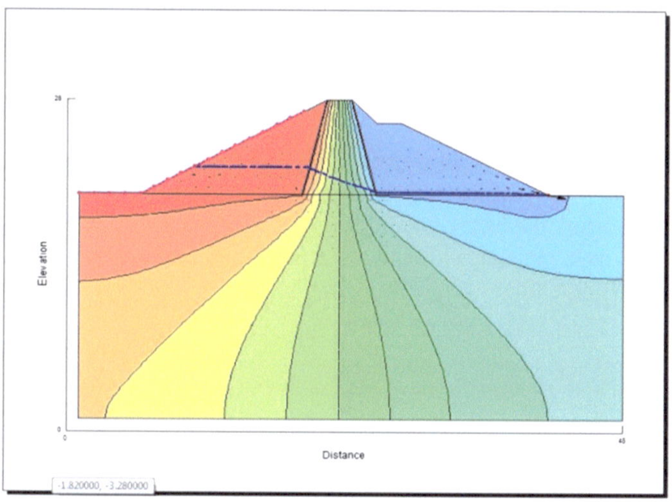

Fig. 4 Earthen dam flow net with clay core for draw down rate = 0.85 m/h and t = 9697 s

From Fig. 3 to Fig. 5 it is clearly observed that the phreatic surface passes through the clay core zone which decreases the exit gradient and therefore increases the factor of safety. From Fig. 5, in all the cases, phreatic surface maintains continuity from the upstream to downstream side of the interface with the increment of time.

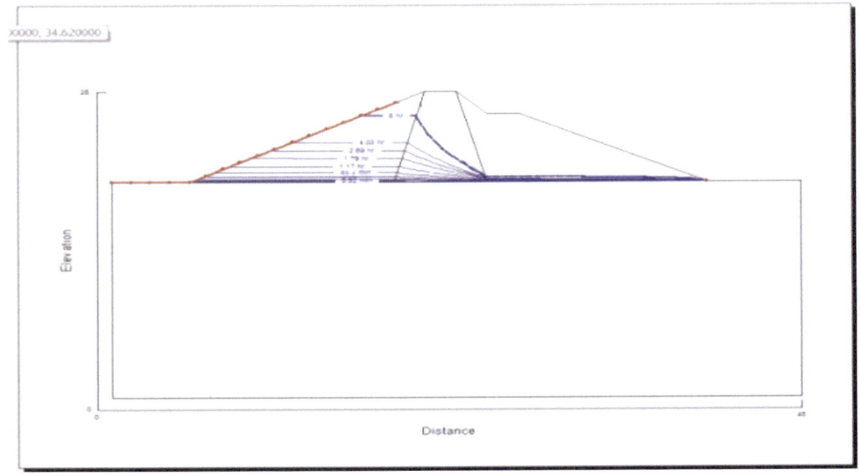

Fig. 5 Iso-line of phreatic surface of earthen dam with clay core for draw down rate = 0.85 m/h for (t) = 0 s to 21,600 s

4.2 *Pore Pressure Variation with Time*

The porewater pressure variation for t = 9697 h with and without clay core represented in Figs. 6 and 7. Figure 8 represent the porewater pressure variation with clay core 5 m below the base of dam for time = 0 s to 21,600 s. It is observed that in case of the embankment with clay core pore pressure is reduced at the downstream side in single tidal cycle development.

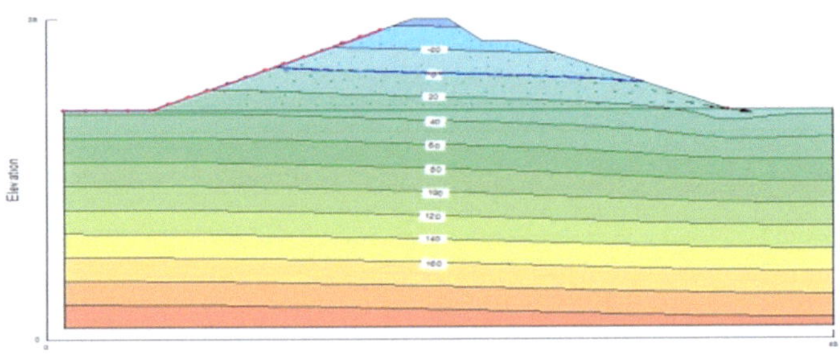

Fig. 6 Porewater pressure variation of earthen dam with out clay core for drawdown rate = 0.85 m/h for time (t) = 9697 s

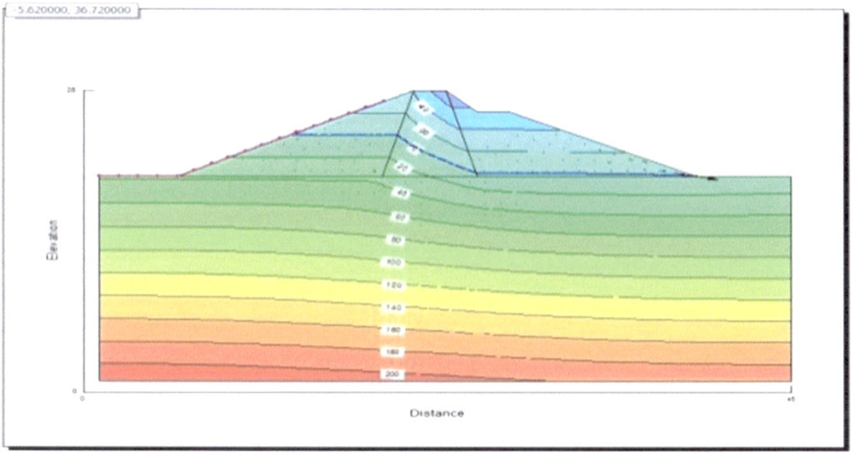

Fig. 7 Earthen dam with clay core for drawdown rate = 0.85 m/h for time (t) = 9697 s

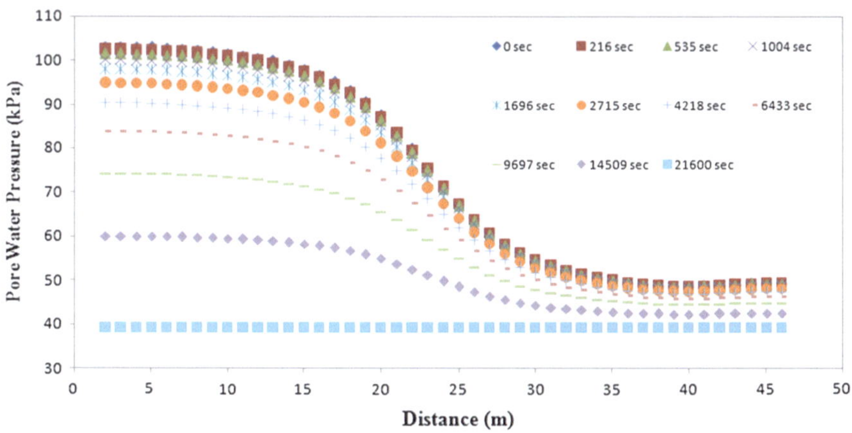

Fig. 8 Porewater pressure variation of earthen dam with clay core for drawdown rate = 0.85 m/h for time (t) = 0 s to 21,600 s (5 m below the embankment)

4.3 Variation of Factor of Safety with Time

Slope Stability analysis has been carried out during the entire tidal cycle using SLOPE/W. The Limit equilibrium method was followed for the uncoupled analysis, with each slice analyzed by the Spencer method. A typical Slip surfaces, F.O.S and Safety map during Draw-down of a single tidal cycle has been shown in Fig. 9. A total of 176 trial slip surfaces were taken and the minimum Factor of Safety was recorded. It is observed from Fig. 10 that with the use of clay core in embankment factor of safety increases up to 17% in transient seepage analysis than embankment

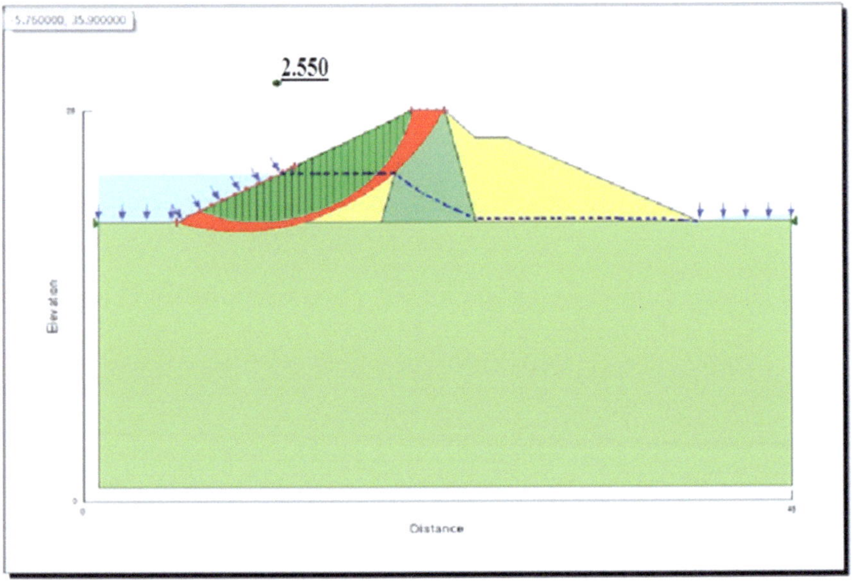

Fig. 9 Factor of safety of earthen dam with clay core for drawdown rate = 0.85 m/h for time (t) = 14,509 s during a single tidal cycle

without clay core in both drawdown conditions. It has been seen that the Factor of Safety (F.O.S) decreases from t = 6 h to t = 12 h, for a single tidal cycle. It is observed from Fig. 11 that approximately factor of safety increases up to 28% due to increment of slope of core material. For a slope under slow drawdown or drawdown with a constant drop in water levels, the factor of safety decreases to the minimum and then increases slightly. The reason for this phenomenon may be due to the balance achieved between the rate of internal porewater pressure and the external water pressure on the submerged boundary.

Fig. 10 Plot of factor of safety of earthen dam with and with out clay core against time during a single tidal cycle

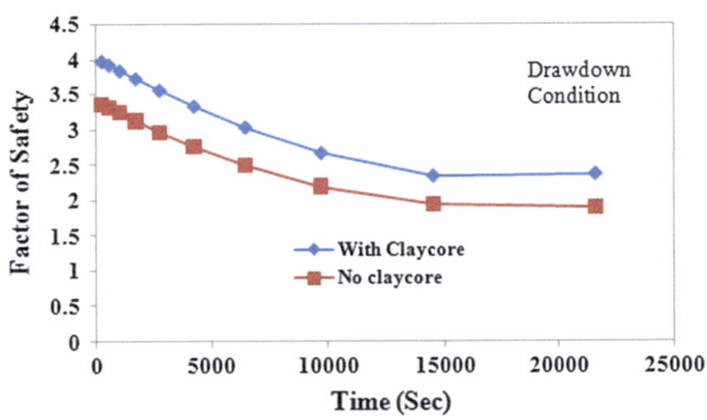

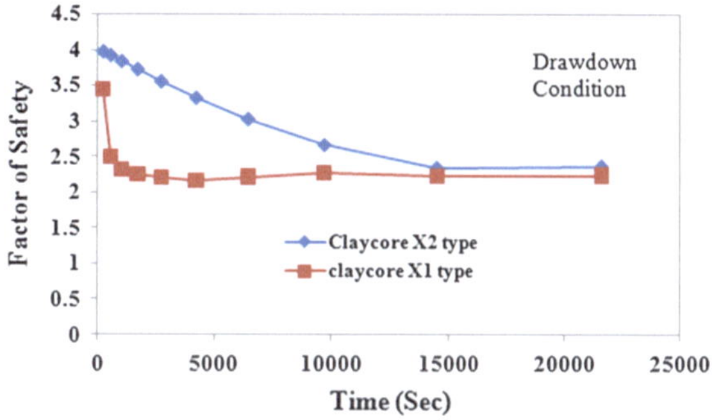

Fig. 11 Plot of factor of safety of earthen dam for different clay core sizes against time during a single tidal cycle

5 Conclusions

The following conclusions may be drawn from the present study:

1. Clay core within embankment effectively change the position of phreatic surface.
2. Development of pore pressure within the embankment and its foundation effectively reduces by using clay core under drawdown condition.
3. With use of clay core factor of safety increases up to 17% in transient seepage analysis than embankment without clay core in tidal cycle.
4. Factor of safety increases up to 28% due to increment of size of core up to 25%.

References

Dai, F. C., Deng, J. H., Tham, L. G., Law, K. T., & Lee, C. F. (2004). A large landslide in Zigui County, Three Gorges area. *Canadian Geotechnical Journal, 41*(6), 1233–1240.

Gao, Y., Zhu, D., Zhang, F., Lei, G. H., Qin, H. (2014). Stability analysis of three-dimensional slopes under water drawdown conditions. *Canadian Geotechnical Journal, 51*, 1355–1364.

GEO-SLOPE International. (2012). *Seepage modeling with SEEP/W, an engineering methodology*.

Morgenstern, N. (1963). Stability charts for earth slopes during rapid drawdown. *Geotechnique, 13*(2), 121–131.

Pauls, G. J., Sauer, E. K., Christiansen, E. A., & Wigder, R. A. (1999). A transient analysis of slope stability following drawdown after flooding of highly plastic cla. *Canadian Geotechnical Journal, 36*(6), 1151–1171.

Reduction of Surcharge Induced Earth Pressure on Rigid Non-yielding Retaining Wall Using Relief Shelves

Vinay Bhushan Chauhan, Rizwan Khan and Satyanarayana Murty Dasaka

Abstract The present study is conducted to investigate the effectiveness of relief shelf to reduce the lateral thrust on rigid non-yielding retaining wall, through a small-scale physical model test in laboratory. The height of the model retaining wall was 700 mm, and five earth pressure sensors were used to measure the lateral pressures along the height of the retaining wall with and without relief shelves. A maximum surcharge pressure of 50 kPa was applied on the backfill. From the results of the study, it is found that relief shelves are effective in reducing total thrust on wall. A parametric study is also carried out on 8 m high wall having 3 relief shelves by varying the width of shelves, through numerical analyses using $FLAC^{3D}$. The study reveals that relief shelves having width factor ranging 0.3–0.8 can reduce total thrust on the wall in range of 11–26%. Among all the cases of retaining wall with relief shelves analysed in the present study, retaining walls with shelf width of 1.4 m (width factor of 0.7) exhibited substantial reduction in lateral thrust, without leading to excessive deflection of relief shelves and backfill surface settlement.

Keywords Lateral earth pressure · Retaining wall · Relief shelf $FLAC^{3D}$

V. B. Chauhan (✉) · R. Khan · S. M. Dasaka
Department of Civil Engineering, Indian Institute of Technology Bombay, Mumbai 400076, India
e-mail: chauhan.vinaybhushan@gmail.com

R. Khan
e-mail: rizwancivil99@gmail.com

S. M. Dasaka
e-mail: dasaka@civil.iitb.ac.in

A. I. V. and V. B. Maji (eds.), *Geotechnical Applications*, Lecture Notes in Civil Engineering 13, https://doi.org/10.1007/978-981-13-0368-5_23

1 Introduction

Assessment of magnitude and distribution of lateral earth pressure on retaining wall under various loading conditions had been an important area of research for geotechnical engineers, as the total thrust on wall is the key factor to decide the sectional dimensions of retaining wall. By reducing the lateral earth pressure on wall, sectional dimensions of wall can be significantly reduced, which would lead to overall economy in construction of retaining wall. Among various measures to reduce total thrust on wall, such as use of expanded polystyrene (EPS) geofoam, geo-boards, rubber tyre chips etc., construction of retaining walls with relief shelves is the least explored technique. However, many retaining walls have been constructed with relief shelves in countries like India, China and Korea (Chauhan et al. 2016). A similar study was also conducted to understand the possible reasons behind the failure of a cantilever retaining wall with relief shelves, which is located in the heart of Hyderabad city, India (Chauhan et al. 2016). Although such type of walls have already been constructed for more than a decade, the mechanism behind the reduction of lateral pressures and design parameter estimation are still in immature state. Hence, present study is aimed to understand the behaviour of such walls and to evaluate the effectiveness of relief shelves in lateral earth pressure reduction. Rigid non-yielding retaining walls with and without relief shelves are considered in the present study and modelled through small-scale physical model tests and full scale numerical model study.

2 Physical Model Tests

To evaluate the effectiveness of relief shelf, small-scale physical model tests are carried out with and without relief shelves with various combinations of location and width of relief shelves. To study the influence of relief shelf on lateral earth pressure distribution on an at-rest wall, 1-g small-scale physical model tests are carried out in a stainless steel tank having dimensions of 1.2 m length, 0.31 m width and 0.7 m depth. Details of the experimental setup are shown in Fig. 1. Six diaphragm type earth pressure cells (EPCs) are fixed along the height of model retaining wall to get the lateral earth pressure distribution on wall. A mechanized travelling pluviator (Gade and Dasaka 2016) is used to prepare uniform sand bed of 80% relative density (friction angle 39° and bulk unit weight 16.5 kN/m^3), while maintaining a height of fall 0.3 m. A uniformly distributed static loading of maximum 50 kPa in 10 kPa increments is applied on the surface of backfill using a hydraulic actuator, and the corresponding load distribution system is shown in Fig. 1. The set up facilitates uniform application of surcharge on the backfill.

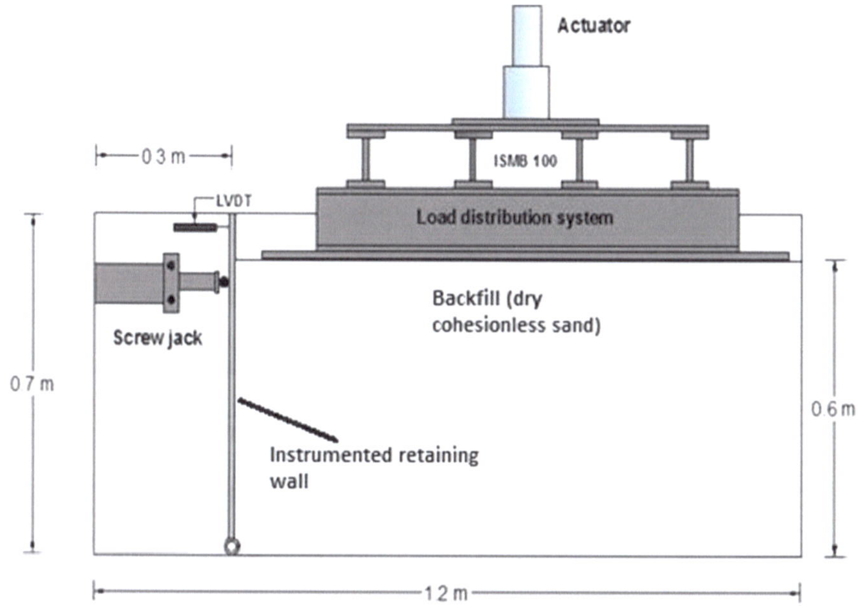

Fig. 1 Detailed experimental setup used in model tests

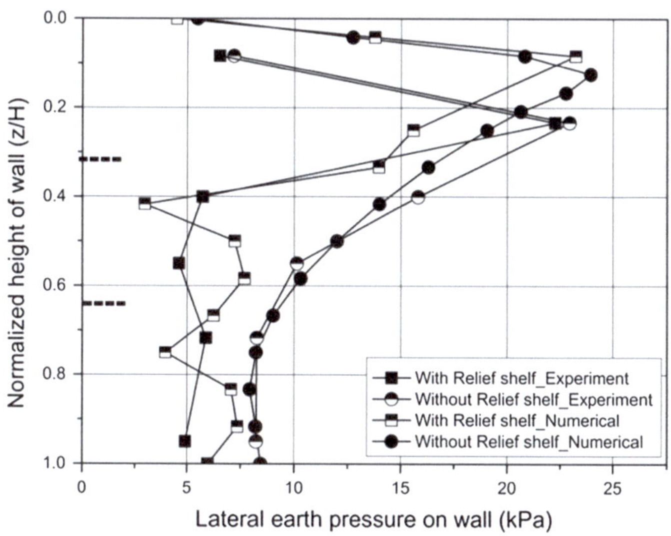

Fig. 2 Comparison of physical model and numerical analysis with and without relief shelves

Figure 2 shows the experimental findings of lateral earth pressure distribution on 0.6 high wall with 2 relief shelves (width 10 cm each placed at z/H = 0.33 and 0.66) and static surcharge of 50 kPa. The earth pressure variation on walls with without relief shelves is also shown in the figure. From the results, it is noted that lateral earth pressure below the relief shelves gets reduced substantially.

3 Modelling of Retaining Wall with Relief Shelves

To evaluate the influence of relief shelf location and width, numerical simulations are carried out with a 8 m high wall (H), having three relief shelves of same widths and located at different position factors, α (z/H = 0.27, 0.56 and 0.84) along the wall. The thickness of relief shelves is kept constant at 0.3 m. Figure 3 shows the numerical model used in the present study.

Physical properties and chosen model for the backfill (dry cohesionless soil) and retaining wall are selected from a similar study on retaining wall with relief shelves (Chauhan et al. 2016). In the present study, width factor, β, of relief shelf is varied from 0.3 to 0.8, to examine the distribution of lateral earth pressure at various sections of wall and total thrust reduction, where β is defined as ratio of B/h (B is width of relief shelf and h represents intermediate height of section of wall between two consecutive relief shelves, which is considered as 2 m in the present study). With this convention of notation, conventional retaining wall without relief shelves is referred to as the wall with β equal to zero. Fixed boundary condition at bottom of the model and roller boundary condition at vertical end of soils are chosen to simulate the field conditions of rigid retaining wall-backfill system. Length of retaining wall (across the plane) is considered as 1.0 m in the present analyses. The rigid wall is modelled as elastic material and not allowed to move away from its initial position to simulate non-yielding condition (at-rest) of wall. The interface between wall and soil is modelled as linear spring-slider system with interface shear

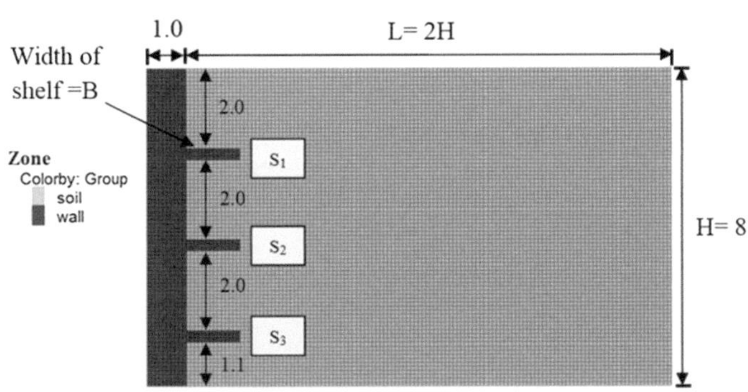

Fig. 3 Numerical model of rigid retaining wall with relief shelves, all dimensions in m (not to scale)

strength defined by the Mohr–Coulomb failure criterion (FLAC^{3D} 2011). Once the model reaches equilibrium condition, a static surcharge of 50 kPa is applied in the form of strip loading on the backfill surface starting at 0.4 m (*H*/20) away from the edge of wall. The numerical model used in the present study is validated using lateral earth pressure profiles obtained from experimental studies on wall with two relief shelves and without relief shelf, as presented in Fig. 2.

4 Results and Discussion

In the present study, rigid retaining walls with three relief shelves positioned at $\alpha = 0.27$, 0.56 and 0.84, and β in the range of 0.3–0.8 are analysed with FLAC^{3D}, to evaluate the influence of β on lateral earth pressure distribution, total lateral thrust, backfill settlement and deflection of relief shelves.

Earth pressure distribution on all walls with and without relief shelves are studied and shown in Fig. 4. It can be observed that lateral earth pressure in topmost segment of wall increases with the increase in width factor of relief shelf. This behaviour may be attributed to the fraction of applied surcharge load carried by topmost relief shelf. As the width of relief shelf increases, a greater portion of

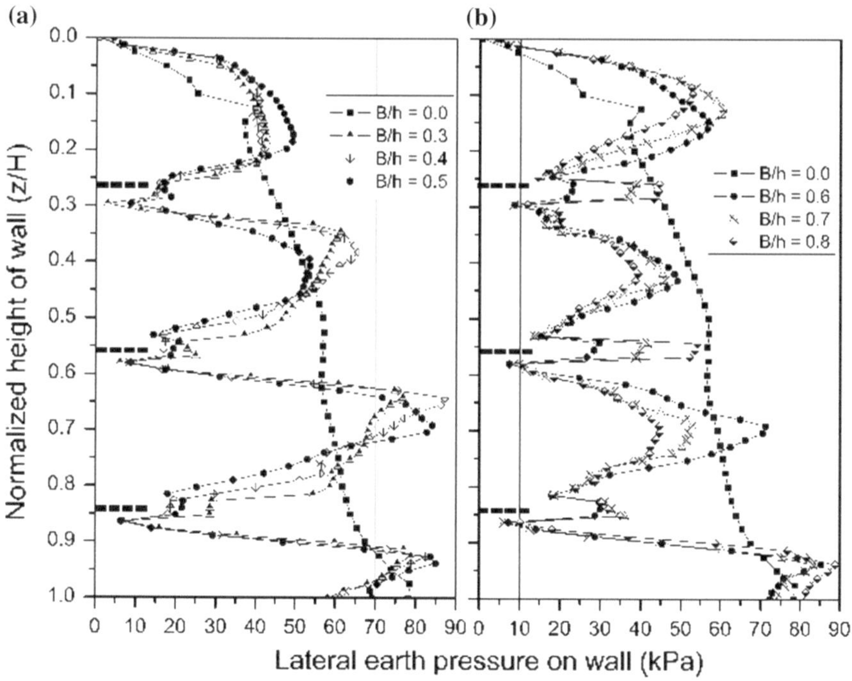

Fig. 4 **a**, **b** Comparison of lateral earth pressure on the wall with and without relief shelves

surcharge is carried by the topmost relief shelf itself. Once loading on relief shelf increases, vertical pressure in the soil overlying a relief shelf also increases, leading to increase in the lateral earth pressure on wall. However, in lower sections of wall, lateral earth pressure has reduced significantly below the relief shelf (Fig. 4a) compared to wall without relief shelf for $\beta > 0.5$.

This behaviour of lateral earth pressure profile may be attributed to the surcharge above the relief shelf is being carried by relief shelves itself, and soil overburden and static surcharge is not getting transferred to the soil below the relief shelf, unlike in the case of wall without relief shelf. It is noted that when $\beta \geq 0.6$, significant reduction in lateral earth pressure is observed in the sections of wall lying between any two relief shelves (Fig. 4b). A substantial reduction of lateral thrust, in the range of 11–26%, is noticed by provision of relief shelves of various widths, as shown in Fig. 5. Although, for the relief shelves having $\beta = 0.3–0.5$, a reduction of lateral thrust in the range of 11–17% is observed, this reduction is very significant for walls with $0.6 \leq \beta \geq 0.7$. With further increase in β, no further increase in reduction of lateral thrust is noted. Maximum deflection of all relief shelves from top to bottom (S_1, S_2 and S_3) are compared and summarized in Fig. 6. For a given wall, maximum deflection of relief shelf is found maximum for topmost relief shelf and it decreases from top to bottom relief shelves for all retaining walls. Maximum deflection of relief shelves has immensely increased when $\beta > 0.7$. This behaviour may be attributed to greater part of applied surcharge is being supported by higher width of relief shelf and thickness of such relief shelves (0.3 m) is not significant to support that much of surcharge. Surface settlement of backfill is an important serviceability criterion for retaining walls. Excessive backfill settlement leads to collapse of backfill soil and leads to failure of structures founding on it or in the close proximity.

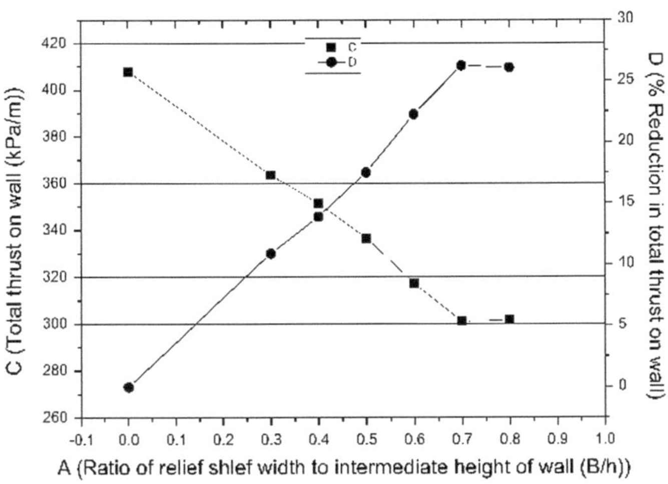

Fig. 5 Total thrust and reduction in thrust on retaining walls

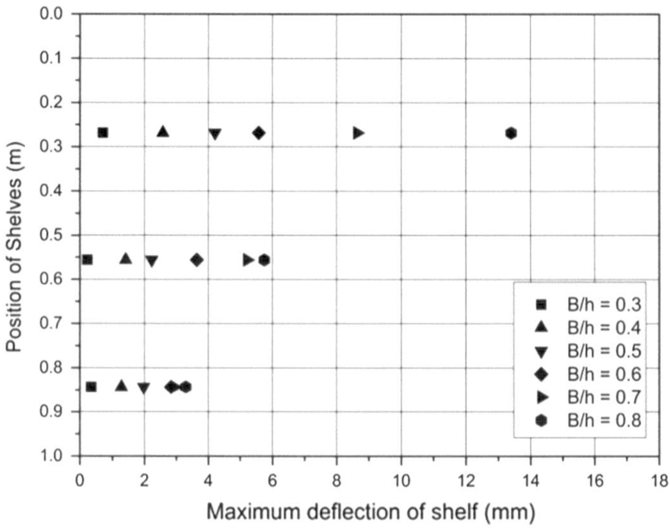

Fig. 6 Maximum vertical deflection (mm) profile of relief shelves

Backfill settlement near the wall is small (1–5 mm) and it increases for walls with $\beta > 0.5$, compared to wall without relief shelf (Fig. 7). Although, with the provision of relief shelf, backfill settlement has reduced near the wall stem compared to wall without relief shelf ($\beta \leq 0.5$), but for walls having wider relief shelves ($\beta > 0.5$), surface settlement near the wall is higher compared to that of wall without relief shelf. As discussed earlier, with increase in width of relief shelf, deflection of relief shelf increases, which leads to higher backfill settlement near the wall. When $\beta > 0.7$, rapid increase in backfill surface settlement is observed near the wall stem, which is due to the higher deflection of relief shelf. Effect of provision of relief shelves on backfill surface settlement has continuously been diminished with increasing distance from stem and achieved the same profile, as that of walls without relief shelves beyond $z/H \geq 1$. Among all the cases of retaining wall with relief shelves analysed in the present study, wall with $\beta = 0.7$ provides maximum benefit in terms of reduction in total thrust, without leading to excessive deflection of relief shelves and backfill surface settlement. Maximum width of relief shelf should be restricted, as relief shelves having higher width experience large deflection leading to higher backfill surface settlement, which may affect the serviceability of nearby structures. Although, it is noteworthy that with increase in width of relief shelf, reduction in total thrust also increases but for a given height of wall and surcharge loading, there exists a certain upper value for width of relief shelf, which provides maximum reduction of total thrust on wall while satisfying the criteria for serviceability within limits. This maximum value of width of relief shelf depends on factors like height of wall, number of shelves and

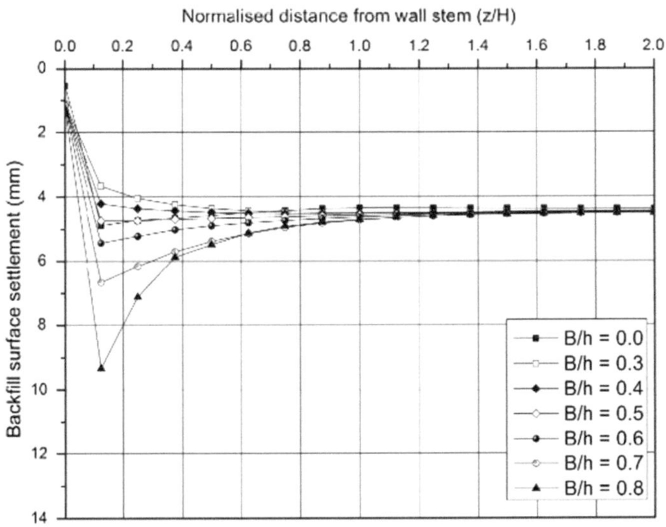

Fig. 7 Backfill surface settlement profile of backfill

thickness of relief shelves, magnitude of surcharge loading and position of surcharge loading from face of the stem, etc. So, it is customary to examine aforementioned factors before deciding the number, position, width and thickness of relief shelf to be provided for any retaining wall, to achieve maximum benefit from the retaining walls with relief shelves.

5 Conclusions

Present study evaluates the effectiveness of rigid non-yielding retaining walls with relief shelves in reducing the lateral thrust acting on the wall. Small-scale physical model studies and numerical model studies on 8 m high retaining wall are carried out in the study. A surcharge load of 50 kPa on the backfill is considered in the study. Influence of width factor of relief shelf having three relief shelves of same width provided at different position factors, $\alpha = z/H$, of 0.27, 0.56 and 0.84, is studied. The results obtained from the analyses revealed that provision of relief shelves on non-yielding rigid retaining wall provides significant reduction in total thrust on wall. For 8 m high retaining wall with three relief shelves, the reduction in lateral thrust is observed in the range of 11–26%. It is also noted that backfill surface settlement reduces due to relief shelves near the wall up to $\beta = 0.5$. However, for walls with $\beta > 0.5$, the backfill surface settlement increases with increase in width of relief. Also, it is observed that deflection of relief shelf is proportional to the width of relief shelf, and it also decreases from top shelf to bottom shelf for a given retaining wall with relief shelves. Among all the walls analysed in the present study, retaining wall with 1.4 m wide relief shelves ($\beta = 0.7$) proves ideal, as it substantially reduces total lateral thrust, without leading to excessive deflection of relief shelves and backfill surface settlement.

References

Chauhan, V. B., Dasaka, S. M., & Gade, V. K. (2016). Investigation of failure of a rigid retaining wall with relief shelves. *Japanese Geotechnical Society*. https://doi.org/10.3208/JGSSP.TC302-02.

FLAC^{3D}, 5.0. (2011). *3-dimensional finite difference code*, Itasca, Minneapolis, USA.

Gade, V. K., & Dasaka, S. M. (2016). Development of a mechanized traveling pluviator to prepare reconstituted uniform sand specimens. *Journal of Materials in Civil Engineering*. https://doi.org/10.1061/(asce)mt.1943-5533.0001396,04015117.

Stability Analysis of Non-homogeneous Soil Slopes Using Numerical Techniques

D. Chatterjee and A. Murali Krishna

Abstract This paper presents the stability analyses of non-homogeneous slopes under different loading conditions. A non-homogeneous soil slope with two different soil layers is considered. Special cases are created, varying the height of the layers, to account for the non-homogeneity of the earth slope. A water table is considered in this study to account for the seepage forces. Pseudo-static earthquake force is considered, taking into account both the forces in the horizontal and vertical directions. A rigorous limit equilibrium method of slices, i.e. Morgenstern-Price method is used to analyze the stability of the slope. Finite element shear strength reduction technique is also used for displacement calculations and comparison with limit equilibrium method. Safety factor, critical slip surfaces and displacements with different loading conditions are studied and compared.

Keywords Numerical model · Non-homogeneous slope · Displacements
Critical surface

1 Introduction

Slope stability is a major area of concern for practicing engineers for filled slopes or cut slopes. Factor of safety is an important parameter in stability evaluation of slopes but the total displacements occurring can give a better idea about the zone of failure. Position of critical slip surface evaluation is another important issue in slope stability. Slopes can be man-made or natural and many external natural forces act on these earth structures making it vulnerable to failure. Out of these external

D. Chatterjee (✉) · A. Murali Krishna (✉)
Department of Civil Engineering, Indian Institute of Technology Guwahati,
Guwahati 781039, Assam, India
e-mail: dooradarshi@iitg.ernet.in

A. Murali Krishna
e-mail: amurali@iitg.ernet.in

A. I. V. and V. B. Maji (eds.), *Geotechnical Applications*, Lecture Notes in Civil
Engineering 13, https://doi.org/10.1007/978-981-13-0368-5_24

forces, the effects of hydrology and seismicity arc thc most common. Rainfall affects the stability of slopes in numerous ways like the increase in height of water table in an area due to heavy downpour or the formation of phreatic line. In high seismic zones slopes suffer greater dynamic loadings which may eventually lead to instability. Varying soil properties make the slope non-homogeneous in nature. Many studies are available in literature like the effect of variation of cohesion with depth has been investigated by Koppula (1984). Kim et al. (2002) compared stability of complex soil slopes using upper bound and lower bound limit analysis. Hammouri et al. (2008) studied stability of layered slopes and showed effects of drawdown, tension crack and undrained clay soils. Qian et al. (2014) proposed stability charts for two-layered cohesive slopes. This paper deals with a two-layered non-homogeneous slope whose layer height is varied, water table is introduced, pseudo-static forces are applied and then the stability of the slope is evaluated.

2 Methodology

Various methods are available for stability determination of earth slopes, they have their own advantages and limitations. Limit equilibrium method gives the factor of safety of slope and position of critical slip surface. It does not give information on deformations occurring within the slope so to overcome this, finite element method is selected as well.

2.1 *Limit Equilibrium Method*

It is a statically indeterminate method, hence for the solution; some assumptions are required to make it determinate. Therefore, anon-circular failure surface is assumed

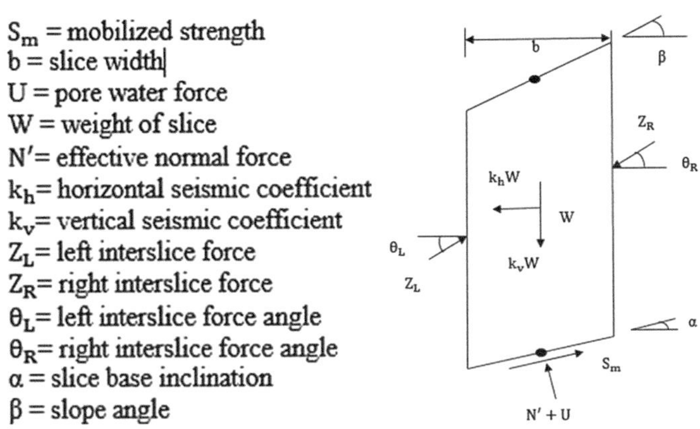

Fig. 1 Slice representing various forces

and the soil mass is divided into a number of slices and each slice is checked for equilibrium of forces and moments. Figure 1 depicts the various forces acting on a typical slice.

Depending upon the type of equilibrium satisfied, various methods are formulated for the factor of safety of slope such as ordinary method of slices (Fellenius 1936), Bishop's Simplified method (Bishop 1955) and Janbu's generalized method of slices (Janbu 1968). Complete equilibrium is followed by Spencer (1967) and Morgenstern and Price (1965) methods. Then a search procedure is used to find the critical slip surface giving the minimum factor of safety of the slope. Morgenstern and Price (1965) method with a half sine function has been used throughout this study. Circular slip surface is assumed for homogeneous profiles while non-circular is taken for non-homogeneous profiles. The number of slices is fixed at 20 while the tolerance is taken to be 0.005.

2.2 *Finite Element Method*

Zienkiewicz et al. (1975) studied c-ϕ slopes using the finite element method showing good comparison with slip circle results. With more use and confidence gained, researchers like Matsui and San (1992), Ugai and Leshchinsky (1995), Griffiths and Lane (1999) and Cheng et al. (2007) used it for stability analysis of slopes. The safety factor, F is defined as the factor by which the strength parameters need to be reduced to bring the slope to a failure point.

$$c'_f = \frac{c'}{F} \tag{1}$$

$$\tan \phi'_f = \frac{\tan \phi'}{F} \tag{2}$$

This method is known as the shear strength reduction technique. Here failure is taken as that point where the solutions fail to converge within a specified number of iterations. The stress distribution fails which implies that the Mohr–Coulomb failure criterion and the global equilibrium is not satisfied. Failure of slopes is accompanied by a sudden rise of displacements.

3 Development of Numerical Model

A numerical model (Fig. 2) has been developed into represent the homogeneous and non-homogeneous slope in Rocscience software suite using Phase and Slide modules (Rocscience 2016). Soil-1 is silty clay while soil-2 is silty sand in nature. The slope angle is taken as 2:1 (H:V) and the slope height is 20 m and depth factor

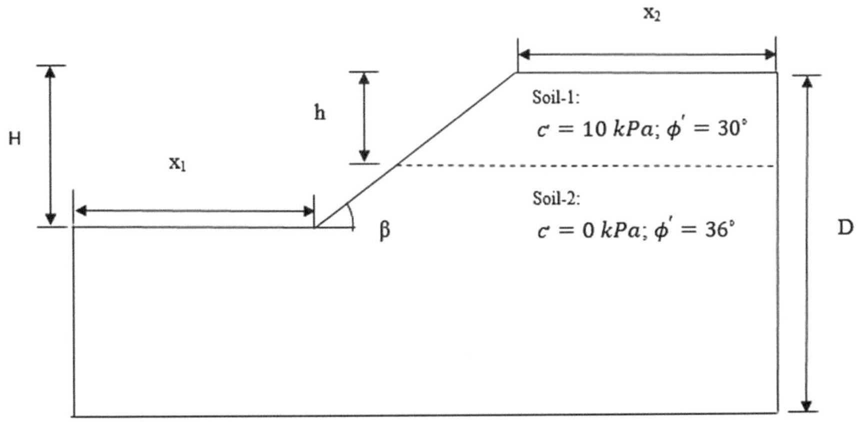

Fig. 2 Non-homogeneous slope model

D is 2. The Mohr–Coulomb failure criterion is used for modeling the soils. Drained values are selected for shear strength parameters of the two soils. Young's modulus value for the two soils is taken as 5 MPa while the Poisson's ratio is 0.3. Total unit weight of 16 kN/m^3 is applied to the layered soil model. Cohesion values for soil-1 and 2 is 10 and 0 kPa, respectively. Friction angle values are 30° and 36°, respectively. The permeability values of the two soils are 4×10^{-7} and 8.2×10^{-6} m/s, respectively.

Stability of slopes being an unconfined problem, the selection of the dilation angle doesn't make much of a difference in the analysis (Griffiths and Lane 1999). A non-associated flow rule has been used throughout this study. The soil model consists of 2700 uniform triangular six-noded elements and 5500 nodes. The height of the top layer is varied from zero to total depth, *DH*. In essence the first and the last case becomes a homogeneous slope with either of the soil layers. Table 1 shows the variation of the layer height. The *h* values shown in table represent the depth of layer from the ground surface. When the ratio is increasing the soil in the upper region is actually forming the major portion of the slope. A pseudo-static seismic coefficient of 0.2 is taken representing violent earthquakes. Both horizontal and vertical pseudo-static acceleration are taken equal as suggested by Shukha and Baker (2007). A water table is also introduced and classified as case-2 while the other case, i.e. case-1 is free of pseudo-static force and water table.

Table 1 Variation of *h*/*D* ratio

h/*D*	0	0.2	0.4	0.6	0.8	1
h (m)	0	8	16	24	32	40

4 Results

The slope has been analyzed using two different approaches, their results and comparison is shown in Tables 2 and 3 for homogeneous and non-homogeneous cases, respectively. Difference in safety factor values between the two methods range from 0.19 to 12.45% for the non-homogeneous slope.

The factor of safety increases with layer height and becomes constant after $h = 0.6D$. The increase in content of the silty clay layer within the slope increases its safety factor value. In terms of the two cases considered, the factor of safety from case-2 show a huge decrease from case-1. Figure 3 depicts the critical slip surfaces for various layer heights corresponding to case-1. Critical surface for homogeneous slope with silty clay soil show a toe type of failure. Shallow slope failure is observed for layer heights $h = 0.2D$, $0.4D$ and homogeneous slope with silty sand soil. Critical surface for layer heights $h = 0.6D$ and $0.8D$ coincide with each other having a deeper slope failure. The increase in the silty clay layer shifts the critical slip surface deeper into the slope simultaneously increasing the factor of safety as well.

Critical slip surfaces for various layer heights corresponding to case 2 is shown in Fig. 4. For $h = 0.2D$ and $h = 0.4D$ the slip surface is shallow and parallel to the slope while for the other two heights they coincide with each other having a deep slip surface. From the figure it is evident that the slope with more silty sand soil has a critical slip surface which is shallow and parallel to slope while with more silty clay soil the failure is a deep slope failure. The total displacement contours and the

Table 2 Homogeneous slope with two different soils

Slope model	Factor of safety					
	LEM		FEM		Difference (%)	
	Case-1	Case-2	Case-1	Case-2	Case-1	Case-2
Soil-1 $h = D$	1.687	1.114	1.68	1.05	0.41	5.74
Soil-2 $h = 0D$	1.453	0.999	1.48	0.98	1.85	1.90

Table 3 Non-homogeneous slope with two soil layers

Layer height	Factor of safety					
	LEM		FEM		Difference (%)	
	Case-1	Case-2	Case-1	Case-2	Case-1	Case-2
$h = 0.2D$	1.478	1.018	1.55	1.02	4.87	0.19
$h = 0.4D$	1.578	1.088	1.64	1.05	3.92	3.49
$h = 0.6D$	1.736	1.188	1.68	1.f05	3.22	11.61
$h = 0.8D$	1.736	1.188	1.68	1.05	3.22	11.61

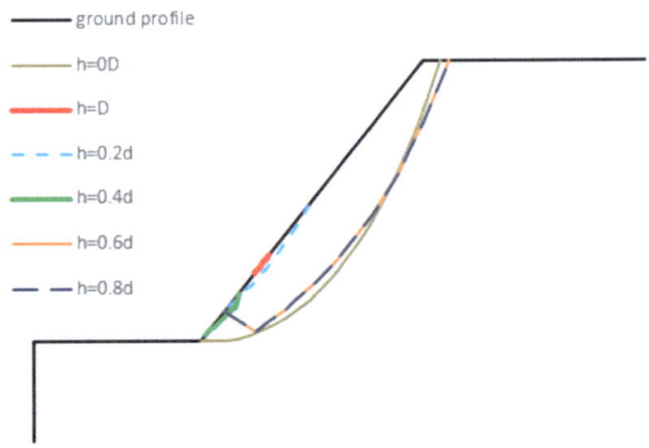

Fig. 3 Critical surfaces for various layer heights for case-1using the limit equilibrium method

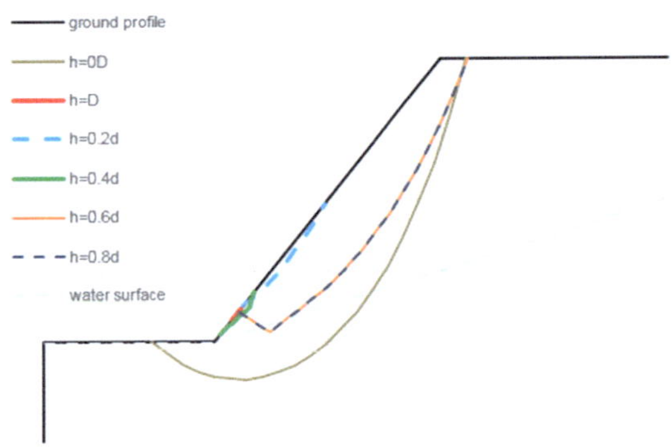

Fig. 4 Critical surfaces for various layer heights for case-2 using the limit equilibrium method

critical slip surface from the limit equilibrium method for case-2 and four layer heights are shown in Fig. 5. The displacement increases with increase in layer height, i.e., the zone of failure increases and parallel to the slope while the slip surface from limit equilibrium method is small and confined to the toe region.

The maximum displacement is 0.066 m. The same is observed for $h = 0.4D$ but here the zone of failure increases but the slip surface from limit equilibrium method is much smaller. For this height the maximum displacement is 0.108 m. For $h = 0.6D$ and $h = 0.8D$ the zone of failure increases and it is seen spreading outwards.

For the last two-layer heights, the critical slip surface from limit equilibrium method is same and they coincide while finite element method depicts that their displacements are not the same, layer height $h = 0.8D$ has more displacement than height $h = 0.6D$. The maximum displacement for the last two layer heights are 0.107 and 0.126 m respectively. The variation of factor of safety with layer height to depth factor ratio for case-1 and case-2 is shown in Fig. 6. Out of the two cases considered it is seen that for case-1 the factor of safety values are higher than that of case-2. The factor of safety increases with layer height to depth factor ratio up to height 0.6D and stays constant after that.

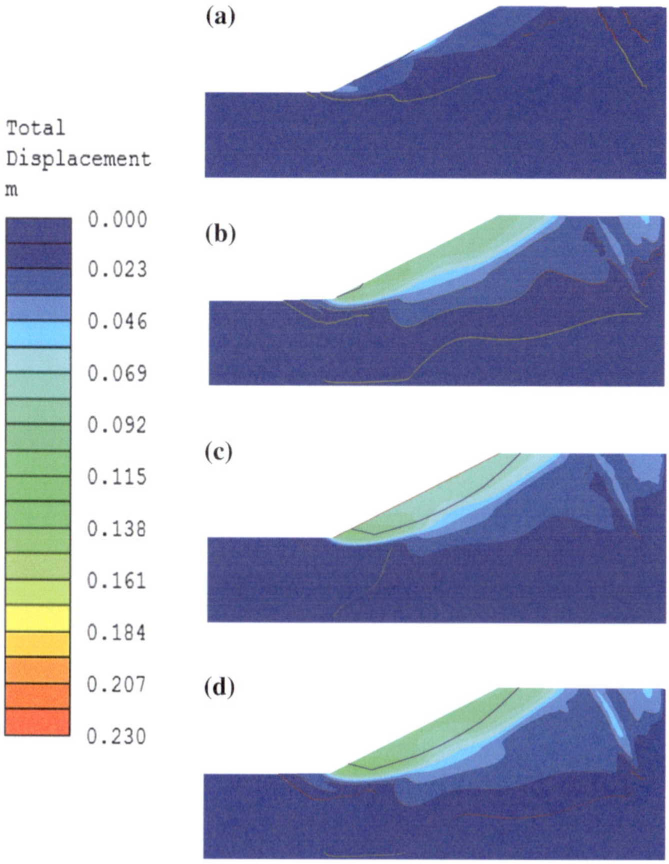

Fig. 5 Contours of total displacements from FEM and critical surface from LEM for case-2 and layer height **a** $h = 0.2D$, **b** $h = 0.4D$, **c** $h = 0.6D$ and **d** $h = 0.8D$

Fig. 6 Variation of safety factor with height ratio for case-1 and 2

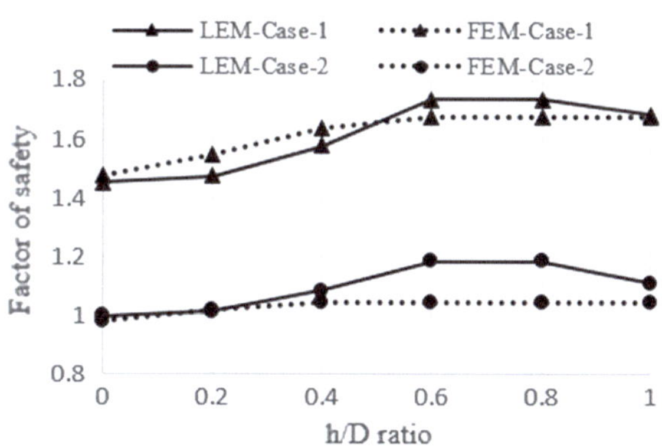

5 Conclusions

A two-layered non-homogeneous slope with a silty clay layer at the top and silty sand layer at the bottom is prepared. Two cases are considered here of which case-2 is more critical where a water table and pseudo-static force has been introduced. Considering only the non-homogeneous profiles, the factor of safety and strength reduction factor is getting increased with increase in layer height from $0.2D$ to $0.6D$ then it stays constant. The minimum safety factor is observed for layer height $h = 0.2D$. It is clear that the silty clay soil imparts strength to the two-layered slope and makes it more stable. For layer height $0.2D$ and $0.4D$ the critical surfaces are shallow and parallel to slope while the surface coincides for heights $0.6D$ and $0.8D$ which has a deeper surface depicting a slope failure. Hence it is found that slopes with sandy soil have a shallow slope failure parallel to slope surface and slopes with clay soil have a deeper slope failure. Displacements show an increasing trend with increasing layer height, i.e. with increase in layer height the silty clay soil in the slope increases and with it the differential displacement also increases indicating a deeper zone of failure.

References

Bishop, A. W. (1955). The use of the slip circle in the stability analysis of slopes. *Geotechnique, 5*(1), 7–17.

Cheng, Y. M., Lansivaara, T., & Wei, W. B. (2007). Two-dimensional slope stability analysis by limit equilibrium and strength reduction methods. *Computers and Geotechnics, 34,* 137–150.

Fellenius, W. (1936). Calculation of the stability of earthdams. In *Proceedings of 2nd congress on large dams*, Washington, D.C. 4.

Griffiths, D. V., & Lane, P. A. (1999). Slope stability analysis by finite elements. *Geotechnique, 49* (3), 387–403.

Hammouri, A., Malkawi, A. I. H., & Yamin, M. M. A. (2008). Stability analysis of slopes using the finite element method and limit equilibrium approach. *Bulletin of Engineering Geology and the Environment, 67,* 471–478.

Janbu, N. (1968). Slope stability computations. In *Soil mechanics foundation engineering report.* Trondheim: Technical University of Norway.

Kim, J., Salgado, R., & Lee, J. (2002). Stability analysis of complex soil slopes using limit analysis. *Journal of Geotechnical and Geoenvironmental Engineering, 128*(7), 546–557.

Koppula, S. D. (1984). Pseudo-static analysis of clay slopes subjected to earthquakes. *Geotechnique, 34*(1), 71–79.

Matsui, T., & San, K. C. (1992). Finite element slope stability analysis by shear strength reduction technique. *Soils and Foundations, 32*(1), 59–70.

Morgenstern, N. R., & Price, V. E. (1965). The analysis of the stability of general slip surfaces. *Geotechnique, 15*(1), 79–93.

Qian, Z. G., Li, A. J., Merifield, R. S., & Lyamin, A. V. (2014). Slope stability charts for two-layered purely cohesive soils based on finite-element limit analysis methods. *International Journal of Geomechanics, ASCE.*

Rocscience Inc. (2016). RS2 v9. Toronto, Canada: Rocscience Inc.

Shukha, R., & Baker, R. (2007). Design implications of the vertical pseudo-static coefficient in slope analysis. *Computers and Geotechnics, 35,* 86–96.

Spencer, E. (1967). A method of analysis of the stability of embankments assuming parallel inter-slice forces. *Geotechnique, 17*(1), 11–26.

Ugai, K., & Leshchinsky, D. (1995). Three-dimensional limit equilibrium and finite element analysis: a comparison of results. *Soils and Foundations, 35*(4), 1–7.

Zienkiewicz, O. C., Humpheson, C., & Lewis, R. W. (1975). Associated and nonassociated visco-plasticity and plasticity in soil mechanics. *Geotechnique, 25*(4), 671–689.

Part III
Theme 7: Rock Engineering, Tunneling and Underground Constructions

Static and Dynamic Slope Stability Assessment of a Himalayan Rock Slope

Amalesh Jana, Mithresh Pushpan, Arindam Dey, S. Sreedeep and A. Murali Krishna

Abstract Stability of rock slope along the stretch of Badrinath National Highway (NH-58) of Garhwal Himalayas in Uttarakhand, India is very much essential to protect the infrastructure and livelihood in the area nearby. Every year this part of the slope has been facing severe landslides due to intensive rainfall causing substantial damages. The chosen area is also highly sensitive to earthquake according to Geological survey of India. Due to the severe earthquake and intensive rainfall, stability of rock mass is under continuous depletion. Rock slope in this area is highly jointed and possess a joint orientation which is highly vulnerable to seismic force. In this study, finite element slope stability analysis of this rock slope has been carried out in commercial software $PHASE^2$. As per the earthquake zonation map of India, the site is located in seismic zone V, and hence, dynamic stability of slope has been performed considering pseudo static and time-response method. Initially, the rock slope is considered to be devoid of any joints and is modeled as a continuum mass using equivalent Mohr–Coulomb shear strength criteria. In the second model, the discontinuities are applied in the continuum model by providing interface elements as joints in between the rock walls. Results from the analysis accentuate the fact that the of rock slopes is stable under high seismic force.

Keywords Rockslope stability · Static and dynamic analysis · Continuum and discontinuum modeling · $PHASE^2$

A. Jana (✉) · M. Pushpan · A. Dey · S. Sreedeep · A. Murali Krishna
Department of Civil Engineering, Indian Institute of Technology Guwahati, Guwahati 781039, India
e-mail: j.amalesh@iitg.ernet.in

M. Pushpan
e-mail: mithresh@iitg.ernet.in

A. Dey
e-mail: arindamdeyiitk@gmail.com

S. Sreedeep
e-mail: srees@iitg.ernet.in

A. Murali Krishna
e-mail: amurali@iitg.ernet.in

A. I. V. and V. B. Maji (eds.), *Geotechnical Applications*, Lecture Notes in Civil Engineering 13, https://doi.org/10.1007/978-981-13-0368-5_25

1 Introduction

In rock mass, discontinuities can be present in the form of joints, faults, fissures, bedding, etc. The mechanical behavior of jointed rock masses is strongly affected by the properties and geometry of the joints (Cai and Horri 1992). The characterisation of rock slope which is located at a distance of 18 km from Rishikesh towards Vyasi National Highway (NH-58) is mainly quartzite, dolomites with slates. The static stability analysis of this rock slope is reported by Pain et al. (2014) and it is mentioned that rock slope is safe under static condition. However, no dynamic analysis is done to investigate the vulnerability of the rock slope in seismic condition. Every year minor to major landslide triggered by incessant rainfall in all six hilly districts of Uttarakhand causes death of many lives and enormous loss of properties. This area is also located in a high seismic prone zone and experienced a number of earthquakes in past and in recent times. Uttarkashi earthquake in 1991 and Chamoli earthquake in 1999 caused significant damage of life and property. So, it is essential to study the behavior of landslide and the stability of rock slope under different destructive environment condition which will be benevolent for planning and implementing mitigation measures.

1.1 Geological Description of the Area

The study area is located in the Alakananda valley of Garhwal Himalayas in Uttarakhand. This region consists of several tectonic faults, fold and overthrusts which are responsible for the crushing and shearing of rocks. In this case study, three set of parallel joints are present in the rock slope in different orientations, having different joint strength and stiffness, which influence the stability of the rock slope. Persistence of joints plays an important role in stability which indicates the length of joint. A fully persistent joint is more vulnerable than low persistent joints. It is very cumbersome process to consider all the persistence in the numerical model because measuring the length of persistence from the field is difficult for high slope. So the continuous parallel set of joints is considered for conservative solution. The dip direction of slope face and Joint set J1 are same (180°) and the dip angle of J1 (27°) is less than dip angle of slope face (54°). Joint set J1 is dipping out of the slope face and striking parallel to the slope face. This means the joint set J1 daylights in the slope face which indicates planar failure along joint set J1. The dip direction of other joint set J2 and J3 is unfavorable to sliding. Because the dip angle of both J2 (86°) and J3 (75°) are more than the dip angle of slope face (54°). Dip direction and dip angle of the joints are unfavorable to wedge and toppling failure.

2 Numerical Model

PHASE^2 is a powerful 2D elasto-plastic finite element program used to calculate stresses and displacement in soil and rock applications and can even generate discrete joint or fracture networks in the rock model. In this study, the analysis is carried out in two parts. First, the rock slope is modeled considering the continuum mass in finite element model using equivalent Mohr–Coulomb shear strength criteria without having any joints. However, the effect of discontinuity is considered by reducing the strength parameter of the intact rock (Table 3) to get equivalent Mohr–Coulomb strength parameter (Pain et al. 2014) shown in Table 2. In the second model, the discontinuities are applied in the continuum model by providing interface elements of zero thickness as joints in between the rock walls. The generalised Hoek–Brown criteria (Hoek et al. 2002) for rock mass shear strength and Barton-Bandis constitutive model to represent the joint strength are used in the model. Static and dynamic stability analysis is implemented in both part of study. In the numerical model, the left boundary of the slope is restrained against horizontal movement but allowed to move in vertical direction (roller type). The base of the rock slope is considered to be restrained in both horizontal and vertical direction. In this study, the in situ horizontal stress is considered to be half of vertical stress as reported in Pain et al. (2014) (Table 1).

2.1 Material Properties

To perform the stability analysis material properties reported in Pain et al. (2014) is used in the numerical simulation.

Table 1 Joint characteristics

Joint set	Dip/dip direction	In situ block size (m)	Normal stiffness (GPa/m)	Joint shear stiffness (GPa/m)	JRC	JCS	Residual friction angle (Ø)
J1	27°/N 180°	0.525	15.99	1.599	4.9	29	28
J2	86°/N 120°	0.225	37.31	3.731	5.37	76	30
J3	75°/N 242	0.375	22.38	2.238	5	40	30
Slope	54°/N 180						

Table 2 Rock mass properties

Unit weight (kN/m^3)	Elastic modulus (GPa)	Poisson's ratio	Cohesion (kPa)	Tension (kPa)	Angle of internal friction (Ø)	GSI
26	6.12	0.30	763	113	57.17	48

Table 3 Intact rock properties

Unit weight (kN/m^3)	Elastic modulus (GPa)	Poisson's ratio	mi	s	a
26	22.53	0.30	15.311	1	0.5

3 Stability Analysis

First, entire domain of the model is discretized into six-noded triangular finite element mesh. PHASE^2 solves the problem using Gaussian elimination technique using properties defined at each node of mesh. The shear and normal stiffness data is provided in the joint interface element to simulate shear and normal displacement at the interface. Three joint sets are oriented according to their dip angle and the dip direction (Table 1). A uniform distribution of joints is incorporated to simulate the random variation of joint spacing. Barton and Bandis (1990) model is utilized in the discontinuum analysis to incorporate the joint strength and deformability at the joint interface. This model is given by equation.

$$\tau = \sigma_n \tan\left(\phi_{\mathrm{r}} + \mathrm{JRC}\log_{10}\left(\frac{\mathrm{JCS}}{\sigma_n}\right)\right), \tag{1}$$

where, τ is the shear strength of joint, ϕ_{r} is the residual friction angle, JRC is the joint roughness coefficient, JCS is the joint wall compressive strength and σ_n is the normal stress acting across the discontinuity (Fig. 1).

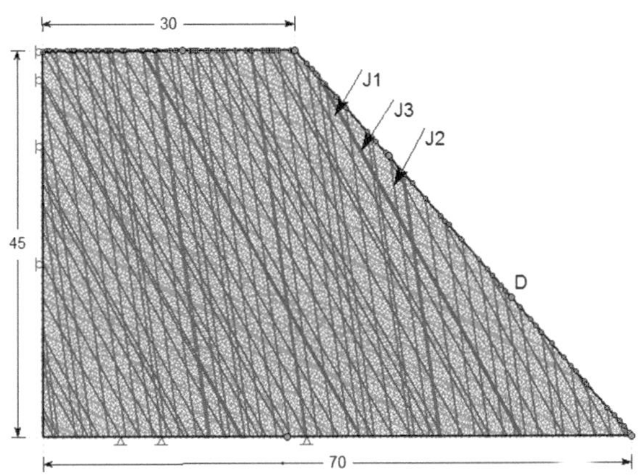

Fig. 1 Jointed rock slope for discontinuum analysis

3.1 Static Stability Analysis

Factor of safety is calculated using shear strength reduction method to compute the critical strength reduction factor in $PHASE^2$. In this method, the strength of the material is reduced by some factor and the finite element solution is calculated to obtain the critical SRF when model become unstable. The SRF value from the continuum slope stability analysis is 4.47 (Fig. 2) which is very high value than the result from the discontinuum model (Fig. 3). The factor of safety of the discontinuum model is presented in Table 4.

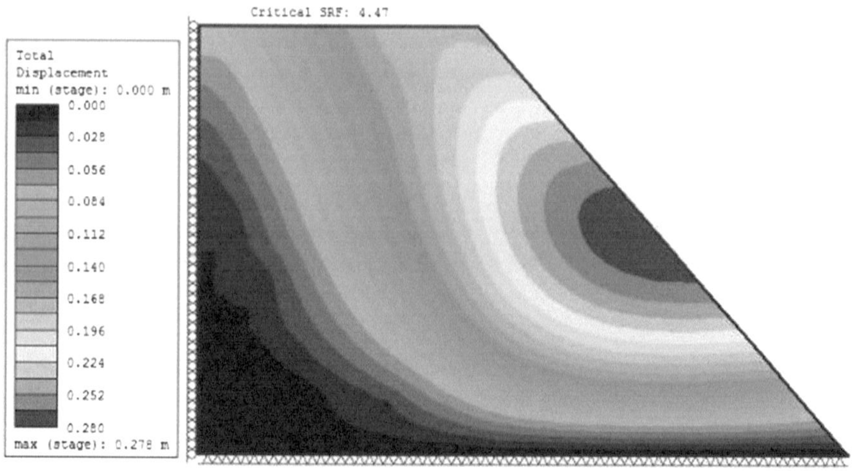

Fig. 2 Total displacement contour of continuum model showing critical SRF 4.47

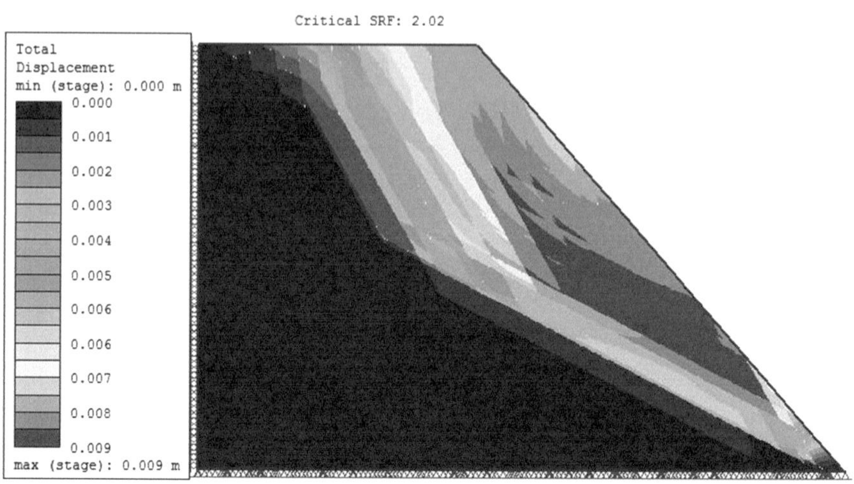

Fig. 3 Displacement contours of discontinuum model for all joint set (J1, J2 and J3), showing critical SRF 2

Table 4 Comparison of SRF values for jointed rocks

Case	Slope with joint	SRF (Pain et al. 2014)	SRF present study
1	J1	2.62	2.15
2	J2 and J3	1.96	2.23
3	J1, J2 and J3	1.74	2.02

Results show small deviation of SRF from Pain et al. (2014) due to difference in the assumption maximum spacing of the joint, although the mean spacing is maintained same. The discontinuum modeling for rock slope with all the three joint sets showed an SRF value of 2.02. This can be explained by the fact that more unstable rock blocks are formed when all the three joint sets are incorporated.

It is important to note that in all the three cases of discontinuum modeling, the rock slope failure occurred along the joint set J1 which further establishes the criticality of joint set J1. Both studies have manifested that the rock slope is stable under static condition where factor of safety is more than one.

3.2 Application of Dynamic Loading: Seismic Condition

The Chamoli earthquake (1999), magnitude of 6.8 is used in the time response analysis to predict the vulnerability of the slope. The acceleration time history which is used in the dynamic analysis is shown in Fig. 4. To provide infinite boundary condition at the left and at bottom face of slope, Lysmer-Kuhlemeyer dashpot boundary condition is used (Lysmer and Kuhlemeyer 1969). The absorbent boundary condition at the base which will absorb incoming pressure and shear waves at this boundary and transmitting boundary condition at the left face is provided which will

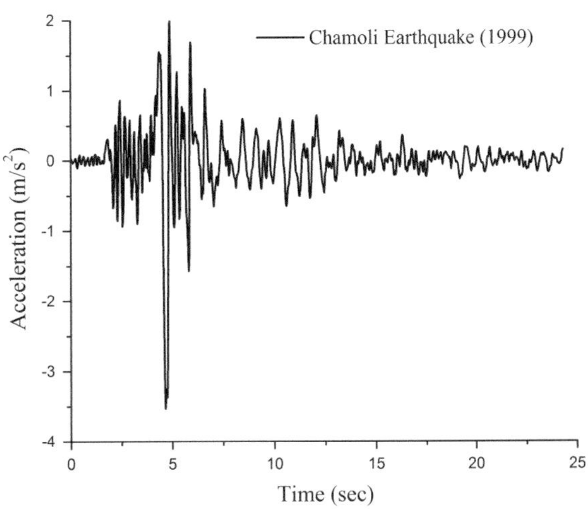

Fig. 4 Corrected transverse component of acceleration–time history of Chamoli earthquake, 1999, taken from USGS used in the study

ensure no reflection of outgoing wave at this boundary. The transverse time acceleration data is provided at the base of the model for dynamic simulation. Rayleigh damping of 5% is considered in this study. To provide the Rayleigh damping, the model is first solved without damping to obtain the significant frequency bandwidth from the output power spectrum. This frequency range is further used to estimate Rayleigh damping constants $\alpha_m = 4.71$ and $\beta_k = 3.9 * 10^{-4}$ is obtained for 5% damping. Using this constant, the damped vibration analysis is done. Pseudo-static analysis is also carried out considering different range of horizontal seismic coefficient (k_{H}).

3.2.1 Dynamic Stability Analysis

After the application of dynamic loading at the base of the slope, the time response analysis is performed in discontinuum model having three sets of joints. Vertical and horizontal displacement at the slope face (point D Fig. 1) is observed throughout the motion. Maximum residual horizontal displacement is observed 86 mm. The residual vertical displacement is 26 mm. Figure 5 inferred that the slope is stable because after the application of earthquake slope is not failing due to earthquake. Though, the rock slope has faced residual deformation due to some joints failure. The pseudo static analysis results (Fig. 6) manifested that the jointed slope is stable below $k_{\mathrm{H}} = 0.36$ g. SRF value is found to be more than one below maximum credible earthquake in the Uttarakhand region.

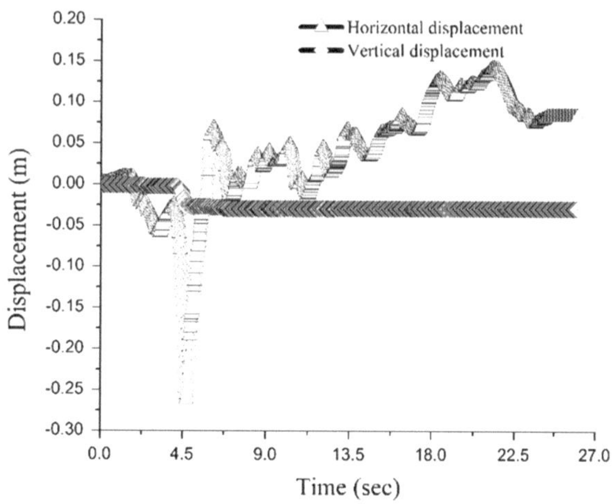

Fig. 5 Horizontal and vertical displacement at face of slope after application of seismic excitation

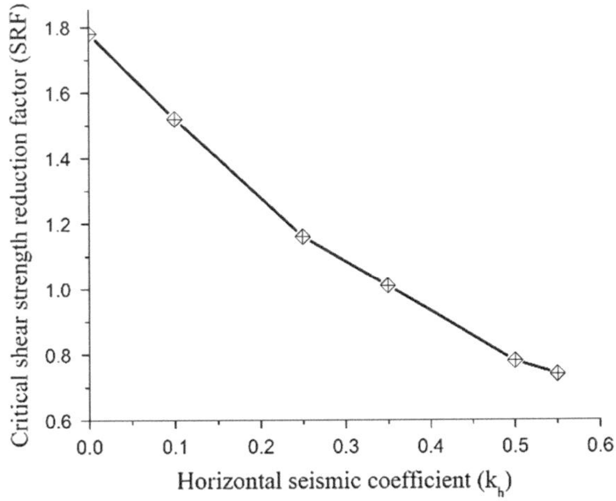

Fig. 6 SRF of pseudo static analysis for discontinuum model

4 Conclusion

The rock slope along the Rishikesh-Badrinath National Highway (NH-58) in Garhwal Himalayas, India is considered. The numerical analysis in PHASE^2 manifested that the discontinuum modeling gives a better insight of the in situ conditions as it takes consideration of the effect of joints. From the discontinuum analysis results, the influence of joints has been studied and it showed that joint orientation (dip angle and dip direction) have significant effect on the stability of rock slopes.

The concerned rock slope is practically stable which is manifested by the stability analysis results. The rock slope is found stable subjected to Chamoli earthquake motion (1999-03-29). The peak acceleration of 3.53 m/s^2 and low frequency content is not causing severe damage to the slope but joints faced severe stress reversal. Pseudo static analysis results indicated that the rock slope is vulnerable to high seismic event at k_H more than 0.36 g.

References

Barton, N. R., & Bandis, S. C. (1990). Review of predictive capabilities of JRC-JCS model in engineering practice. In: *Rock joints'. Proceedings of the International Symposium on Rock Joints*, June 4–6, 1990, Loen, Norway (pp. 603–610). Boca Raton, FL: CRC Press.

Cai, M., & Horri, H. (1992). A constitutive model of highly jointed rock mass. *Mechanics of Materials, 13*(3), 217–246.

Pain, A., Kanungo, D. P., & Sarkar, S. (2014). Rock slope stability assessment using finite element based modelling—examples from the Indian Himalayas. *Geomechanics and Geoengineering, 9*(3), 215–230.

PHASE^2. (2012). A 2D finite element program for calculating stresses and estimating support around the underground excavations. *Geomechanics Software and Research*. Toronto, ON: Rocscience Inc.

Hoek, E., Carranza-Torres, C., & Corkum, B. (2002). Hoek-Brown criterion—2002 edition. In *Proceedings of the NARMS-TAC Conference*, July 10, 2002, Toronto, Canada (Vol. 1, pp. 267–273). University of Toronto Press.

Lysmer, J., & Kuhlemeyer, R. L. (1969). Finite dynamic model for infinite media. *Journal of Engineering Mechanics, ASCE, 95*(4), 859–877.

Rock Failure Pattern Under Uniaxial, Triaxial Compression and Brazilian Loading Conditions

Tarun Singh, Ashwani Jain and K. S. Rao

Abstract Failure of rock is an important problem arising during underground construction of shafts, tunnels etc. It has been determined that the peripheral stress developed around a circular tunnel causes rock failure which makes it essential to recognize these failures patterns under these circumstances. Prediction and quantification of failure mode is very complex and difficult. Therefore, it is important to carry out comprehensive study in laboratory to determine rock failure pattern which is significant to stabilize engineering rock masses and to recognize the suitability of support system as per nature of engineering work. Uniaxial compressive strength, Brazilian tensile strength and triaxial strength tests were conducted on three different rock samples namely Migmatitic Gneiss (MG), Phyllitic Quartzite (PQ), and Quarzitic Phyllite (QP), collected from the vicinity of Rohtang tunnel in Higher Himalaya, to understand the nature of failure patterns. Test results shows axial splitting along with shearing along plane and multiple fracture patterns in phyllites and gneiss whereas gneiss also shows multiple fracture failure patterns with increase in comprehensive strength of rock under triaxial conditions. In case of Brazilian tensile strength tests failure pattern are generally of central multiple type. This paper highlights the failure modes of gneiss and phyllite rocks under different loading conditions using ISRM and BIS suggested methods.

Keywords Failure pattern · Uniaxial compressive strength
Triaxial compressive strength · Brazilian tensile strength

T. Singh (✉) · A. Jain · K. S. Rao
Indian Institute of Technology Delhi, Hauz Khaz, New Delhi 110016, India
e-mail: tarun.iitd@outlook.com

A. Jain
e-mail: ashwanijain5@rediffmail.com

K. S. Rao
e-mail: raoks@iitd.ac.in

A. I. V. and V. B. Maji (eds.), *Geotechnical Applications*, Lecture Notes in Civil Engineering 13, https://doi.org/10.1007/978-981-13-0368-5_26

1 Introduction

Rock failure is very significant in rock engineering. The mineralogy, particle arrangements, void ratio, microcracks play an important role to control the mechanical behavior of rock, microfaults are often seen dominant in these failures (Sammis and Ashby 1986; Akesson et al. 2004; Szwedzicki 2007; Basu et al. 2009).

Rock materials under uniaxial, triaxial or other compression conditions undergo complex process of initial damage like crack initiation, crack growth, crack closure, and finally failure of rock (Brace et al. 1966; Bieniawski 1967; Scholz 1968; Lajtai and Lajtai 1974; Martin 1993; Eberhardt et al. 1998; Li et al. 2003; Jaeger et al. 2007). Prediction and quantification of failure mode are complex and difficult. Even failure theories of Mohr–Coloumb and Irwin–Griffith cannot predict the development of fracture in rock materials (Santarelli and Brown 1989; Peng and Johnson 1972). Manifestation of rock failure under low to high confining pressure depends on rock type and microstructure of rock. Specimens of similar lithological composition observes various failure pattern and range of uniaxial compressive strength which is attributed to microcracks developed due to microstructural variations (Szwedzicki 2007; Basu et al. 2009).

Hence to apprehend rock failure in tunnels it is essential to understand failure pattern under circumstances developed due to peripheral stresses around a tunnel (Hudson 1989). As no analytical or numerical method can determine the nature of fracture development in rock materials the laboratory test can ascertained information about failure patterns (Bieniawski 1967). Therefore, it is important to have a comprehensive study on rock failure pattern to identify the suitability of the design support. In the past, several researches have been carried but our understanding of rock failure is still uncertain and inadequate. This study aims to understand the failure patterns of Migmatitic gneiss (crystalline metamorphic rock), Quartzitic phyllite and Phyllitic quartzite (anisotropic metamorphic rock) under uniaxial, triaxial compression and Brazilian tensile condition.

2 Description of Rock Types and Laboratory Investigations

2.1 Rock Types Investigated

Three rock types, Migmatic gneiss, Quartzitic phyllite and Phyllitic quartzite were collected from Rohtang tunnel's south portal area. The lump samples were collected from site and specimen of size 38 mm were drilled and prepared after cutting and grinding as per IS: 9179 (1979) standard. Core sample photographs and corresponding scanning electron micrographs (SEM) of the collected rocks are given in Fig. 1. Migmatic gneiss is essentially bearing polygonised quartz, biotite, plagioclase feldspar, microcline, clinozoisite, zircon, sphene, tourmaline, and apatite.

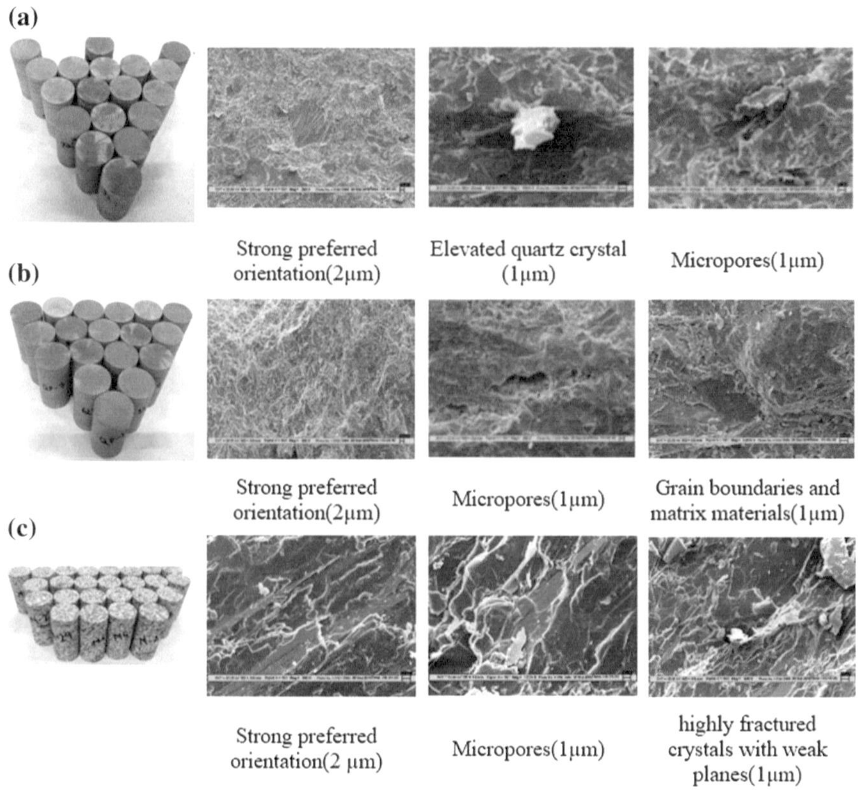

Fig. 1 Core samples and scanning electron micrographs of **a** phyllitic quartzite, **b** quartzitic phyllite and **c** migmatite gneiss

Phyllitic quartzite contains quartz (both with or without ribbon quartz), plagioclase feldspar, sanidine, ilmenite along with sigma type prophyroclast of quartz and feldspar. The foliation in Migmatic gneiss and Phyllitic quartzite is defined by biotite. Quartzitic phyllite mainly comprises of calcite, diopside, quartz, tourmaline, zircon and biotite.

2.2 *Specimen Preparation*

The failure mode of rock specimen is greatly affected by the end surface of specimen and constraints offered by the platens used during testing (Vutukuri et al. 1974; Szwedzicki 2007), hence the specimens were prepared as per IS: 9179 (1979) standards with length to diameter ratio 2 for triaxial and uniaxial compression test where as it was 0.5 for Brazilian test.

3 Failure Pattern Under Uniaxial Compression

A total of 18 specimens (six each of Migmatic gneiss, Quartzitic phyllite and Phyllitic quartzite) were tested under uniaxial compression in this investigation. A constant loading rate of 0.25 MPa/s is used and the failure patterns of each were observed and tabulated in Table 1 along with its failure strength (UCS) value.

In case of Migmatic gneiss axial and shearing failure pattern were observed. The specimen showing axial pattern has more UCS value than shearing pattern. With uniaxial stress the cracks are developed from the tips of the microcracks parallel to the maximum principal stress resulting in axial pattern. The presence of highly fractured crystals with weak planes as shown in micrographs has constrained the propagation of cracks developed along the maximum principal stress hence suitably oriented microcracks have developed to release the strain energy in the form of shear fracture. Highly fractured crystal and weak planes have reduced the value of UCS of specimen showing shear fracture pattern. In many cases the presence of strong preferred orientation, elevated quartz crystals and micropores leads the Quartzitic phyllite specimen to fail under axial pattern predominantly showing multiple failure for UCS value less than 80 MPa.

4 Failure Under Triaxial Test Conditions

A total of 9 specimens (three each of Migmatic gneiss, Quartzitic phyllite and Phyllitic quartzite) were tested under triaxial compression in this investigation. The failure patterns of each were observed and their corresponding failure values have been presented in Table 2.

In triaxial compression the Migmatic gneiss initially shows shear splitting at σ_3 value of 8 MPa, which on increase in confining pressure (σ_3) to 12 MPa, results in multiple fracturing of rock, on further increasing σ_3 value to 16 MPa, specimen get completely crushed at TCS value of 172.86 MPa; whereas in both phyllites (i.e. Phyllitic quartzite and Quartzitic phyllite) predominately shearing and foliation splitting were observed, their corresponding failure values are presented in Table 2.

5 Failure Under Brazilian Test Conditions

In the present investigation, a total of 15 specimens (five each from MG, PQ and QP) were tested under Brazilian test conditions. Failure pattern of each was recorded and its corresponding Brazilian strength is presented in Table 3.

Central and central multiple patterns were observed generally for all types of rocks specimens.

Table 1 UCS values and corresponding failure pattern of migmatic gneiss, quartzitic phyllite and phyllitic quartzite specimens

S. No.	Specimen No.	UCS (MPa)	Failure mode	S. No.	Specimen No.	UCS (MPa)	Failure mode	S. No.	Specimen No.	UCS (MPa)	Failure mode
1	MG1	43.03	Axial with shearing	7	QP1	71.67	Axial	13	PQ1	83.16	Shearing
2	MG2	45.22	Shearing	8	QP2	95.09	Axial	14	PQ2	101.09	Axial
3	MG3	46.04	Axial with shearing	9	QP3	75.41	Axial and Foliation	15	PQ3	94.20	Multiple
4	MG4	48.68	Axial	10	QP4	87.07	Axial	16	PQ4	91.05	Axial
5	MG5	46.15	Axial	11	QP5	92.16	Multiple	17	PQ5	119.69	Multiple
6	MG6	44.33	Axial	12	QP6	82.18	Axial	18	PQ6	97.40	Axial

Table 2 TCS values and corresponding failure pattern under triaxial test

S. No	Specimen No.	TCS (MPa)	Failure mode	S. No.	Specimen No.	TCS (MPa)	Failure mode	S. No.	Specimen No.	TCS (MPa)	Failure mode
1	MGT1	110.45	Shearing	4	QPT1	135.78	Axial and foliation	7	PQT1	143.68	Shearing and foliation
2	MGT2	148.65	Shearing	5	QPT2	176.41	Shearing	8	PQT2	180.12	Shearing and foliation
3	MGT3	172.86	Multiple	6	QPT3	202.89	Axial	9	PQT3	212.24	Shearing

Table 3 Brazilian tensile strength values and corresponding failure pattern under Brazilian test

S. No.	Specimen No.	σ_t (MPa)	Failure mode	S. No.	Specimen No.	σ_t (MPa)	Failure mode	S. No.	Specimen No.	σ_t (MPa)	Failure mode
1	MGB1	5.79	Central multiple	6	QPB1	10.78	Central	10	PQB1	12.63	Central
2	MGB2	6.27	Central	7	QPB2	10.77	Central multiple	11	PQB2	11.87	Central multiple
3	MGB3	5.66	Central multiple	8	QPB3	11.13	Central multiple	12	PQB3	11.61	Central multiple
4	MGB4	6.41	Central	9	QPB4	11.16	Central multiple	13	PQB4	10.52	Central
5	MGB5	6.12	Central	10	QPB5	Test Failed		14	PQB5	12.32	Central multiple

6 Conclusions

Failure modes of Migmatic gneiss, Quartzitic phyllite and Phyllitic quartzite under Uniaxial, Triaxial compression and Brazilian tensile conditions were examined.

Axial failure in uniaxial compression is mainly controlled by micro factures present in the rock along with strong preferred orientation of minerals. In Migmatic gneiss UCS value of specimen is more in case of axial failure than shearing failure. For UCS value higher than 90 MPa axial failure pattern is observed in Phyllitic quartzite, whereas axial failure pattern was observed in Quartzitic phyllite for UCS value between 71 and 95 MPa. Broadly the failure of all three rocks under uniaxial compression and its relation with the corresponding UCS values can broadly be viewed in terms of cracks growth and damage of the rocks.

In all three types of rock the principal nature of failure pattern is axial. A total of five discrete failure patterns (i) Axial splitting, (ii) Shearing along plane, (iii) Axial splitting with shearing, (iv) Multiple fracturing and (v) Failure along foliation, and sketches of these pattern is shown in Fig. 2. While in the Brazilian test, the distinctive central fracture pattern is observed along with multiple central fracture, for all types of rocks specimens (i.e. MG, PQ and QP) indicating development of high strain energy to initiate failure in tension.

The findings of above study can be helpful during engineering design of underground structures. These types of investigations can provide useful information about probable rock failure during excavation and to ascertain support design. Moreover research is required in this direction with various rock types to understand rock failure pattern and its relation with existing engineering situations.

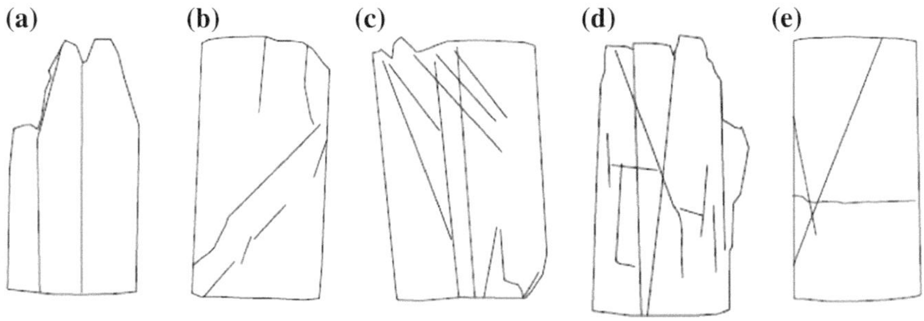

Fig. 2 Failure patterns observed under uniaxial and triaxial conditions **a** axial splitting, **b** shearing along plane, **c** axial splitting with shearing, **d** multiple fracturing and **e** along foliation

References

Akesson, U., Hansson, J., & Stigh, J. (2004). Characterization of micro-cracks in the Bohus granite, western Sweden, caused by uniaxial cyclic loading. *Engineering Geology, 72,* 131–142.

Basu, A., Celestino, T. B., & Bortolucci, A. A. (2009). Evaluation of rock mechanical behaviours under uniaxial compression for different weathering grades. *Rock Mechanics and Rock Engineering, 42,* 73–93.

Bieniawski, Z. T. (1967). Mechanism of brittle fracture of rock: Part I—theory of the fracture process. *International Journal of Rock Mechanics and Mining Sciences & Geomechanics Abstracts, 4,* 395–406.

Brace, W. F., Paulding, B. W., & Scholz, C. H. (1966). Dilatancy in the fracture of crystalline rocks. *Journal of Geophysical Research, 71,* 3939–3953.

Eberhardt, E., Stead, D., Stimpson, B., & Read, R. S. (1998). Identifying crack initiation and propagation thresholds in brittle rock. *Canadian Geotechnical Journal, 35,* 222–233.

Hudson, J. A. (1989). *Rock Mechanics principles in engineering practice*. Butterworths, London: CIRIA Report.

IS 9179. (1979). Method for the preparation of rock specimen for laboratory testing.

Jaeger, J. C., Cook, N. G. W., & Zimmerman, R. W. (2007). *Fundamentals of rock mechanics* (4th ed.). Oxford: Blackwell.

Lajtai, E. Z., & Lajtai, V. N. (1974). The evolution of brittle fracture in rocks. *Journal of the Geological Society London, 130,* 1–16.

Li, L., Lee, P. K. K., Tsui, Y., Tham, L. G., & Tang, C. A. (2003). Failure process of granite. *International Journal of Geomechanics, 3,* 84–98.

Martin, C. D. (1993). *The strength of massive Lac du granite around underground openings* (Ph.D. thesis, University of Manitoba).

Peng, S., & Johnson, A. M. (1972). Crack growth and faulting in cylindrical specimen of chelmsford granite. *International Journal of Rock Mechanics and Mining Sciences & Geomechanics Abstracts, 9,* 37–86.

Sammis, C. G., & Ashby, M. F. (1986). The failure of brittle porous solids under compressive stress state. *Acta Metallurgica, 30,* 511–526.

Santarelli, F. J., & Brown, E. T. (1989). Failure of three sedimentary rocks in triaxial and hollow cylinder compression tests. *International Journal of Rock Mechanics and Mining Sciences & Geomechanics Abstracts, 26,* 401–413.

Scholz, C. (1968). Experimental study of the fracturing process in brittle rock. *Journal of Geophysical Research, 73,* 1447–1454.

Szwedzicki, T. A. (2007). A hypothesis on modes of failure of rock samples tested in uniaxial compression. Technical note. *Rock Mechanics and Rock Engineering, 40*, 97–104.

Vutukuri, V. S., Lama, R. D., & Saluja, S. S. (1974). *Handbook on mechanical properties of rocks*. Clausthal: Trans Tech Publications.

Analysis of a Diaphragm Wall Panel After Leakage During Deep Excavation

Murli Iyengar

Abstract Instances of leakage from the joints between the panels of a diaphragm wall, during construction, are not uncommon and several cases have been reported in literature, along with measures taken for repair and rectification. This paper deals with one such case wherein, following repairs to stop the leakage, a parametric study was carried out using PLAXIS, considering various scenarios of subsoil disturbance, both the decrease in the subsoil strength as well as likely extent of zone of disturbance. The back-analysis was based on the data from the inclinometers, installed in the diaphragm wall panels and cone penetration tests carried out to establish the subsoil condition, pre and post leakage. This enabled to establish the most probable combination for assessment of the BM developed in the wall for ascertaining its structural adequacy. Based on the study it was concluded that the disturbed zone is limited to about 5 m beyond the face of the D'Wall, defined by Case A. The computed deformation for the lower Es values (Cases B and C) and 10 m disturbed zone (Case D), is much in excess of the observed deformation, hence not applicable.

Keywords Diaphragm wall · Water leakage · Deep excavation

1 Introduction

1.1 Diaphragm Walls with RCC Slabs

Diaphragm walls with RCC slabs at the Roof level, Concourse level and the Base level constitute a typical Station box for a Metro Rail Station. The Diaphragm walls are installed using a segmental installation technique, comprising of Primary and Secondary wall panels. Rubber bars are installed in the joints between the panels with the help of stop-ends. Several publications have reported instances of leakage

M. Iyengar (✉)
Engineers India Ltd. (Rtd.), Chennai, India
e-mail: miyengar.in@gmail.com

A. I. V. and V. B. Maji (eds.), *Geotechnical Applications*, Lecture Notes in Civil Engineering 13, https://doi.org/10.1007/978-981-13-0368-5_27

from the joints between the wall panels (Horodecki et al. 2004; Likitlersuang et al. 2013; Mair 1998). Measures taken for assessment, repair and rectification are described in these publications. In the case described in this paper, an adequacy check of the wall was carried out, after taking steps to seal the leakage. Using the data from the instrumentation installed, a parametric study was carried out, considering the possible scenarios of disturbed soil.

1.2 Instrumentation and Monitoring

Extensive Instrumentation and monitoring system was provided both in the diaphragm walls panels as well as in the ground and buildings in the vicinity of the excavation. These included D'wall Inclinometers, Soil Inclinometers, Strain gauges and Load cells for structural struts, where provided. In the buildings in the immediate vicinity, Ground and Building settlement markers, Crackmeters and Tiltmeters were installed. Piezometers for Ground water table recording were installed all around the excavation. All instrumentations were monitored on a daily basis to provide early warning of any unexpected behaviour. They also provided indication whether the response of the structure was on predicted lines.

2 Subsoil Conditions

2.1 Configuration of Station and Subsoil

The subsoil at the site comprises of predominantly Silty Sands (SM) layers underlain by Charnockite rock. Ground water table is within 4–5 m of the GL. Figure 1 depicts the subsoil at site along with the main structural configuration. High water table called for a well-designed dewatering system for ground water lowering, during the excavation process.

3 The Event

3.1 Excavation for Station Box

While the excavation was proceeding towards the concourse slab level, water started leaking from a Diaphragm wall joint, about 6 m below GL. Immediately a steel plate was fixed on the joint and sand bags were stacked against the leaking joint from inside the station to arrest sand coming out of the joint. Cement grout was pumped in up to the ground level in the joint from the outside. This was followed by PU grouting from inside, till the water flow was completely arrested.

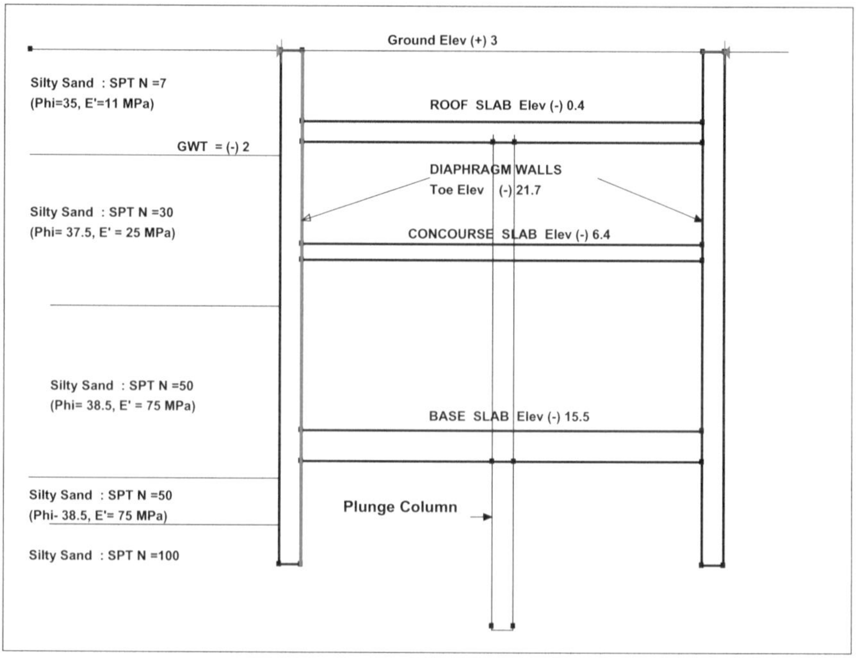

Fig. 1 Section of station box and subsoil details

Thereafter, TAM grouting was carried out from the outer side to increase the water tightness of the joint.

3.2 D'Wall Inclinometer Movement

During the above leakage, the D'Wall inclinometer showed a movement towards the excavation of the order of 11 mm. Figure 2 presents the D'Wall inclinometer behaviour, both before and after the leakage.

4 Assessment of the Subsoil Conditions After the Event

4.1 DCPT Tests

It was apprehended that subsoil in the vicinity of the leaking joint could have been affected. In order to assess the disturbance to the subsoil, DCPT (Dynamic Cone Penetration Tests), were conducted using "Geotools". To facilitate comparison with

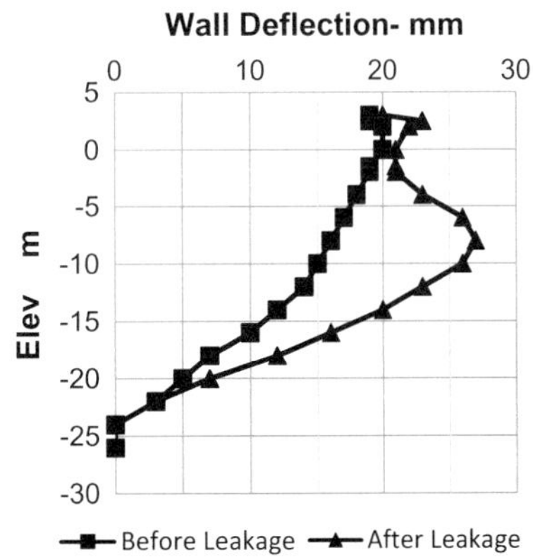

Fig. 2 Deformation of wall versus elevation

a "base reading", tests were also carried out in the unaffected area, nearest to the site, where the subsoil was relatively undisturbed by the event.

4.2 Geotools Tests

The Geotools equipment uses a automatic trip device to release a "hammer" falling on a jar block, in order to drive a cone, attached at the end of the string of rods, into the subsoil. A small engine provides the power for the lifting of the hammer each time before its free fall on the Jar Block. Blow counts are recorded for each metre of the penetration of the Cone. Though these are not directly correlated to SPT or CPT values, they, nevertheless, give a comparative assessment of the subsoil strength.

4.3 Blow Counts Before and After Disturbance

Figure 3 presents the blow counts recorded in the tests, both before and after the soil disturbance had taken place. Tests A, B and C were carried out in the affected area and the blow counts were lower than those carried out at the corresponding depths beyond the affected area (Tests D and E). Consequently, it was apprehended that the area in the vicinity of leakage might have developed zones of loose soil.

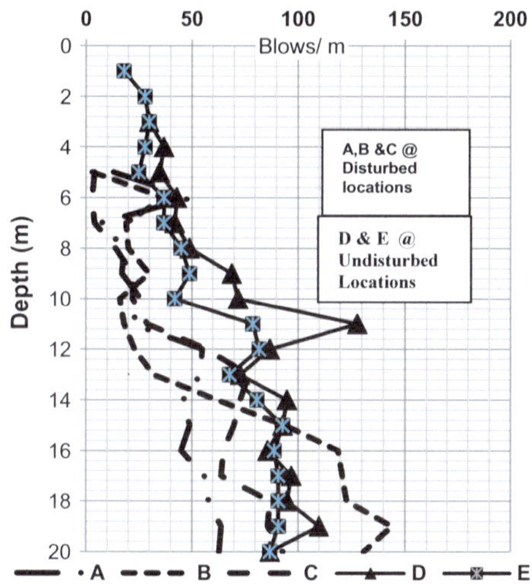

Fig. 3 Geotools test—blow counts versus depth

5 Effects of Subsoil Strength Reduction

5.1 Assessment of Subsoil Disturbance

It was important to assess both the extent and magnitude of reduction in soil parameters due to leakage and the consequent implications on the Diaphragm wall safety. The DCPT tests carried out are indicative of subsoil strength reduction in zone affected by leakage. However, quantification of the same cannot be done solely by these tests. Therefore, it was decided to carry out the assessment by a back-analysis of the Instrumentation data, primarily the D'Wall Inclinometer and link it with the subsoil conditions.

For this purpose a parametric study was carried out, using a range of subsoil parameters, representative of the disturbed soil conditions, both in terms of reduced strength as well as zone of disturbed soil.

5.2 PLAXIS Modelling

Use of numerical modelling to predict wall displacements during excavations is widely adopted (Mohammed et al. (2013) and other referenced papers). Models were developed in PLAXIS representing various likely subsoil scenarios after disturbance, represented by reduced soil parameters such as Modulus of Elasticity Es, Phi and Density. Several cases representing disturbed soil zone were considered and analysed, extending 5 and 10 m from the face of the Diaphragm wall. The analysis were carried out for the stage of Concourse slab excavation.

5.3 *Outputs of Analysis*

Representative outputs of D'Wall deflections based on analysis carried out are presented in Fig. 4. This exercise enabled a comparison to be made between the computed deflections with the actual wall deflections, recorded by the Inclinometer. A match between the computed and actually observed deflection was considered to represent the likely subsoil condition after disturbance. The BM and SF values for the D'Wall corresponding to this scenario, was considered to represent the actual values developed in the wall after soil disturbance.

6 Results of Analysis and Discussions

The results of the analysis performed are presented in Table 1.

From a study of Figs. 2 and 4; Table 1, it is seen that during the leakage, the maximum deflection of the Diaphragm wall increased from 16 mm to a value of 27 mm at El. −8.0 m. Above this level also the same trend is noted up to El. −2, although the increase in deflection is smaller. At the time of leakage, the Roof slab was already in position.

In Fig. 4, the actual deflection of the Inclinometer, installed in the wall panel, is also shown. It is seen that the recorded Inclinometer readings immediately after leakage, match the plot Case A representing the projected values for reduced Es value of 3000 kPa and disturbance zone of 5 m.

For Cases B and C representing lower Es values, i.e., higher soil disturbance conditions, it is seen that computed deflection values of the Inclinometer, exceed the actually recorded values. So is the case for Case D, which considers a 10 disturbed zone. Based on this, it is concluded that the soil disturbance has not spread beyond 5 m from D'Wall face.

Since Case A matches closest to the actual inclinometer readings, it is considered most representative to reflect BM and SF values developed in D'Wall after leakage. As per Table 1, the BM in D'Wall has increased from 858 to 952 kN m/m. However, it is still well within its design capacity. Hence no additional structural modifications were necessary, other than improvement of surrounding subsoil by grouting.

It appears that the "Water Bars" provided in the panel joints may have got disturbed during construction, leading to water leakage. Therefore, as an additional measure, it was decided to carry out "Jet grouting" behind the joints for all remaining panel joints, prior to proceeding with excavation. No such leakage issues were reported thereafter. The Station box is now fully completed and ready to be put into service.

Fig. 4 PLAXIS analysis outputs and actual deflection

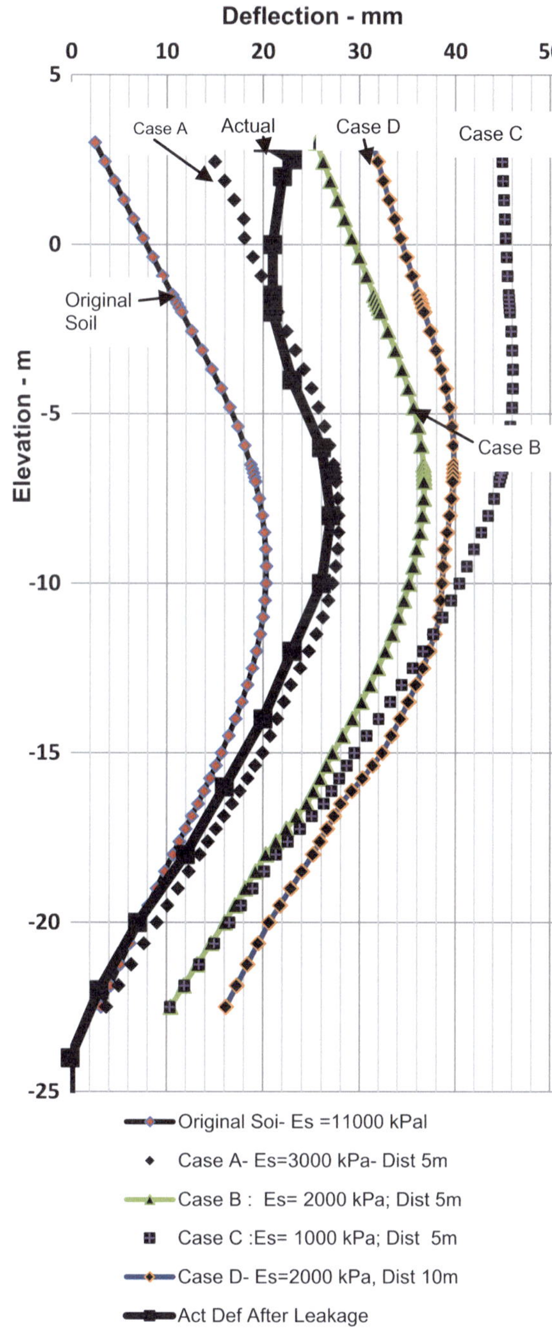

Table 1 Summary of PLAXIS analysis results

Cases	Es[a] (kPa)	Max D'Wall def. (mm)	BM (kN m/m)	Remarks
Original soil	11,000	20.5	858	Case A is closest to the actual inclinometer readings
A-Es = 3000 disturbance up to 5 m	3000	28.9	922	
B-Es = 2000 disturbance up to 5 m	2000	36.7	1020	
C-Es = 1000 disturbance up to 5 m	1000	46	900	
D-Es = 2000 disturbance up to 10 m	2000	40	1040	
Actually measured values		27		

[a]Es—Mod. of elasticity of soil

7 Conclusions

- The observed values for the D'Wall inclinometer matches well with the computed deformation for Case A.
- As per Case A the BM in D'Wall has increased from 858 to 952 kN m/m on account of leakage but remains well within its capacity.
- The disturbed zone is limited to about 5 m beyond the face of the D'Wall.
- The computed deformation for the lower Es values (Cases B and C) and 10 m disturbed zone (Case D), is in excess of the observed deformation, hence not considered applicable.
- The reduction in Subsoil strength has a more pronounced effect on deformation values of the D'Wall compared to the effect of the extent of zone of disturbance.

References

Horodecki, et al. (2004). Deep excavations braced by diaphragm walls. In *Proceedings Fifth International Conference on Case Histories in Geotechnical Engineering*, New York, April 13–17, 2004.

Likitlersuang, S., et al. (2013). Finite element analysis of a deep excavation: A case study from Bankok MRT. *Soils and Foundations, 53*(5), 756–773.

Mair, R. J. J. (1998). Recent experiences of tunneling and deep excavations in London SOA. In *Proceeding Fourth International Conference on Case histories in Geotechnical Engineering.* Paper 9. March 1998.

Mohammad, et al. (2013). Characteristics of diaphragm wall lateral deformations and ground surface settlements. Case study in Iran-Ahwaz metro. *Tunnelling and Underground Space Technology, 35,* 109–121.

Response of Single Pile Due to Deep Excavation and Underground Openings

Akhil Ambooken, R. K. Madhumathi and K. Ilamparuthi

Abstract One of the important issues of excavation and tunneling in urban areas is the assessment of its impact on foundations of neighboring structures due to ground movements in particular on pile foundations. Excavation and tunneling operations cause lateral soil movements which induce additional lateral loads on piles and they are known as passive piles. They are subjected to additional bending moment and shear force which can lead to serviceability problems or even failure of piles itself. The present study focuses on 1 g model tests to determine the response of single pile adjacent to a tunnel and a combined effect of excavation and tunneling. The effect of various parameters related to the pile, tunnel, and soil on the response of pile to passive loading was analyzed in this study. The following parameters viz. pile length to diameter ratio, distance between pile and retaining wall, distance between pile and tunnel springline, position of pile with respect to retaining wall and tunnel are varied to understand the response of piles and also to identify the factor which influences the response most. The lateral deflection of pile head was measured. Analysis of the test results showed that the shorter pile ($L/d = 10$) deflects more when compared to a long pile ($L/d = 20$) in case of excavation, tunneling and combined case.

Keywords Pile · Tunnel · Excavation · Lateral soil movement

A. Ambooken · R. K. Madhumathi · K. Ilamparuthi (✉)
Anna University, Guindy, Chennai 600025, India
e-mail: kanniilam@gmail.com

A. Ambooken
e-mail: akhilambs@gmail.com

R. K. Madhumathi
e-mail: madhumathi17@gmail.com

A. I. V. and V. B. Maji (eds.), *Geotechnical Applications*, Lecture Notes in Civil Engineering 13, https://doi.org/10.1007/978-981-13-0368-5_28

1 Introduction

Excavation below ground level causes relaxation of in situ stress due to ground movement. Tunneling is one form of excavation which causes ground movement. These ground movements affect the performance of structures located nearby such underground activity particularly the foundations of structures. The various field conditions in which piles are subjected to lateral soil movements are piles located adjacent to deep excavations, and tunnel construction, embankment construction, and piles adjacent to pile installation operations. The lateral forces induced on piles due to soil movements as a result of situations stated above are termed as "passive loads" and the piles are termed as passive piles.

The lateral forces acting on piles due to soil movement may induce additional stresses and excessive deflection on the piles. In critical situations, they might damage the pile and compromise the stability and serviceability of supported structures. Thus, it is essential to evaluate pile responses due to lateral soil movements. Reasonably good number of research works is reported in literature for the prediction of ground movement due to excavation as well as tunneling and response of pile foundations towards these ground movements independently. Both 1 g model tests (Chen et al. 1995; Ilamparuthi and Madhumathi 2011; Meguid and Mattar 2009; etc.,) and centrifuge model tests (Leung et al. 2000) have been performed. Analysis were also carried out on numerically simulated models by researchers (Poulos and Chen 1997; Chen et al. 2000; Min et al. 2011; Basile 2014; etc.,) to understand the response of piles due to excavation and tunneling. However, the response of piles due to combined effect of excavation and tunneling has not been investigated adequately. Thus the combined effect of ground movement due to excavation and tunneling on pile is investigated in this research study through 1 g model tests. The objective of the present study is to investigate the response of single pile to lateral soil movement by varying the pile length to diameter ratio, distance between pile and retaining wall, distance between the pile and tunnel springline, depth of tunnel and location of pile with respect to retaining wall and tunnel.

2 Experimental Facility

The experimental facility includes model tank, laboratory models of pile, retaining wall and tunnel and necessary instrumentation.

2.1 Experimental Procedure

The response of single pile due to tunneling is studied using a model tank of dimension 0.7 m × 0.29 m × 0.44 m made of steel. Aluminum hollow tubes of

19 mm outer diameter with 1 mm wall thickness were used to fabricate the piles with length to diameter ratio of 10 and 20. Retaining wall is made out of aluminum sheet 1.2 mm thick. Poorly graded sand with specific gravity 2.65 was used for experiments under medium dense condition with density 16.5 kN/m^3. Tunneling operation is performed along the shorter side of the tank. The tunnel axis depth is normalized by taking the ratio of depth of tunnel axis to diameter of tunnel. Two normalized depths having H/D ratio 2.2 and 4.2 are considered in the experiments where H is the depth of tunnel axis from surface of sand bed and D is the tunnel diameter. The properties of the materials used in the experiment are given in Table 1.

The tunneling operation is simulated using a helical auger which is rotated at a uniform rate to cut through the soil. A hollow tube made of tin with outer diameter 74 mm and thickness of 0.5 mm is used as the liner. The auger was manually rotated at a uniform rate with a handle along with the liner. As the auger cuts the sand and moves forward, thrust was applied to the walls of the liner while operating the auger. Thus the liner moves along with the auger as in tunnel boring machine. The liner prevents the flow of sand during tunnel boring and the helical auger brings out the cuttings through the space between liner and auger. The experimental investigation on response of single pile due to excavation and tunneling is observed by simulating an excavation operation first followed by tunneling. The location of pile is changed in such a way that in first case, pile is located equidistant from retaining wall and tunnel springline. Schematic arrangement of test setup is as shown in Fig. 1.

In the second case pile was located at a distance of 10d from the retaining wall. Sand in front of the retaining wall was removed manually in layers of 20 mm each.

Table 1 Properties of the materials used in the experiments

Material	Parameter	Value
Soil	γ_{soil}	16.5 kN/m^3
	Angle of internal friction, Φ	37°
	μ_{soil}	0.30
	E_{soil}	9000 kPa
	c_{soil}	1 kPa
	Angle of dilatancy, ψ	7°
Pile	d (outer diameter of pile)	19.05 mm
	Thickness of pile	1 mm
	E_{pile}	70 GPa
	γ_{pile}	27.5 kN/m^3
Retaining wall	E_{wall}	70 GPa
	Thickness of retaining wall	1.2 mm
Tunnel	D (outer diameter of tunnel)	74 mm
	Thickness of liner	0.5 mm
	E_{tun}	50 GPa
	γ_{tun}	72 kN/m^3

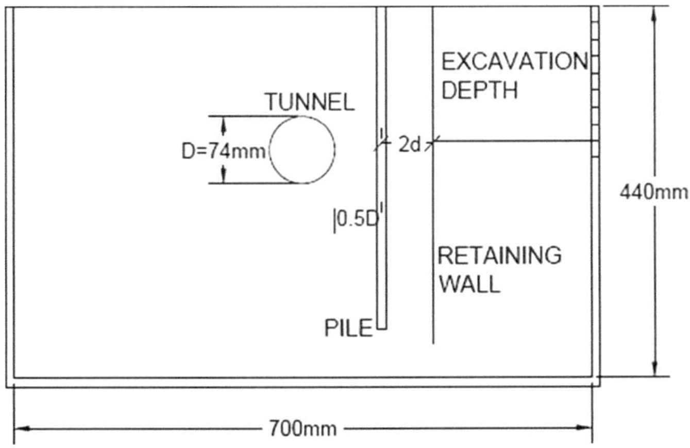

Fig. 1 Schematic arrangement of test setup for excavation and tunneling

This will induce movements in retaining wall and pile due to the lateral pressure exerted by the soil behind it. The excavation is continued up to depths of 200 mm. Deflections were noted for every 20 mm depth of excavation. Once the excavation activity was completed, the tunnel was bored to a length of 250 mm by adopting procedure as explained in previous para. Pile deflections for various length of penetration of tunnel were observed.

3 Results and Discussion

In all the tests pile was located at a distance of 145 mm from the edge of the tank (mid-width of the tank). The deflection of the pile head are observed for two different tunnel axis depths and six different pile tip locations as shown in Fig. 2.

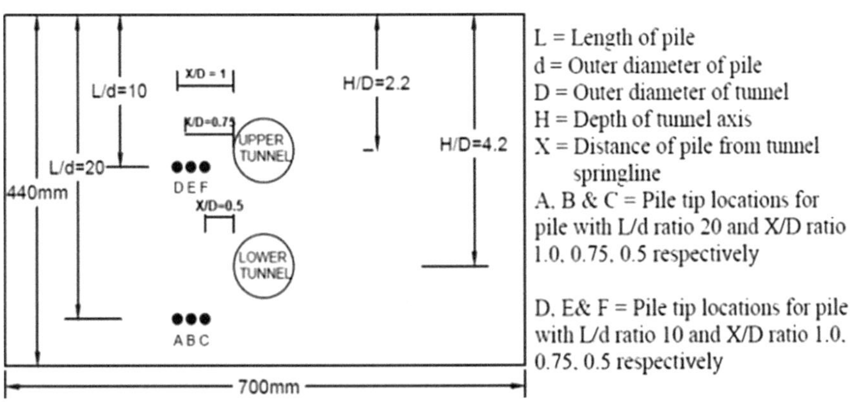

Fig. 2 Location of pile tip and tunnel axis

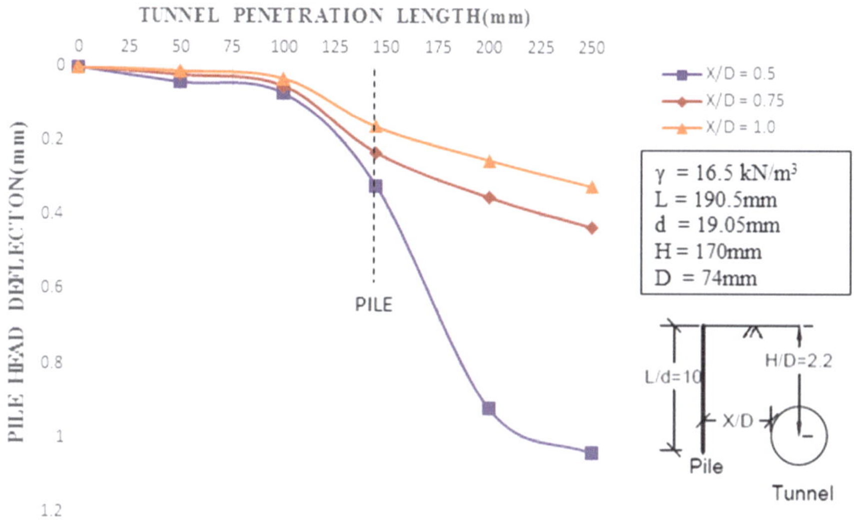

Fig. 3 Pile head deflection versus penetration depth for *L*/*d* = 10, *H*/*D* = 2.2

The variation of pile head deflection with respect to depth of penetration of tunnel for *H*/*D* = 2.2 is shown in Fig. 3. The maximum pile head deflection decreases when the distance of pile from tunnel springline increases for the condition of the sand bed and *L*/*d* ratio of pile.

The rate of pile head deflections shows steady increase after 100 mm of tunnel penetration irrespective of the distance of pile from the springline. This confirms the fact that, as the tunnel face approaches the pile, rapid soil movement occurs in the vicinity of the pile causing increase in pile head deflection. Figure 4 shows variation of pile head deflection versus penetration depth for *H*/*D* = 4.2. The maximum value of pile head deflection is lesser compared to the previous case (*H*/*D* = 2.2). As the tunneling operation is performed at a deeper depth (i.e., *H*/*D* = 4.2) the soil around the entire length of pile is subjected to movement which resulted in overall vertical movement of pile. Hence the lateral deflection of pile head has reduced for shorter pile when tunneling is performed at deeper depth. The location of pile is indicated with a dashed line. A comparison of Figs. 3 and 4 also confirms that the rate of pile head deflection is higher within a distance of 0.75 times diameter of tunnel from the center of the pile.

Figure 5 shows the comparison of maximum pile head deflection for pile with *L*/*d* ratio 20 located at a distance of *X*/*D* = 0.5 from tunnel springline under two depths (*H*/*D* = 2.2, 4.2) of tunneling. The dashed line in indicates the position of pile which is at 145 mm from the edge of the tank. Maximum deflection in the pile for tunnel at *H*/*D* = 2.2 was 4.5 times lesser when compared to the tunnel axis at *H*/*D* = 4.2. When the length of pile extending below the tunnel axis is more, pile

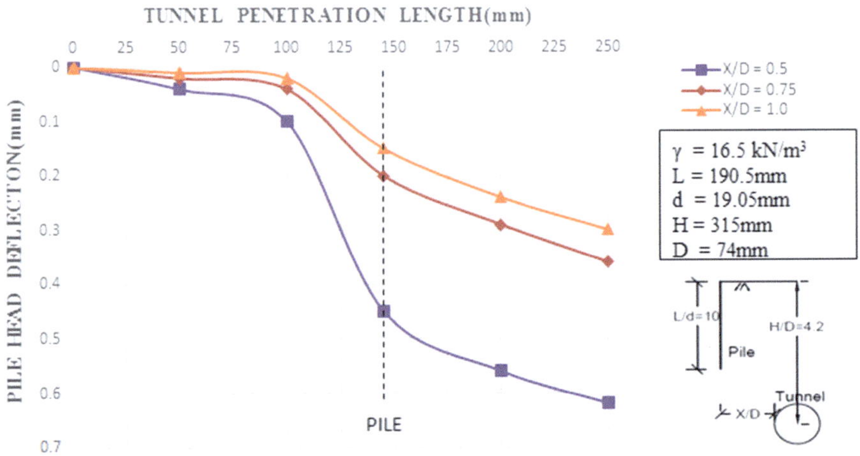

Fig. 4 Pile head deflection versus penetration depth for L/d = 10, H/D = 4.2

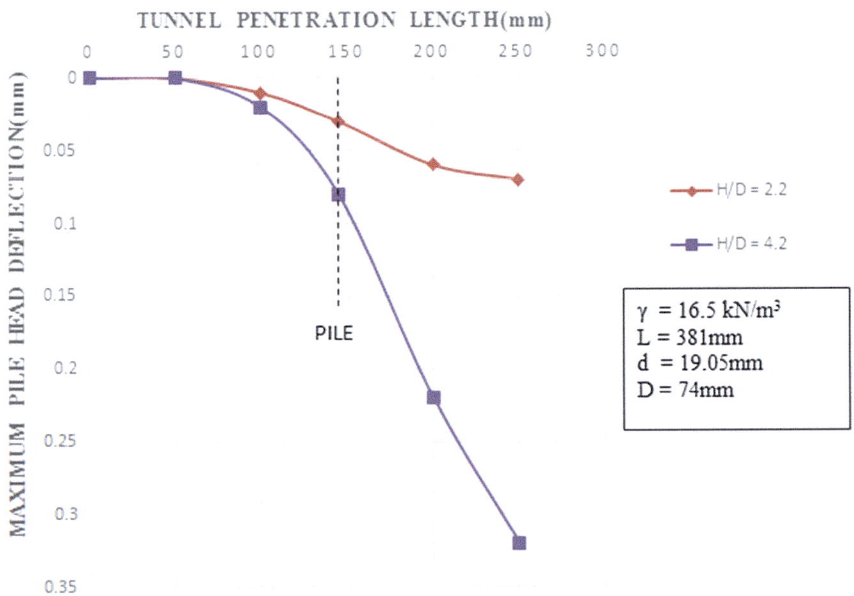

Fig. 5 Maximum pile head deflections for tunnels with H/D = 2.2 and 4.2

head deflection decreases because a significant part of the pile is embedded in sand which does not undergo movement due to tunneling. The rate of pile head deflection is very high between 100 and 200 mm penetration. This implies that within a distance of 0.75 times diameter of the tunnel from the center of pile, the rate of increase in pile head deflections are maximum.

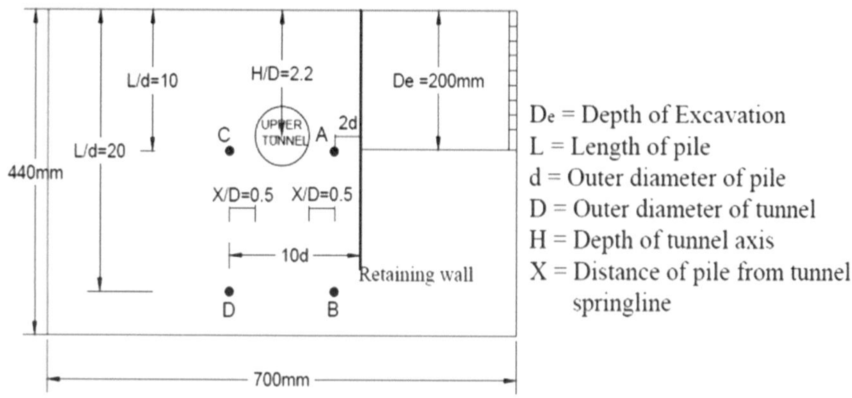

Fig. 6 Location of pile tips for excavation and tunneling case

The variation of pile head deflection for combined condition of excavation and tunneling is studied at four different tip locations of the pile as shown in Fig. 6, where *A*, *B*, *C* and *D* are the tip locations of the pile. Tunneling was performed at *H*/*D* ratio 2.2.

Pile head deflection for every 20 mm excavation is recorded and continued up to 200 mm of excavation. Once excavation operation was completed tunneling process was commenced. The deflections of the pile for case *A*, *B*, *C* and *D* are shown in Table 2. For pile tip at A the pile head deflection was 5.3 mm (towards retaining wall) for the pile at 2*d* distance from the excavation face and for the excavation depth of 200 mm. Thereafter the pile head deflection remained constant till a tunnel penetration length of 100 mm was achieved. Beyond 100 mm of penetration the pile deflected slightly towards the tunnel. As the face of the tunnel approached the pile, the soil movements around the pile increased causing deflection of pile head towards the tunnel side. Thus, deflection reduced to 5.23 mm when the maximum penetration distance of 250 mm is reached. During boring the tunneling the pile

Table 2 Pile head deflection due to excavation followed by tunneling for pile Tip at *A*, *B*, *C* and *D*

Depth of excavation (mm)	Tunnel penetration distance (mm)	Pile head deflection (mm)			
		Pile tip at *A*	Pile tip at *B*	Pile tip at *C*	Pile tip at *D*
40	0	0	0	0	0
80	0	0.03	0.02	0	0
160	0	2.80	0.25	0	0
200	0	5.30	0.60	0	0
200	100	5.30	0.60	0.05	0.01
200	200	5.23	0.60	0.90	0.06
200	250	5.23	0.60	1.0	0.07

head deflection remains the same without any change in deflection till the distance between the tunnel face and pile is 0.6*D* approximately.

Unlike the case of short pile, long pile at *B* did not undergo deflection due to tunneling. It is observed that the stress condition in the part of pile extending below the tunnel is not affected significantly. Thus embedment of pile in this region provided sufficient resistance against deflection. When the pile tip is located at *C* and *D* zero deflection is recorded which shows that there is no influence of excavation when pile is located at 10 times the diameter of pile from the face of the excavation.

4 Conclusion

- The maximum lateral pile head deflection decreases when the distance of pile from tunnel springline increases for the condition of the sand bed and *L*/*d* ratio of pile.
- Lateral deflection of pile head is observed to be higher for short pile with less embedment depth when compared to a long pile.
- The magnitude of pile head deflection increased when the normalized depth of tunneling is increased from 2.2 to 4.2 for long pile ($L/d = 20$). Whereas, for a short pile with $L/d = 10$ the magnitude of lateral deflection of pile decreased with increase in normalized depth of tunneling due to vertical settlement of pile.
- The rate of increase in pile head deflection is maximum when the advancing tunnel face is at a distance of 0.75 times diameter of the tunnel from the center of the pile.
- The net deflection of short pile when excavation and tunneling are performed on either side of the pile at equal distance is observed to be lesser than the total deflection when excavation alone is performed.
- In case of long pile with $L/d = 20$, the deflection due to tunneling is observed to be zero. Hence the net deflection is same as the deflection due to excavation alone.
- Zero pile head deflection due to excavation was observed for a pile at 10*d* from the retaining wall and 0.5*D* from the tunnel springline.

References

Basile, F. (2014). Effects of tunneling on pile foundations. *Soils and Foundations, 54*(3), 280–295.

Chen, L. T., Poulos, H. G., & Hull, T. S. (1995). Model tests on single piles subjected to lateral soil movement. *Soils and Foundations, 37*(1), 1–12.

Chen, L. T., Poulos, H. G., & Loganathan, N. (2000). Pile responses caused by tunneling. *Journal of Geotechnical and Geoenvironmental Engineering, ASCE, 125*(3), 207–215.

Ilamparuthi, K., & Madhumathi, R. K. (2011). Effect of ground movement on the performance of pile foundation. *Proceedings of Indian Geotechnical Conference, Kochi, 1,* 187–190.

Leung, C. F., Chow, Y. K., & Shen, R. F. (2000). Behaviour of pile subjected to excavation-induced soil movement. *Journal of Geotechnical and Geoenvironmental Engineering, ASCE, 126*(11), 947–954.

Meguid, M. A., & Mattar, J. (2009). Investigation of tunnel-soil-pile interaction in cohesive soils. *Journal of Geotechnical and Geoenvironmental Engineering, ASCE, 135*(7), 973–979.

Min, Y., Qing, S., Wei-Chao, L., & Kang, M. (2011). Three-dimensional finite element analysis on effects of tunnel construction on nearby pile foundation. *Journal for Central South University and Technology, 18*(3), 909–916.

Poulos, H. G., & Chen, L. T. (1997). Pile response due to excavation-induced lateral soil movements. *Journal of Geotechnical and Geoenvironmental Engineering, ASCE, 123*(2), 94–99.

Part IV
Theme 10: Forensic Investigations and Case Histories

Support of Deep Excavation Using Contiguous Pile—A Case Study

M. Vinoth and S. M. Ghan

Abstract The design of deep excavations requires careful consideration of the strength and stability of the various structural elements and soil properties at all stages during the construction process. In addition, the ground movements induced by the excavation need to be carefully studied and controlled, to ensure no damage to nearby properties and services and to keep these within serviceable limits. The performance of a deep excavation depends on the method of construction as well as precise assessment of the local ground conditions. Making reliable predictions of performance often presents a considerable challenge looking at various factors and uncertainties involved. This paper presents a case history on the use of temporary support of excavation using contiguous bored piled wall technique for deep basement excavation in Chennai. Case history presents an excavation of up to 9.3 m deep for the construction of a basement for a commercial building. The site stratification is sedimentary formation (Sandy Clay underlain by Stiff Clay) and the rock is at very deep depths beyond even shoring pile socketing extents. The geotechnical challenge is to design and construct a deep basement up to 9.3 m depth where SPT '*N*' values are less than 10 up to 15 m below Existing Ground Level (EGL), high ground water table, close proximity of high-rise structures and adjacency of underground metro line to the shoring system. Finite element analysis using PLAXIS was used to analyse this staged excavation. Based on the finite element analyses, temporary strut supports were used to provide lateral support for contiguous piles. Monitoring works all along the shoring line was carried out using total station on regular intervals, and these readings were compared with the analysis results.

M. Vinoth (✉) · S. M. Ghan
CSIR - CRRI, New Delhi 110025, India
e-mail: vinothm.27@gmail.com

S. M. Ghan
e-mail: SANDEEP-GHAN@lntecc.com

A. I. V. and V. B. Maji (eds.), *Geotechnical Applications*, Lecture Notes in Civil Engineering 13, https://doi.org/10.1007/978-981-13-0368-5_29

1 Introduction

The proposed project is a retail development located in Anna Nagar, Chennai. Retail development consists of construction of structure with 3 Basements + Ground + 4 Floors. Plot area of the proposed building is approximately 27,700 m^2. The three basement floors will be used as car parking and remaining all floors are to be used for retail and cinema. Permanent contiguous piles are proposed all along the periphery of the basement as a retaining wall. During the excavation for the basement construction, the contiguous pile will be supported by struts (installed at two levels). After basement construction is over, the Contiguous piles with skin wall will derive horizontal support from the floor slabs which will convert to permanent system, due to which the struts shall be removed and the contiguous piles will be permanently supported by basement floor slabs system including raft. As the deep excavation is made very close to high-rise buildings, underground metro line and arterial roads (which require safeguarding against damage due to soil movement if any), it was decided to adopt observational method where predicted responses are checked by field monitoring during the works and preventive measures implemented as and when necessary. The main purpose of this paper is to assess the performance of the contiguous pile based on the comparison between predicted and measured pile displacements during construction.

2 Subsurface Profile

The subsurface investigation comprised of detailed investigation through 10 boreholes. From all these investigations the average and the generalised soil profile was deduced and is shown in Table 1. The profile of average SPT values observed along the depth of all the bore holes is shown in Fig. 1.

Maximum depth of excavation is 9.3 m from NGL. Profile revealed that soil up to excavation depth is weak. Hence a stiff retaining wall needs to be constructed to enable basement construction activities.

Table 1 Generalized soil profile

Type of layer	Depth from EGL (m)	Avg. SPT '*N*'
Sandy clay-1	0.0–14.0	9
Sandy clay-2	14.0–16.5	20
Stiff clay-1	16.5–20.0	35
Stiff clay-2	20.0–30.0	39

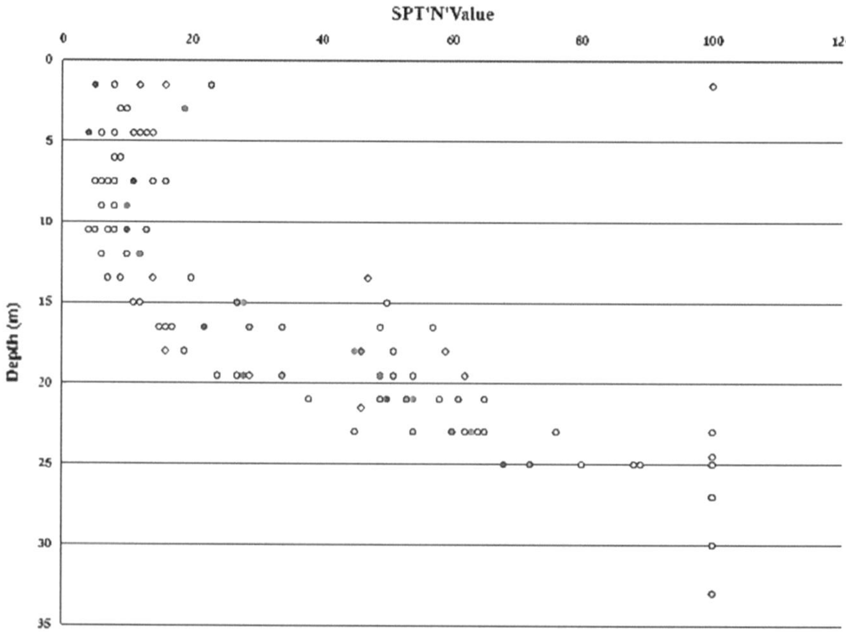

Fig. 1 The profile of SPT values observed along the depth of various bore holes

3 Design

Contiguous pile of 800 mm diameter with embedment depth of 9.2 m and supported with struts is designed for these conditions (and hereafter described in this paper). Contiguous pile was designed in accordance to IS: 456 (2000). As the high capacity anchors could not be designed for such soft soils and the space around the plot was also a constraint, it was decided to design the system with struts inside the construction.

Choice of retaining the soil with contiguous pile was found appropriate for the following features,

- Depth of excavation is less than 10 m.
- Economical and ease in construction.
- Can be used as a permanent retaining wall along with Bentonite Slurry Wall.

Details of contiguous pile are given in Table 2.

Table 2 Details of contiguous pile

Diameter of contiguous pile (mm)	Maximum depth of excavation (m)	Level of struts	Embedment depth (m)
800	9.3	2 levels (2.1 and 5.1 m)	9.2

As contiguous pile is proposed to be used as permanent retaining wall, contiguous piles were analysed for two types of phase namely temporary analysis and permanent analysis. The temporary analysis is carried out till the final excavation level where as the permanent analysis is done considering all the floor slabs in position. For temporary analysis, a surcharge load of 10 kPa is considered in addition to the earth pressure acting on the contiguous pile. The water table was recorded at 8.3 m below NGL. Hence, in the design, the water table was considered at 8.3 m below NGL on retaining side and 9.8 m below NGL on the excavated side for contiguous pile analysis during temporary condition. This was later validated during the excavation process in the proposed site. Inside the excavation area, ground water table was supposed to be lowered by open sump dewatering system. To make the contiguous pile retention system waterproof, on the outer surface (retaining side) 150 mm diameter bentonite slurry walls were installed.

The contiguous pile system is analysed for various stages of excavation for the purpose of installation of struts. The initial ground water table is assumed to be at −8.30 m and lowered to −9.8 m (i.e. 0.5 m below excavation level) below NGL, for the temporary analysis. The excavation was carried out using two levels of struts for supporting the contiguous pile during the basement construction. Hollow steel tubes were considered for struts and are designed in accordance to IS: 806 (1968). The struts are designed for a temporary design life of 2 years.

The proposed excavation scheme for the temporary stage is analysed in PLAXIS-2D software as a plane strain problem. In this analysis, contiguous pile is modelled as plate element, struts are modelled as fixed end anchors and the soil behaviour is modelled using Mohr–Coulomb model. Globally meshes were refined to a relative element size factor (r_e) of 1 and locally where plates, struts and surcharges are located a r_e value of 0.5 was adopted. Properties of materials considered in PLAXIS model are given in Tables 3 and 4.

The stages considered for the temporary analysis for 800 mm diameter contiguous pile is given below,

Stage-1 Installation of contiguous pile and application of surcharge of 10 kPa.
Stage-2 Excavation of soil up to −1.5 m from the NGL.
Stage-3 Excavation of soil from −1.5 to −3.0 m.
Stage-4 Excavation of soil from −3.0 to −4.5 m.
Stage-5 Excavation of soil from −4.5 to −6.0 m.
Stage-6 Excavation of soil from −6.0 to −7.5 m.
Stage-7 Excavation of soil from −7.5 to −9.3 m. Water table inside the plot is lowered from −8.3 to −9.8 m.

Table 3 Soil properties considered for PLAXIS analysis

Type of layer	γ (kN/m^3)	c (kPa)	E (kPa)	μ
Sandy clay-1	18	45	13,500	0.3
Sandy clay-2	18	100	30,000	0.3
Stiff clay-1	19	175	52,500	0.3
Stiff clay-2	20	197	75,000	0.3

Table 4 Pile and strut properties considered for PLAXIS analysis

Description	EI (kN m^2/m)	EA (kN/m)	W (kN/m/m)	μ
Contiguous pile	558,505	13,962,634	13.96	0.15
Strut	–	2,221,600	–	–

Stage-8 Installation of first level of strut at −2.1 m at a horizontal spacing of 1.0 m c/c.
Stage-9 Excavation of soil (acting as passive resistance until the installation of strut) up to second berm level.
Stage-10 Installation of second level of strut at −5.1 m at a horizontal spacing of 1.0 m c/c.
Stage-11 Excavation of soil (maintained at a slope of 1V:1.5H, acting as passive resistance until the installation of strut) till excavation level.

4 Field Instrumentation

The construction and subsequent behaviour of the excavation was monitored by measuring displacements of wall. In the present case study, total station was used to measure the lateral deformation of contiguous pile. Field monitoring gives information on pattern of deformation and effectiveness of construction control measures. It helps in monitoring behaviour after construction and indicates potentially dangerous conditions that may adversely affect stability of the structure, its foundation and surrounding area. It also provides basic data for design improvement that will promote safer and economical design and construction. Around 30 points were identified in each stretch of the contiguous pile line.

5 Comparison of Design Calculations and Site Observations

Design calculations have been compared with site observations measured by the instrumentation carried out within the site during the period of construction.

5.1 *Site Observations of Horizontal Displacements of Contiguous Pile at Top*

Figures 2, 3 and 4 shows the horizontal displacement at top of contiguous pile obtained from total station readings for 800 mm diameter contiguous pile. Positive

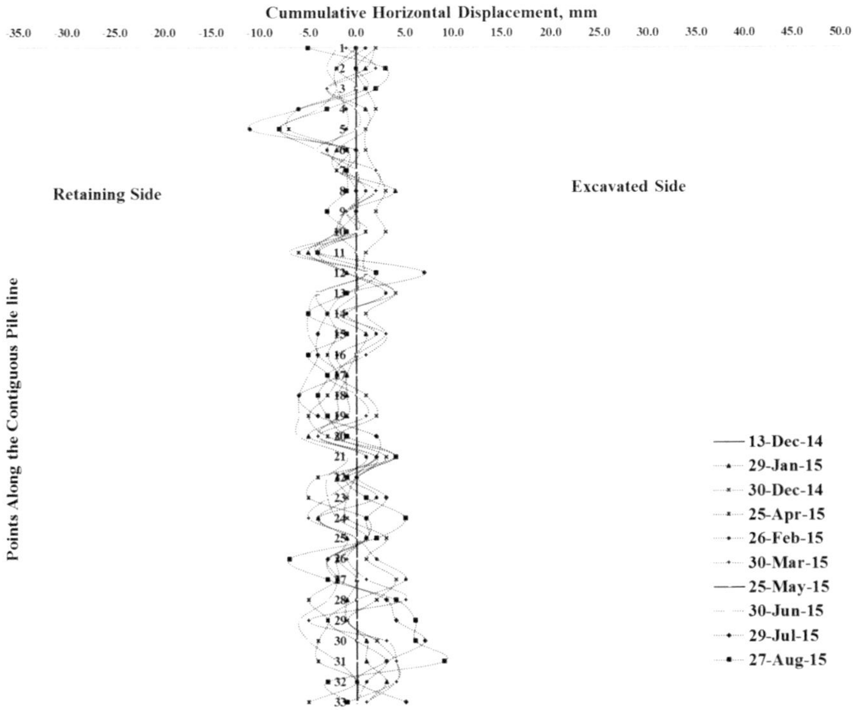

Fig. 2 Horizontal displacement at top of contiguous pile along east side of the plot

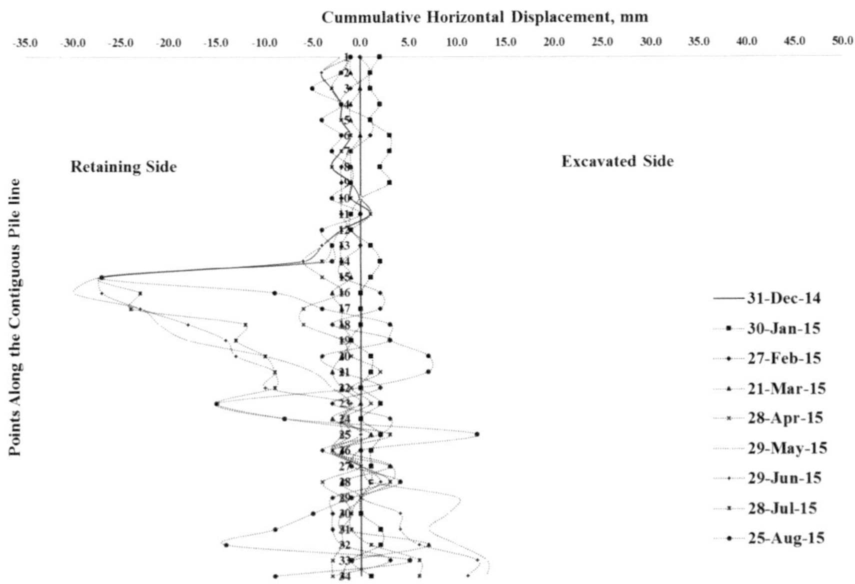

Fig. 3 Horizontal displacement at top of contiguous pile along south side of the plot

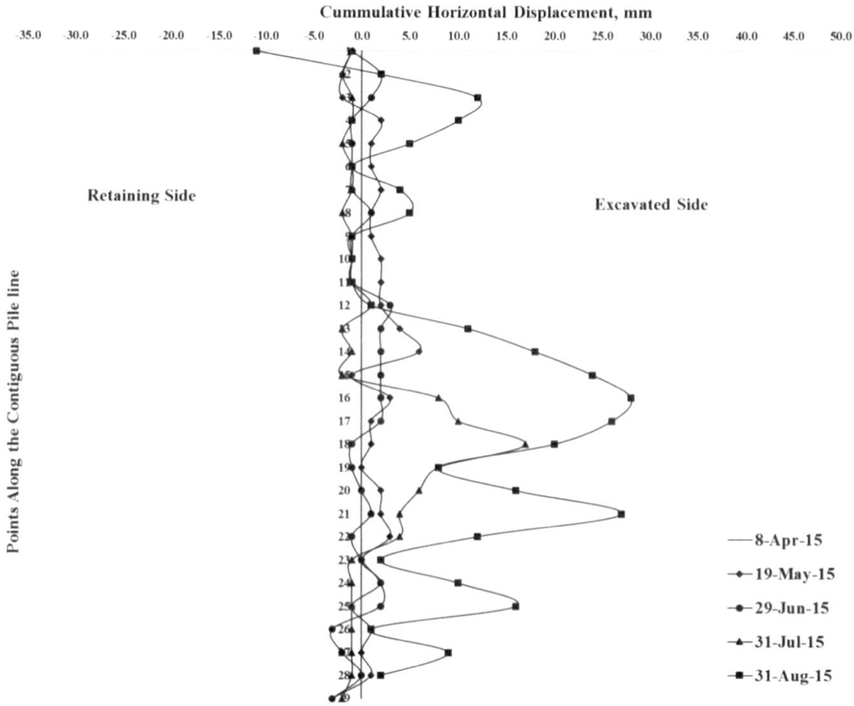

Fig. 4 Horizontal displacement at top of contiguous pile along north side of the plot

and negative signs in the horizontal displacement represent, respectively, inward and outward displacements relative to the contiguous pile.

On the west, retaining side soil was removed for basement construction activities of the adjacent party. As no active pressure is going to be transferred to the contiguous piles in this stretch, field monitoring was not carried out.

By the end of excavation works for east side, the maximum deflection was about 9 mm. Displacements observed on south side was more, with a maximum of about 13 mm. Maximum horizontal displacement at top in the north side was still of greater magnitude (28 mm) against theoretical predicted value of 42.5 mm.

The actual realised movements are consequently below the predicted values. This may be because of various reasons like presence of higher cohesion values leading to deeper tension zones thereby reducing the active thrust on the wall, assumed design surcharge not actually experienced in the site by the walls, caused by the exclusion and horizontal resistance provided by the capping beam, which is difficult to quantify in a two-dimensional model.

Figure 5 shows the comparison plot between the theoretical and obtained horizontal displacement at top for 800 mm diameter contiguous pile after final level of excavation has reached.

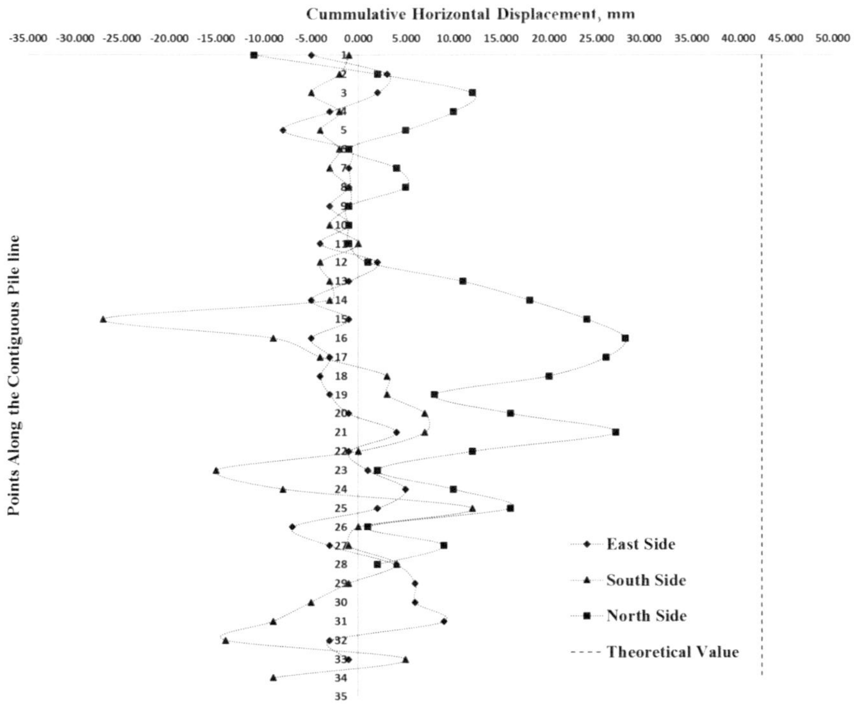

Fig. 5 Comparative plot for theoretical versus actual lateral deflection at top of contiguous pile after final level of excavation has reached

5.2 *Comparison of Field Performance and Design Calculations*

There were some key differences between field observations and design predictions.

- It is noticed that in majority of locations, after the installation of strut, contiguous pile moved towards retaining side (negative deflection). This confirms that tension zones have been developed behind the contiguous pile wall.
- The maximum measured horizontal wall displacements were approximately 65% of the predicted design values.

This information confirmed that the design was adequate.

6 Conclusions

This study focuses on the effect of construction of a deep basement on contiguous pile for Retail building. The contiguous piles were monitored during the works using total station. The results from the observations have been used to analyse the contiguous pile response to various construction and excavation activities. Observations from the monitoring results and the deduced contiguous pile responses can be summarised as follows.

- Total station measurements at the top of contiguous pile indicated that contiguous pile moved towards retaining side (negative deflection) due to development of tension zones behind the contiguous pile.
- In general, reasonable agreement with the measured top deflections of the contiguous pile was obtained till both the struts are installed.
- Total station data indicated that the maximum measured horizontal displacements at the top of contiguous pile varied from 21 to 65% of the theoretical values.
- It was observed that temporary struts were satisfying in design aspect but on the other hand it had a major disadvantage. By obstructing the free movements of machineries and equipment's within the site during execution. Thus, in locations where tie back anchors are possible, shall be preferred in place of struts, so as to provide hindrance free area for carrying out basement construction activities.

Extensive monitoring carried out during the construction has given valuable insight into the performance of the contiguous pile and the behaviour of the soil strata.

References

IS 456. (2000). Plain and Reinforced Concrete—Code of Practice.
IS 806. (1968). Code of Practice for Use of Steel Tubes in General Building Construction.

Punch-Through Analysis of Jack-Up Rig at a Site Off the East Coast of India—A Case Study

Rupam Mahanta

Abstract Jack-up rigs are primarily used for carrying out offshore operations related to drilling for hydrocarbons. Geotechnical assessment is a pre-requisite for deployment of any jack-up rig at a new offshore site. When deployed for operations, jack-ups are supported by their legs which initially penetrate below the seafloor on loading. Majority of jack-up units have individual/independent foundation, called ‘spudcans’, at the bottom of their legs. This paper presents a case of deployment of a jack-up at a site off the east coast of India where the jack-up had a ‘punch-through’ during preloading. Punch-through is a sudden and uncontrolled penetration of the spudcan often causing structural damage of the unit. It may occur when the load becomes equal to bearing capacity for the spudcan while within a soil layer having relatively higher bearing capacity than the soil underlying that layer. The deployment of the rig was carried out on the basis of preliminary soil investigation report from the consultant on-board the geotechnical vessel deployed for soil investigation. The case is analysed and discussed. Computer program ‘MAHAJACK’ developed by the author for carrying out leg-penetration and punch-through analysis for foundation of jack-up rigs has been used for the analysis.

Keywords Jack-up · Punch-through · Offshore

1 Introduction

Jack-up units having independent foundations at their bottom are termed as ‘independent-legged’ or ‘spudcan-type’ jack-up rigs. Most of the modern jack-ups have three legs. The case presented in this paper is related to a three-legged spudcan-type jack-up rig. Geotechnical investigation was carried out at the site for assessment of safety of the jack-up before its deployment. A jack-up, after getting

R. Mahanta (✉)
Institute of Engineering and Ocean Technology (IEOT), ONGC, Navi Mumbai, India
e-mail: mahanta_rupam@ongc.co.in

A. I. V. and V. B. Maji (eds.), *Geotechnical Applications*, Lecture Notes in Civil Engineering 13, https://doi.org/10.1007/978-981-13-0368-5_30

positioned at a location, is preloaded and its legs are subjected to penetration below the seafloor before the hull is fully raised to higher elevated position above the water surface. Preloading operation is to apply loads on the legs of the jack-up to a pre-determined load level to ensure that the soil resistance is sufficient for the foundation to resist the maximum design loads (includes environmental loads) for operational in that location. Once the maximum penetration of the spudcan below the seafloor is achieved after application of designated preload on the legs, the extra load (generally seawater ballast) is subsequently dumped, before elevating the hull to its final position. A careful geotechnical analysis needs to be carried out to predict the load-penetration behaviour of the spudcan in advance in order to avoid the possibility of an undesired situation of sudden and uncontrolled leg penetration that may prove to be too costly in terms of time, money and in the worst case injury/loss of human resource. In the present case, there was sudden penetration of the spudcan by about 5 m during preloading operation for the rig. The analysis of such incidents is very helpful for understanding the behaviour of combinations of soil layers in response to loading from large foundation of jack-ups in the offshore and may help in adopting improved approach in future for prevention of such cases. All depths of foundation and soil layers mentioned in the paper are referenced to the seafloor. Depth of the spudcan means the depth of the tip of the spudcan below the seafloor.

2 Leg-Penetration Analysis

Leg-penetration analysis is a bearing capacity analysis for the spudcan with respect to penetration depth of the spudcan. Spudcans are generally approximated as circular in shape. Its diameter is typically in the range of 8–20 m and it normally has a conical base with a pointed tip. Significant depth of soil below the spudcan is affected, often involving more than one layer of soil below the spudcan due to its large size. Bearing capacity at a foundation depth depends mainly on soil strength and its shear behaviour within the significant stress zone, geometry of the spudcan and sequence and combinations of layers of soil. Leg-penetration analysis shows the vertical load carrying capacity of a spudcan at depths below the seafloor. The final penetration depth of the spudcan is where the maximum preload is equal to the bearing capacity of soil. From this analysis, the risk of sudden, uncontrolled penetration for the spudcan, if any, is also found out. Guidelines recommended by SNAME (Society of Naval Architects and Marine Engineers) and ISO (International standard organisation) are generally followed for the analysis.

3 Preloading

The purpose of preloading is already described. The preloading of a jack-up is carried out in fair weather. Initially, the legs are lowered to touch the seabed. As the hull is gradually raised from water, vertical compressive load, i.e., self-weight of the jack-up is transferred to the foundations. The load is generally increased gradually up to the required maximum preload by pumping seawater into large tanks in the hull of the jack-up unit. For a typical jack-up rig, the required preload is about 45 MN per leg of the unit. However, it may exceed 150 MN per leg in case of some heavy jack-ups. The penetration of legs generally increases with increasing loads. During preloading, the spudcan penetrates to a depth where the mobilised bearing capacity is sufficient against the applied preload.

Since preloading is carried out in fair weather, the forces due to wave and current are relatively insignificant during preloading and the load is predominantly vertical. After completion of preloading, the water ballast if offloaded from the hull. The hull is then raised to a safe and suitable elevation with sufficient air gap between the hull and water surface.

4 Foundation Details

The water depth at the location was about 20.5 m. The details of foundation of the jack-up are given below

Maximum preload = 76.5 MN
Equivalent diameter of spudcan = 16.1 m
Height of spudcan = 3.8 m
Volume of spudcan = 570 m^3
Tip extension = 0.9 m
Still water load = 47.5 MN.

5 Punch-Through

The deployment of the rig in the present case was carried out by the operator of the rig based on preliminary soil report from the on-board consultant. During the preloading operation, before the maximum preload was applied, punch-through had occurred. The spudcan suddenly penetrated from a depth of 18.7–24 m. Final penetration of spudcan with maximum preload was 27 m (average for the three spudcans). No damage to the rig was reported, but the incident was a cause for serious concern. The author was subsequently involved in field investigation, analysis and deliberations for deciding on suitable locations for further deployment of jack-ups in the area.

6 Soil Condition

Soil investigation, before deployment of jack-up, included in situ Cone Penetration Test (CPT) and on-board laboratory test of soil samples. The soil condition was predominantly silty clay with intermittent layers of silt and sand having small

Table 1 Soil parameters from on-board report

Depth (m)	Soil type	ϕ' (°)	S_u (kN/m^2)	γ' (kN/m^3)
0–2.6	Clay	–	5	7
2.6–6.7	Clay	–	12–20	9
6.7–13.0	Sandy-silt	25	–	9
13.0–13.8	Clay	–	20	9
13.8–15.5	Sandy-silt	25	–	9
15.5–17.6	Clayey-silt	–	20–30	9
17.6–18.7	Sandy-silt	25	–	9
18.7–28.5	Clay	–	35–40	9
28.5–41.8	Clay	–	40–60	9

Note ϕ'—angle of internal friction; s_u—undrained shear strength and γ'—effective unit weight of soil

thickness. Based on the soil properties, a leg-penetration analysis for jack-up rig was carried out by the on-board consultant. However, possibility of punch-through was not indicated in the report. The soil properties and thickness of layers (from the on-board report) is presented in Table 1.

Clay occurs in very soft to soft condition up to about 17 m. The increase of shear strength of the clay with depth is at a very low rate. Analyses in the present study were carried out using both the original soil parameters (as recommended in the on-board soil report) and also with revision of the soil parameters from 6.7 to 15.5 m. The revised soil parameters are presented in Table 2. However, the revised parameters within this depth have no analytical effect on the actual punch-through depth that occurred in the field when the spudcan depth was 17.6 m below seafloor. But, this modification helped in matching the Field experience for the relevant range of depths and also it is recommended by codes to examine both drained and

Table 2 Soil parameters with re-interpretation/ modification (from 6.7 to 15.5 m)

Depth (m)	Soil type	ϕ' (°)	S_u kN/m^2)	γ' (kN/m^3)
0–2.6	Clay	–	5	7
2.6–6.7	Clay	–	12–20	9
6.7–7.8	Clay	–	20	9
7.8–8.3	Sandy-silt	20	–	9
8.3–9.2	Clay	–	20	9
9.2–10.2	Sandy-silt	25	–	9
10.2–12	Clay	–	20	9
12–12.8	Sandy-silt	25	–	9
12.8–17.6	Clay	–	20–30	9
17.6–18.7	Sandy-silt	25	–	9
18.7–28.5	Clay	–	35–40	9
28.5–41.8	Clay	–	40–60	9

undrained behaviour for silts in bearing capacity analysis for jack-ups. Data of CPT cone resistance are shown in Fig. 1. Unfortunately, the digital data of CPT is not available for the location. Therefore, the available plot of CPT cone resistance for the relevant soil is presented in Fig. 1 (CPT data for higher depths and laboratory test results are not presented due to paucity of space). Mixed nature of soil in the location can be observed especially below 6.7 m.

7 Analysis, Results and Discussion

Conventional limit equilibrium analysis was carried out by using both the original (on-board) soil parameters and also with some revision of the soil parameters (for 6.7–15.5 m). International codes SNAME and ISO, which are the current industry standards, have been followed for the analysis.

The results of analysis are presented in Fig. 2. Both the international guidelines broadly recommend similar procedures of analysis of bearing capacity involving multiple layers. For bearing capacity of clays, Skempton's method (ISO) has been used for all the curves. For squeezing mode of failure, recommendations of ISO/ SNAME and when sandy soil is overlying clayey soil, SNAME's recommendation is followed for the analysis.

Curve-1 (Fig. 2) is based on the soil data of Table 1 (on-board) with a simplified analysis involving maximum up to 2 layers for bearing capacity analysis at any depth. This pattern of analysis does not follow the principle of bearing capacity for multiple layers as recommended by SNAME or ISO. Curve-2 is based on the on-board soil parameters (Table 1) following SNAME/ISO guidelines for bearing capacity involving multiple layers. The analysis as per Curve-2 with on-board soil report indicates a major punch-through starting at 7.5 m, which did not actually happen.

Further analysis was carried out with revision of soil parameters for the depth from 6.7 to 15.5 m. Curve-3 is the analysis with soil parameters modified for the depth from 6.7 to 15.5 m (Table 2) following SNAME/ISO guidelines.

It may be noted that results of all the 3 (three) analyses have significant differences at penetration depths from about 7 to 16 m due to difference in relevant soil parameters (from 6.7 to 15.5 m)and different approach for calculating bearing capacity for Curve 1.

Curve-1 (simplified analysis with original soil parameters) predicts the depth of initiation of punch-through during preload to start at 17.6 m but the predicted distance of uncontrolled penetration is much more than that what actually occurred. Final penetration of spudcan is also predicted to be somewhat higher. This analysis also predicts relatively lesser bearing capacity at depth 7.5 m penetration.

Curve-2 (analysis following SNAME/ISO with original soil parameters) indicates the punch-through exactly at the level where it actually started (17.6 m) and the distance of uncontrolled penetration. It also indicates almost accurately the final spudcan penetration. However, it also indicates a major punch-through starting at

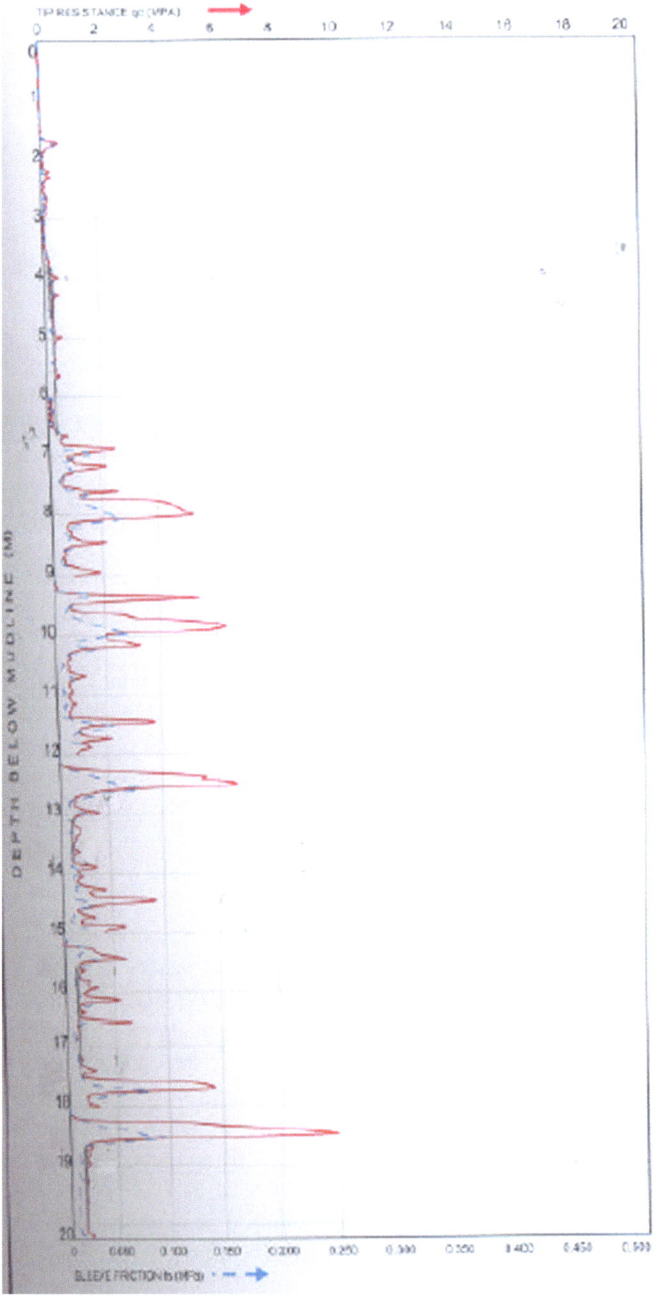

Fig. 1 CPT cone resistance and sleeve friction

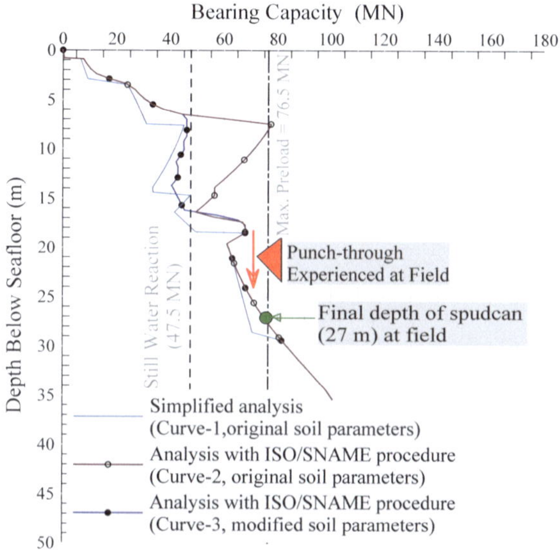

Fig. 2 Leg penetration/punch-through analysis

about 7.5 m which did not occur, indicating the actual bearing capacity as weaker than predicted by using original soil parameters for 6.7–15.5 m.

Curve-3 (analysis following SNAME/ISO with revision of soil parameters from 6.7 to 15.5 m) matches the actual behaviour both for the punch-through and final depth of penetration of the spudcan.

Observations from the study about the leg-penetration behaviour are listed below:

- ISO/SNAME-recommended analysis has proved to be quite useful in the present case for making a prediction closer to actual behaviour in the field for the depth of initiation of punch-through, vertical distance of punch-through penetration and the final depth of penetration of the spudcan. The recommendation on bearing capacity approach for multiple layers of soil need to be followed.
- It also proves the usefulness of the recommendation by codes that silts should be analysed as both drained and undrained materials. In the present case, revision of soils between 6.7 and 15.5 m, along with multi-layer bearing capacity analysis matched the actual result in Field.
- Soil parameters presented in the on-board report was assessed reasonably well. Analysis following existing international guidelines and original soil data could predict the actual punch-through and final depth of penetration.
- The simplified 2 (two) layer analysis (Curve-1 of Fig. 2) did not result in making good predictions for the case having multiple layers within the significant zone below the spudcan. Both punch-through movement and the final depth of penetration are predicted higher than reality by this analysis in the present case. Further, punch-through or rapid penetration may not be indicated or may be wrongly indicated with this kind of analysis.

8 Conclusion

It is observed that occurrence of thin layers of silty/sandy soil within soft/firm clays in the soil zones of relevance, requires careful interpretation of soil and subsequent analysis for prediction of leg penetration and punch-through possibility. From the case study, it was evident that there was an obvious sign of impending punch-through. Unfortunately, the same was not detected in advance.

Although recent research has made progress on mechanism-based approach for bearing capacity analysis simulating the actual process of leg-penetration, analysis carried out in the present case using 'wish in-place' model for the foundation, recommended by SNAME and ISO was found to produce useful results. It is also found from the analysis of the case that Skempton's method for general bearing capacity for clay, as suggested by ISO code, is applicable for clayey soils for this area for analysis of leg penetration for jack-ups.

Acknowledgements The author is grateful to the management of ONGC for granting permission to publish the paper.

References

ISO 19905-1:2012(E), International Standard ISO 19905-1, First edition, Petroleum and natural gas industries—*Site-specific assessment of mobile offshore units, (with errata and supplements), Part 1: Jack-ups.*

Mahanta, R., Prakasha, K. S., & Mishra, A. K. (2010). Geotechnical problems of jack-up rig deployment in offshore West Bengal and Mahanadi Area of Indian Offshore—case studies. In *Proceedings of Indian Geotechnical Conference*, IIT Bombay, Mumbai.

Randolph, M., & Gourvenec, S. (2011). *Offshore geotechnical engineering*. CRC Press.

On-Board Report of Geotechnical Investigation for the Site. (2004).

The Society of Naval Architects and Marine Engineers. (2002). *Guidelines for site specific assessment of mobile jack-up units* (1st ed.). Technical & Research Bulletin 5-5A.

Part V
Theme 11: Reliability in Geotechnical Engineering

Reliability Analysis of Slopes in Soils with Strain-Softening Behaviour

S. Metya, S. Dey, G. Bhattacharya and R. Chowdhury

Abstract This article presents a systematic approach for the reliability analysis of slopes in strain-softening soils based on the first order reliability method (FORM). The performance function is based on the Spencer method modified to take strain-softening into account in terms of the average residual factor R_F over a potential slip surface estimated based on a simple progressive failure model available in the literature. The shear strength parameters, peak and residual, are assumed as normally distributed random variables and the reliability analysis is performed on the probabilistic critical slip surface. For the residual factor R_F, bounded by 0 and 1, a generalized beta distribution has been assumed. Results obtained from an illustrative example indicate that a significant reduction (25%) occurs in the value of the minimum reliability index when R_F is considered as a random variable compared to when R_F is considered as a deterministic parameter. A FORM based sensitivity study also reveals that, amongst the five random variables, residual factor has the most dominating influence on the estimated reliability index and thus justifies its inclusion as one of the random variables.

Keywords Slope reliability · Peak and residual strengths · Residual factor Probability distribution

S. Metya (✉) · S. Dey · G. Bhattacharya
Department of Civil Engineering, IIEST Shibpur, Howrah 711103, India
e-mail: subhadeep.metya@gmail.com

S. Dey
e-mail: sukanta.hit15@gmail.com

G. Bhattacharya
e-mail: bhattacharyag@gmail.com

R. Chowdhury
University of Wollongong, Wollongong, Australia
e-mail: robin@uow.edu.au

A. I. V. and V. B. Maji (eds.), *Geotechnical Applications*, Lecture Notes in Civil Engineering 13, https://doi.org/10.1007/978-981-13-0368-5_31

1 Introduction

The conventional limit equilibrium method of slope analysis is based on the assumption of ideal plastic behaviour of soils. Such an assumption implies simultaneous failure of a slope. In contrast, slopes in strain-softening soils are associated with progressive failure. It is well known that in such soils the processes of progressive failure are often associated with a decrease in the values of shear strength parameters (Skempton 1964, 1985). Skempton proposed a definition of residual factor at a point in a soil mass as the extent to which shear strength has decreased from its peak value to its residual value. He also proposed a definition of the average residual factor over a slip surface as the proportion of slip surface length over which shear strength has reduced to a residual value. The average residual factor over a slip surface, denoted by R_F, therefore, deserves to be included as a parameter in the stability analysis of slopes in strain-softening soils associated with progressive failure. In the past, the residual factor, averaged over a potential slip surface, has been included, either directly or indirectly, as a deterministic variable even in probabilistic studies. However, because of uncertainties associated with the residual factor, it is very important to consider it as a random variable in slope reliability analyses within a probabilistic framework of slopes in strain-softening soils.

In this paper, a systematic approach for reliability analysis of a 2-D simple slope in a strain-softening soil considering the average residual factor over a potential slip surface as a random variable, has been developed on the basis of a limit equilibrium model, specifically, the Spencer's circular method. For the purpose of reliability analysis, the first order reliability method (FORM), widely accepted as the most versatile among the methods of reliability analysis based on analytical approximations (Haldar and Mahadevan 2000), has been made use of in this study. An assumption has been made regarding a suitable probability distribution of the average residual factor R_F.

2 Adopted Methodology

2.1 Slope Stability Analysis—Deterministic and Probabilistic

Deterministic slope stability analyses based on the limit equilibrium approach consists of two joint tasks, namely, computation of factor of safety of a given or trial slip surface, and then the search for the critical slip surface having the minimum factor of safety FS_{min} (called the deterministic critical slip surface) using an optimization technique. The Spencer method of slices (Spencer 1967), regarded as one of the rigorous methods, is used for the calculation of factor of safety and the sequential quadratic programming (SQP), rated as a powerful optimization technique (Hong and Roh 2008), is adopted in the MATLAB platform with its optimization toolbox.

Similar to the deterministic analysis, the probabilistic slope stability analysis can be viewed as the problem of locating the slip surface corresponding to the lowest value of reliability index $\beta_{\min}$ (or the highest value of probability of failure, p_{F}) (called the probabilistic critical slip surface of the slope). As already mentioned, for the determination of reliability index, the first order reliability method (FORM), has been adopted in this study. The detailed description of the computational procedure for the deterministic and probabilistic slope stability analysis can found elsewhere (Metya and Bhattacharya 2014) and the same computer programs have been made use of in this study. As a part of that study, these programs were validated with reference to two benchmark slope problems.

2.2 *Modified Expression for Factor of Safety Based on Spencer Method Including Residual Factor*

Based on the original definition by Skempton (1964, 1985), the average residual factor over a slip surface mass, denoted as R_{F}, can be expressed as

$$R_{\mathrm{F}} = \frac{s_{\mathrm{p}} - s}{s_{\mathrm{p}} - s_{\mathrm{r}}}, \tag{1}$$

where, s_{p}, s_{r}, and s denote the average values of the peak shear strength, the residual shear strength, and the current shear strength respectively, over a potential slip surface.

For the analysis of slopes in strain-softening soils, using the above definition of residual factor, the shear strength parameters have been redefined as follows:

$$c'_{rf} = R_{\mathrm{F}} c'_{\mathrm{r}} + (1 - R_{\mathrm{F}}) c'_{\mathrm{p}} \tag{2}$$

$$\tan \phi'_{rf} = R_{\mathrm{F}} \tan \phi'_{\mathrm{r}} + (1 - R_{\mathrm{F}}) \tan \phi'_{\mathrm{p}} \tag{3}$$

The expression for the factor of safety, FS, associated with a curved slip surface of circular shape for a simple slope, based on the Spencer method (Spencer 1967), has been modified for a strain-softening soil, by replacing c′ with c'_{rf} given by Eq. (2), tan ϕ' with tan ϕ'_{rf} given by Eq. (3).

2.3 Method of Estimation of Statistical Properties of the Residual Factor Using an LEM Based Progressive Failure Model (Chowdhury et al. 2010)

A simple model for progressive failure of slopes in strain-softening soils under the framework of the limit equilibrium methods of slices (LEM) has been given by Chowdhury et al. (2010). If the soil is assumed to be perfectly brittle strain-softening, the shear strength parameters of the overstressed slices will reduce to residual values, whereas the remaining segments will still be at the peak shear strength. Based on this model involving an iterative process of distribution excess shear stress from the overstressed slices, one can identify those segments of a potential slip surface for which the shear strength parameters have decreased from the peak to the residual values. Then the mean of the residual factor can be estimated as $R_F = L_r/L$ in which L is the total length of a slip surface of which the length L_r is at the residual shear strength, the remaining length $(L-L_r)$ being still at the peak shear strength.

2.4 Probability Distribution for the Residual Factor

As regards the probability distribution of residual factor, a choice may be made between the assumption of a normal distribution and that of a generalized beta distribution. However, assuming normal distribution, errors will arise as the mean values approach the end points 0 and 1. Moreover, use of normal distribution excludes consideration of skewed distributions. A generalized beta distribution with the end points of 0 and 1 seems more appropriate. Once the mean of R_F is estimated based on the progressive failure model, as described above, the specific beta distribution to be used in a given situation depends on a reasonable assumption regarding the value of the COV of R_F. After making such an assumption, the beta-distribution parameters q and r corresponding to the specific shape of the beta distribution to be considered in the analysis are then calculated using Eqs. (4) and (5) (Harr 1977).

$$E[x] = a + \frac{q}{q+r}(b-a) \tag{4}$$

and,

$$V[x] = \frac{qr(b-a)^2}{(q+r)^2(q+r-1)} \tag{5}$$

where, $E[x]$ and $V[x]$ are the expected value and variance of residual factor as a beta distributed random variable with $a = 0$ and $b = 1$.

3 Illustrative Example

The application of the proposed procedure for reliability analysis of finite slopes in strain-softened soils is elucidated with the help of an example of a simple slope in a strain-softening soil selected from the literature (Chowdhury et al. 2010).

3.1 *Slope Description and Input Data*

A homogeneous slope is considered with height 25 m, inclination 22°, and unit weight of soil 18.8 kN/m^3. The statistical properties of the peak and the residual strength parameters, considered as random variables with normally distribution, are as given in Table 1. The pore pressure ratio r_u is taken as zero.

3.2 *Studies Conducted*

With the help of the illustrative example described above, studies have been conducted for purposes of numerical demonstration of the application of the proposed simplified procedure for 2-D reliability analysis of finite slopes in strain-softening soils. Of special interest is the analysis of the most likely scenario of strain-softening occurring in part of a slip surface (referred to as Case C) in which case the average residual factor has a value between 0 and 1. For such a scenario, it is necessary to quantify the extent to which strain-softening would advance in a given slope situation. As discussed in Sect. 2.3, for this purpose the simplified LEM-based progressive failure model proposed in Chowdhury et al. (2010) has been used. As against this general case, there are two extreme scenarios, namely, (i) the entire slip surface is at peak strength ($R_F = 0$) (Case A), and (ii) the entire slip surface is at residual strength ($R_F = 1$) (Case B). A comparison of results for the three cases will indicate the error introduced as a result of analysing a Case C scenario as either Case A ($R_F = 0$) or Case B ($R_F = 1$).

Further, to bring out the impact of including R_F as one of the random variables, reliability analyses have been performed first considering the residual factor as a deterministic parameter, and then, as a random variable. Moreover, in order to study

Table 1 Statistical properties of the strength parameters

Description of the parameter		Mean	Coefficient of variation
Peak strength parameters	c_p'	30.0 kPa	0.20
	$\tan \phi_p'$	tan(20)	0.10
Residual strength parameters	c_r'	10.0 kPa	0.20
	$\tan \phi_r'$	tan(12)	0.10

the relative influence of residual factor as a random variable on the reliability index, a sensitivity analysis based on the FORM has also been included.

3.3 *Reliability Analyses and Corresponding Results*

For slope reliability analysis, the probabilistic critical slip surface together with the value of the associated minimum reliability index β_{min} has been determined. In order to explore the effect of considering R_F as a random variable, analysis has been performed first (i) considering R_F as a deterministic (Study 1), and then (ii) considering R_F as a random variable (Study 2).

3.3.1 Study 1: Considering R_F as Deterministic

Considering the residual factor as deterministic (i.e. $R_F = 0$ for Case A and $R_F = 1$ for Case B), the probabilistic critical slip surfaces have been determined based on the computational procedure outlined in Sect. 2.1 using the FORM in conjunction with the SQP technique, and are shown in Fig. 1. The minimum reliability index (β_{min}) associated with these two probabilistic critical slip the surfaces (for Case A and Case B) are obtained as 4.244 ($p_F = 1.10 \times 10^{-5}$) and −2.059 ($p_F = 9.80 \times 10^{-1}$) respectively.

Besides the two extreme scenarios presented above, the third or the most likely scenario (Case C) is considered next. For this case, unlike the above two cases, during the search for the probabilistic critical slip surface, the value of R_F is estimated for each trial slip surface based on the Chowdhury et al. (2010) simplified model. Following the same general procedure as used above, the probabilistic critical slip surface has been determined for the Case C as also shown in Fig. 1 in

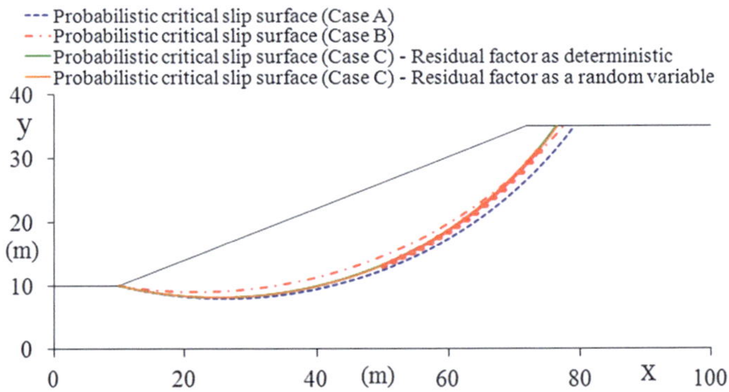

Fig. 1 Probabilistic critical surfaces for case A, case B and case C

which the strain-softened portions (failed slices) of the slip surfaces are highlighted in red bullet points. It is observed that all the three probabilistic critical slip surfaces are rather close to one another except in their upper segments. For Case C, the associated value of β_{min} is obtained as 3.555 ($p_F = 1.89 \times 10^{-4}$) while the associated value of R_F is 0.395. It is thus seen that the β_{min} value for Case C is in between the two extreme cases i.e., β_{min} value for Case A and Case B, which is expected. A comparison of the β_{min} values for the three cases further reveals that assuming peak shear strength for the entire slip surface overestimates the reliability index by nearly 19%, while assuming residual shear strength for the entire slip surface underestimates the reliability index by nearly 158%.

3.3.2 Study 2: Considering R_F as a Random Variable

In this study, the residual factor R_F is considered as a statistically independent random variable. During the search for the probabilistic critical slip surface, for each trial slip surface, its mean value is estimated using the Chowdhury et al. (2010) model, while its COV is assumed to be 0.3. The obtained probabilistic critical slip surface is shown again in Fig. 1. The value of the associated β_{min} is obtained as 2.856 ($p_F = 2.1 \times 10^{-3}$) while the associated value of R_F is 0.396.

3.3.3 Comparison Between Results from Study 1 and Study 2

It is seen that the β_{min} for Case C considering the residual factor R_F as a random variable is significantly lower than that assuming R_F as deterministic and percentage difference is nearly 25%. It is, however, observed from Fig. 1 that these two slip surfaces are rather close to each other.

3.3.4 FORM-Based Sensitivity Analysis

A FORM-based sensitivity analysis comparing the values of direction cosines for various random variables for the probabilistic critical slip surface is shown in Fig. 2. It is observed that the residual factor R_F has the largest influence on the reliability index among the five random variables considered in this study and this observation clearly indicates the consideration of the residual factor as one of the random variables in the reliability analysis of slopes in strain-softening soils.

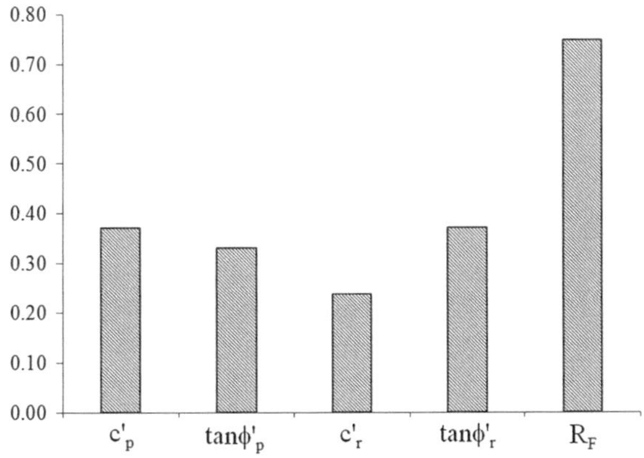

Fig. 2 Results of sensitivity analysis based on the FORM method

4 Conclusions

Based on the studies undertaken in this paper, the following concluding remarks can be made:

1. In this study, three possible scenarios have been studied, namely, the two extreme cases, Case A, when the entire slip surface is at peak strength ($R_F = 0$), and, Case B, when the entire slip surface is at residual strength ($R_F = 1$), and the most likely one, Case C, when strain softening has taken place over part of the slip surface ($0 < R_F < 1$). For Case C, reliability analyses have yielded a β_{min} value of 3.555 for deterministic R_F and 2.856 for random R_F. When R_F is deterministic, Case A overestimates the reliability index by nearly 19%, while Case B underestimates the reliability index by nearly 158%. When R_F is random these differences are even higher (49 and 172% respectively).
2. The effect of including R_F as one of the random variables (with an assumed COV of 0.3) in reliability analysis of a slope in strain-softening soil is substantial; the value of reliability index reduces by a margin of 25% compared to when R_F is considered deterministic. It, however, needs to investigate the effect of the assumption regarding a suitable value of COV for R_F.
3. FORM-based sensitivity studies confirm that, amongst the five random variables, residual factor has the most dominating influence on the estimated reliability index and thus justifies its inclusion as one of the random variables.

References

Chowdhury, R., & Flentje, P., Bhattacharya, G. (2010). *Geotechnical Slope Analysis*. CRC Press/Balkema.

Haldar, A., & Mahadevan, S. (2000). *Probability Reliability and Statistical Methods in Engineering Design*. Wiley.

Harr, M. E. (1977). *Mechanics of Particulate Media—A Probabilistic Approach*. New York: McGraw-Hill.
Hong, H., & Roh, G. (2008). Reliability evaluation of earth slopes. *Journal of Geotechnical and Geoenvironmental Engineering, 134*(12), 1700–1705.
Metya, S., & Bhattacharya, G. (2014). Probabilistic critical slip surface for earth slopes based on the FORM. *Indian Geotechnical Journal, 44*(3), 329–340.
Skempton, A. W. (1964). Long-term stability of clay slopes. *Geotechnique, 14*(2), 77–101.
Skempton, A. W. (1985). Residual strength of clay in landslides, folded strata and the laboratory. *Geotechnique, 35*(1), 3–18.
Spencer, E. (1967). A method of analysis of the stability of embankments assuming parallel inter-slice forces. *Geotechnique, 17*(1), 11–26.

Part VI
Theme 12: Special Topics: Offshore Geotechnics, Remote Sensing and GIS, Geotechnical Education, Codes and Standards

Appraisal of Sensitivity Correlations on Data from Clays of Western Indian Offshore

Balram Nayak, R. K. Ghanekar and A. Ajit

Abstract Sensitivity (S_t) of cohesive soils is defined as the ratio of peak undisturbed shear strength (s_u) to the remoulded shear strength (s_{ur}) at the same water content. In offshore geotechnics, sensitivity is an important design parameter for clays especially where strength of remoulded clays is an important input in foundation analysis and design, for example, in design of some anchors in deep waters, assessment of lateral and vertical pile capacity in the zone disturbed by jack-up spudcan penetration. In many instances in offshore projects, results of laboratory or field tests for remoulded shear strength of clays are not available and in such cases empirical correlations are used which correlate either the remoulded strength or the sensitivity directly with other index parameters measured or derived in laboratory. Presently such correlations have not been derived or investigated for the fields of western Indian offshore and the available correlations in literature may not be applicable for the clayey soils of Western Indian Offshore. This paper discusses the results of evaluation of published correlations, for the soil data from top 30 m of soil profile from various fields of Western Indian Offshore. Results of attempts to develop new correlations for the specific application for the soils of Western Indian offshore are also presented.

Keywords Sensitivity · Remoulded shear strength · Correlations

---

B. Nayak (✉) · R. K. Ghanekar · A. Ajit
Institute of Engineering and Ocean Technology (IEOT), ONGC, Navi Mumbai, India
e-mail: nayak_balram@ongc.co.in

R. K. Ghanekar
e-mail: ghanekar_rk@ongc.co.in

A. Ajit
e-mail: ajit_a@ongc.co.in

A. I. V. and V. B. Maji (eds.), *Geotechnical Applications*, Lecture Notes in Civil Engineering 13, https://doi.org/10.1007/978-981-13-0368-5_32

1 Introduction

Sensitivity (S_t) is defined as ratio of peak undisturbed shear strength (s_u) to the remoulded shear strength (s_{ur}) at the same water content.

In offshore geotechnics, sensitivity is an important design parameter for clays especially where strength of remoulded clays is an important input in foundation analysis and design, e.g., in design of some anchors such as suction and drag anchors in deep waters, assessment of lateral and vertical pile capacity in the zone disturbed by jack-up spudcan penetration.

There are a number of tests in practice for the measurement of remoulded shear strength both in laboratory and in the field. Miniature (motorised) Vane and fall cone test are the most commonly used laboratory tests for the purpose of soil investigation in Western Indian offshore.

In many instances in offshore practice worldwide, results of laboratory or field tests for measuring the remoulded strength of clays are not available and in such cases empirical correlations are used which correlate either the remoulded strength or the sensitivity directly with other index parameters measured in laboratory or parameters measured by cone penetration test (CPT) or piezocone penetration tests (CPTU).

Presently such correlations have not been derived or investigated for western Indian offshore soils.

In the present paper, the available published correlations are evaluated for the soils from various fields of Western Indian Offshore. Efforts made to develop new correlations for the area of Western Indian Offshore, are also presented.

The results of this project will help in deriving sensitivity of clays during the soil investigations, where remoulded shear strengths are not available. The results will also be used as a check where the test results are available.

2 Selected Correlations for Investigation

The following correlations were investigated for the western Indian offshore data:

2.1 Wroth and Wood (1978)

$$s_{ur} = 170 \exp^{(-4.6 I_L)}, \quad (1)$$

where,

s_{ur} remoulded undrained shear strength
I_L Liquidity index of the soil

2.2 Leroueil et al. (1983)

$$s_{ur} = 1/(I_L - 0.21)^2 \quad (2)$$

2.3 NGI (2002)

$$s_{ur} = 4.2(I_L)^{-1.6} \quad (3)$$

2.4 Ramsay (2002)

$$s_{ur} = 2/3 f_s, \quad (4)$$

where f_s = measured value of sleeve friction from CPTU.

2.5 Bjerrum (1954)

Bjerrum proposed a relationship between Sensitivity and Liquidity Index for Norwegian marine clay.

$$S_t = 10^{(0.15 + 0.73 I_L)}. \quad (5)$$

3 The Data

The soil data used in this study have been taken from ONGC's recent Soil Investigation Projects for 27 locations from 8 different Fields of Western Indian Offshore. The selected data are from the top about 30.0 m of soil profile at each location.

The basic intact and remoulded strength data for MV tests are from tests performed on-board the geotechnical vessel. The data for fall cone are from laboratory tests performed at IEOT laboratory. Other laboratory tests data, i.e., water content; liquid limit and plastic limit, at specific depths have been taken from tests performed at IEOT laboratory. CPTU parameter sleeve friction, also taken from the background data of soil investigation reports for the study.

4 Evaluation of Correlations

Figures 1 and 2 show the evaluation of the correlations by Wroth and Wood (1978) and Ramsay (2002) using s_{ur} values determined in the laboratory by the Miniature vane test plotted against the corresponding I_L and CPT f_s respectively.

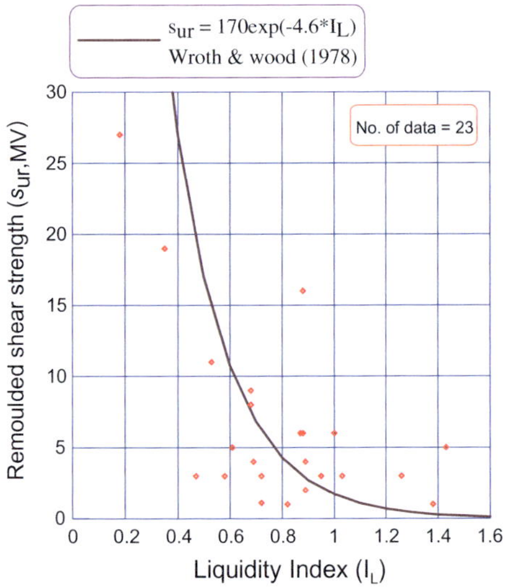

Fig. 1 Evaluation of correlation for remoulded shear strength by Wroth and wood (1978) for the soil data from western Indian offshore (miniature vane)

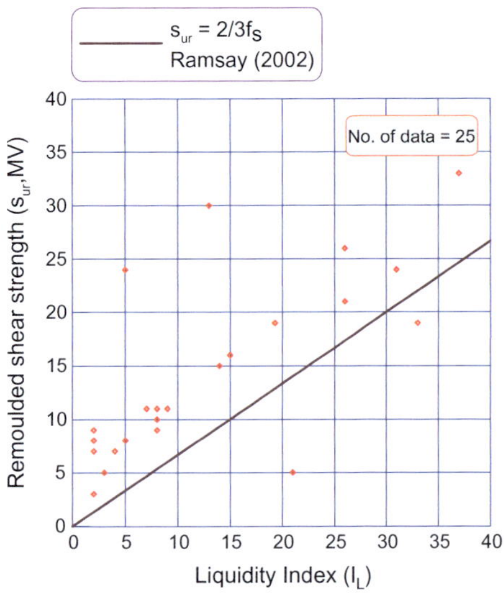

Fig. 2 Evaluation of correlation for remoulded shear strength by Ramsay (2002) for the soil data from western Indian offshore (miniature vane)

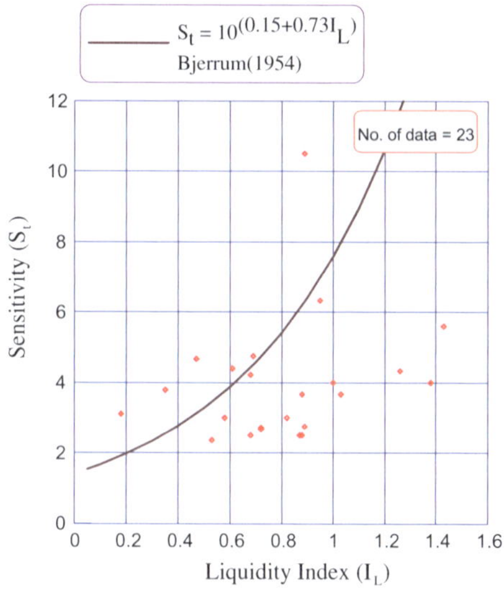

Fig. 3 Evaluation of correlation for sensitivity by Bjerrum (1954) for the soil data from western Indian offshore (miniature vane)

Figure 3 shows the evaluation of the correlation proposed by Bjerrum (1954); using Sensitivity (S_t) values (determined from laboratory Miniature vane test) plotted against the corresponding I_L values. The corresponding correlations are also plotted on the figures for comparison.

From the figures, it is clear that the data are scattered and correlations do not predict well for the western offshore Indian clays. The other existing correlations evaluated using both Miniature vane and Fall Cone data also did not fit the data well.

5 Regression Analysis

Regression Analysis is a statistical process for the investigation of relationships between variables. It includes many techniques for modelling and analysing several variables, when the focus is on the relationship between a dependent (response) variable and one or more independent variables. So, the output of a Regression Analysis is a function that predicts the dependent variable based on independent variables.

Since the predictions by the available published correlation are not very good for the fields of western Indian offshore, development of some new correlations was attempted using Regression Analysis.

Simple Regression Analyses were performed between remoulded shear strength/sensitivity and the Index parameters measured in the laboratory and CPTU measured parameters.

Table 1 Results of regression analyses

Equation	Results of regression analysis	n	R^2	S.E.
A	$\log_{10}(s_{\text{ur,MV}}) = 2.219 - 1.717\left(\frac{w}{w_L}\right)$	17	0.691	0.2016
B	$\log_{10}(s_{\text{ur,MV}}) = 3.359 - 4.694\left(\frac{w}{w_L}\right) + 1.624\left(\frac{w}{w_L}\right)^2$	17	0.724	0.1975
C	$s_{\text{ur,MV}} = 2.254 + 0.6191(f_s)$; (kPa)	144	0.685	2.930
D	$s_{\text{ur,FC}} = 5.211 + 0.6421(f_s)$; (kPa)	25	0.894	2.63

Note: n is number of data points, R^2 is coefficient of determination, S.E. is standard error, w natural water content, w_L water content at liquid limit, $s_{\text{ur,MV}}$ remoulded shear strength from miniature vane test, $s_{\text{ur,FC}}$ remoulded shear strength from fall cone test

The results of the regression analyses are presented in Table 1. Results which show reasonable strength of correlations only are included in the table.

6 Discussion

Due to paucity of space only equation D (refer Table 1) along with the data is presented in Fig. 4, being the strongest correlation statistically. It can be seen that the equation predicts the remoulded shear strength reasonably well although the number of data points are low and some obvious outliers exist. The data is re-plotted in Fig. 5 with 10 and 20% bounds included. Barring a couple of outliers, most of the measured values are within 20% of prediction.

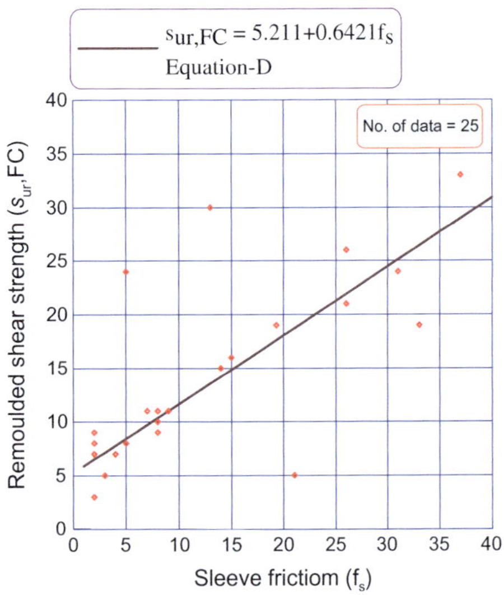

Fig. 4 Evaluation of correlation of remoulded shear strength by regression analysis (Equation D) for the soil data from western Indian offshore (fall cone)

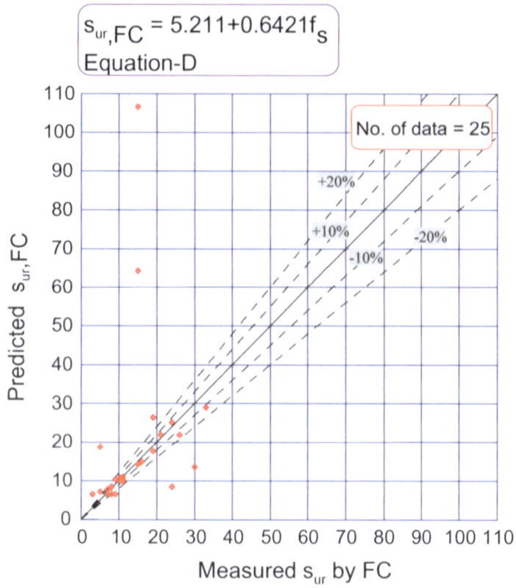

Fig. 5 Predicted remoulded shear strength from Equation D versus measured remoulded shear strength (fall cone)

Since the correlation was not forced through the origin, an anomaly exists that at zero f_s, remoulded strength of about 5 is predicted. With additional data generated during the future soil investigations, all the correlations derived shall be revisited and refined. For the present, it is suggested that the remoulded shear strength shall be derived using the derived equations to arrive at the most reasonable value.

7 Conclusions

Data from 27 offshore locations, from 8 different fields of western Indian offshore have been used in this study to evaluate the applicability of the selected published correlations for deriving remoulded shear strength or Sensitivity of calcareous clays from western Indian offshore from index parameters and CPT parameters. It is found that none of the published correlations are directly applicable for such clays to evaluate the applicability of the selected published correlations for deriving remoulded shear strength or Sensitivity of calcareous clays from western Indian offshore from index parameters and CPT parameters. It is found that none of the published correlations are directly applicable for such clays.

Attempts were hence made to derive some new correlations by performing regression analyses using laboratory measured data and CPTU data. A total of four reasonably strong correlations could be derived for the Remoulded shear strength. Statistically, the best correlation found was between the remoulded shear strength measured with fall cone test and CPT measured.

Acknowledgement Authors are grateful to the management of ONGC for granting permission to publish the paper.

References

Bjerrum, L. (1954). Geotechnical properties of Norwegian marine clays. *Geotechnique, 25*(2), 21–23.

Leroueil, S., Tavenas, F., & Le Bihan, J. P. (1983). Proprittescaracteristiques des argiles de Test du Canada. *Canadian Geotechnical Journal, 20*(4), 681–705.

NGI. (2002). Establishing of soil design parameters from soil index parameters. *Report-III, Statistical Analysis of data and preliminary Recommendations-IEOT*, ONGC, Panvel.

Ramsay, N. (2002). A calibrated model for the interpretation of cone penetration tests (CPTs) in North Sea quaternary soils. In *Proceedings of Offshore Site Investigation and Geotechnics: Diversity and Sustainability*, London, UK (pp. 341–356).

Wroth, C. P., & Wood, D. M. (1978). The correlation of index properties with some basic engineering properties of soils. *Canadian Geotechnical Journal, 15*(2), 13.

Behaviour of Bucket Foundations in Sandy Bed Subjected to Eccentric Lateral Loading

Tanmoy Kumar Deb and Baleshwar Singh

Abstract Monopod bucket foundations supporting offshore wind turbines have to resist high overturning moment due to water waves and wind currents acting laterally on the structure above the seabed level. Hence, the stability of the bucket foundation system is governed by the lateral load-displacement behaviour. In this study, the drained behaviour of monopod bucket foundation under monotonic eccentric lateral loads has been investigated using finite element analysis. The influence of bucket dimensions and wind turbine self-weight on the lateral load response of the foundation system was also studied. The ultimate lateral load capacity of the bucket foundation has been observed to decrease with the increase of load eccentricity, but it increases marginally with self-weight and noticeably with the foundation size. For a preliminary design of the bucket foundation, the permissible loading state is presented as a lateral load-overturning moment interaction diagram for different geometries.

Keywords Bucket foundation · Lateral capacity · Offshore wind turbine
Eccentricity · Sandy site

T. K. Deb (✉) · B. Singh
Department of Civil Engineering, Indian Institute of Technology,
Guwahati 781039, Assam, India
e-mail: tanmoy.deb@iitg.ernet.in

B. Singh
e-mail: baleshwar@iitg.ernet.in

A. I. V. and V. B. Maji (eds.), *Geotechnical Applications*, Lecture Notes in Civil Engineering 13, https://doi.org/10.1007/978-981-13-0368-5_33

1 Introduction

A bucket foundation is a hollow steel or concrete cylinder which is open at the bottom and capped at the top. Generally, a monopod bucket foundation is suitable in water depths near shore up to about 30 m. Under offshore conditions, these monopod bucket foundations have to resist the combined action of vertical and eccentric lateral loads. The overturning moment produced from the eccentric lateral load is a critical aspect in the design of monopod bucket foundation.

Laboratory, field and numerical investigations have been carried out by several researchers in order to assess the behaviour of monopod bucket foundation. The response of bucket foundation under monotonic and cyclic lateral load was investigated by conducting laboratory tests in dry sand bed (Byrne and Houlsby 1999) and in oil-saturated sand bed (Byrne and Houlsby 2004). Kelly et al. (2006) carried out laboratory tests on bucket foundations in sand and clayey soils and compared moment capacity and stiffness with results from their own field tests.

Sun et al. (2009) studied the horizontal bearing capacity of bucket foundation embedded in clay numerically based on three-dimensional finite element method and developed a formulation for horizontal bearing capacity using limit equilibrium method. Achmus et al. (2013) investigated the lateral response of bucket foundations numerically using finite element analysis by varying lateral load eccentricities and superstructure loads for several bucket geometries. Ibsen et al. (2015) carried out laboratory tests in order to determine the effect of embedment on strain hardening behaviour of model buckets founded in saturated dense sands under combined loading condition.

In this study, numerical analyses have been carried out on prototype dimensions of monopod bucket foundations embedded in very dense sand bed under combined loading conditions. Based on the numerical analyses, the response of bucket foundation geometries under eccentric lateral loads and superstructure loads are presented herein.

2 Numerical Analysis

The response of bucket foundation under combined loading is simulated numerically using ABAQUS (2010). The bucket foundation and soil domain have been discretized into 480 and 9568 numbers of 8 noded brick elements respectively, as shown in Fig. 1. Reduced integration scheme has been utilized.

The nonlinear behaviour of soil is simulated using Mohr–Coulomb elastoplastic material model with stress dependent oedometric modulus of elasticity (E_s):

$$\mathrm{E_s} = \kappa.\sigma_{\mathrm{at}}.\left(\frac{\sigma_{\mathrm{m}}}{\sigma_{\mathrm{at}}}\right)^{\lambda}, \tag{1}$$

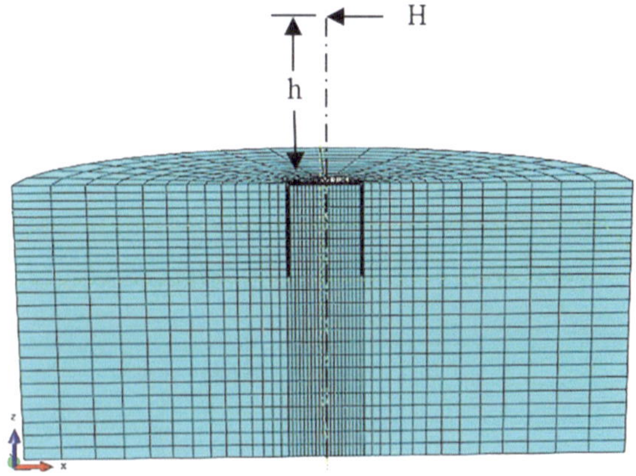

Fig. 1 Discretized bucket foundation soil system

Table 1 Geometric details of bucket foundation considered for the analysis

Geometric details	Superstructure load, V (MN)	Load eccentricity, h (m)
D = 10 m, L = 12 m	30, 20, 10	0, 2.5, 5, 10, 20, 30, 40,100, pure moment
D = 8 m, L = 12 m		
D = 6 m, L = 12 m		

where, σ_{at} is atmospheric pressure, and σ_m is mean principal stress. κ and λ are oedemetric stiffness parameters having values of 600 and 0.55. The submerged unit weight, internal friction angle and dilation angle of the soil are taken as 11 kN/m^3, 40° and 10° respectively. The bucket geometries and loading details are presented in Table 1.

3 Results and Discussion

3.1 Lateral Capacity of Bucket Foundation

In this study, the ultimate lateral capacity of the bucket foundation at any eccentricity is taken as the lateral load which causes displacement of bucket lid equal to 10% of its diameter. Figures 2a, b, c respectively show typical lateral load-displacement response, lateral load-rotation response and overturning moment-rotation response for a bucket foundation of 10 m diameter, up to ultimate condition corresponding to 1 m displacement.

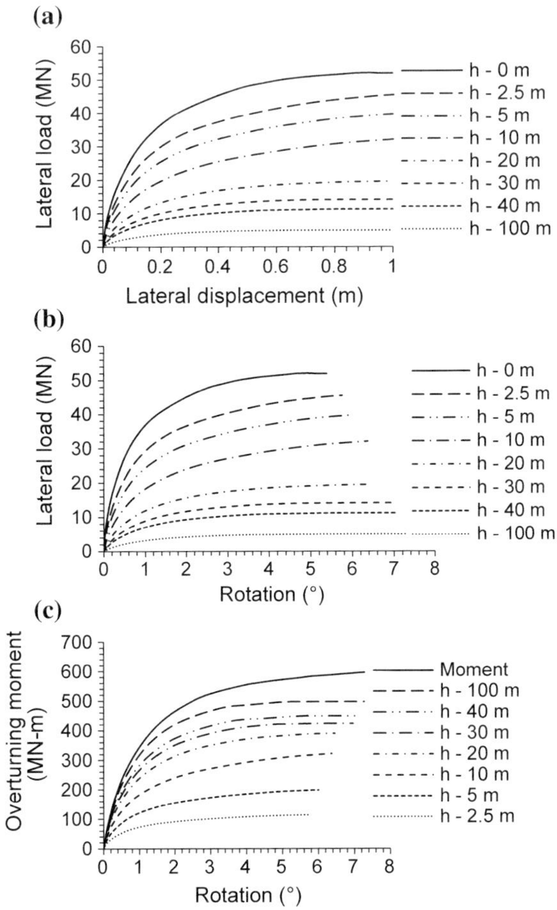

Fig. 2 **a** Lateral load-displacement response of bucket foundation (D = 10 m, L = 12 m and V = 30 MN), **b** lateral load-rotation response of bucket foundation (D = 10 m, L = 12 m and V = 30 MN), **c** overturning moment-rotation of bucket foundation (D = 10 m, L = 12 m and V = 30 MN)

From Fig. 2a, at any particular load eccentricity, the displacement of the bucket lid is noted to increase with applied lateral load. The ultimate lateral load capacity of the bucket foundation is observed to decrease from 51.93 to 4.97 MN with the increase of load eccentricity from h = 0 m (pure lateral load) to h = 100 m.

From Fig. 2c, it is observed that the maximum rotation of the lid of the bucket foundation at ultimate condition is greater for a higher load eccentricity. The overturning moment causing this maximum rotation increases from 112.5 to 497 MNm when the load eccentricity is increased from h = 2.5 to h = 100 m. However, this overturning moment is found to be the highest (596.2 MNm) under pure moment load applied at the centre of the bucket lid.

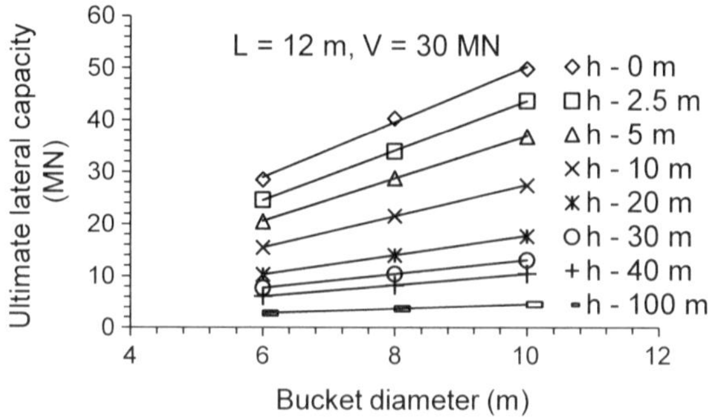

Fig. 3 Ultimate lateral load capacity of bucket foundation

3.2 *Variation of Ultimate Lateral Capacity*

The variation of ultimate lateral load capacity with bucket diameter and superstructure load is shown in Fig. 3 for a superstructure load of 30 MN and all load eccentricities. The percentage increase of ultimate lateral capacity ranges between 50 and 70% when the diameter is increased from 6 to 10 m. The ultimate lateral capacity is found to decrease with superstructure load as presented in Table 2 for load eccentricity of 40 m.

3.3 *Variation of Initial Stiffness*

Determination of the initial stiffness of the bucket foundation is necessary to compute its rotational stiffness, which should be lesser than the prescribed value provided by the wind turbine designer to avoid resonance. From the lateral load-rotation plot in Fig. 2b for a particular load eccentricity, the initial stiffness can be obtained as the slope of the line drawn from the origin to a rotation of 0.5°. Typical plots of variation of initial stiffness with load eccentricity are shown in Fig. 4. The initial stiffness is observed to increase with bucket diameter and superstructure load as presented in Table 3.

Table 2 Variation of ultimate capacity with superstructure load for $h = 40$ m

Geometric details	Ultimate lateral capacity (MN)		
	$V = 30$ MN	20 MN	10 MN
$L = 12$ m, $D = 10$ m	11.22	10.80	10.45
$L = 12$ m, $D = 8$ m	9.97	9.15	8.20
$L = 12$ m, $D = 6$ m	7.14	6.64	6.14

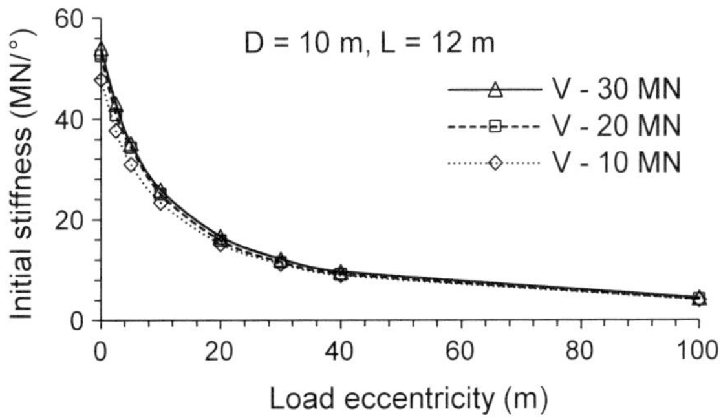

Fig. 4 Initial stiffness of bucket foundation

Table 3 Initial stiffness of bucket foundation for $h = 40$ m

Geometric details	Initial stiffness (MN/°)		
	$V = 30$ MN	20 MN	10 MN
$L = 12$ m, $D = 10$ m	9.64	9.18	8.94
$L = 12$ m, $D = 8$ m	7.66	7.22	6.82
$L = 12$ m, $D = 6$ m	5.24	5.02	4.42

3.4 Variation of Depth of Point of Rotation

The left exterior (LE) and right exterior (RE) sides of the bucket foundation are indicated in Fig. 5, and typical variation of depth of point of rotation of bucket foundation with load eccentricity is shown in Fig. 6. For the superstructure load of 30 MN, with an increase in eccentricity from 0 to 10 m, there is a decrease in the depth of point of rotation from about 10.98 m (0.91 L) to 9.41 m (0.78 L). The variations of lateral earth pressure at the ultimate condition along the depth for both sides are shown in Fig. 7.

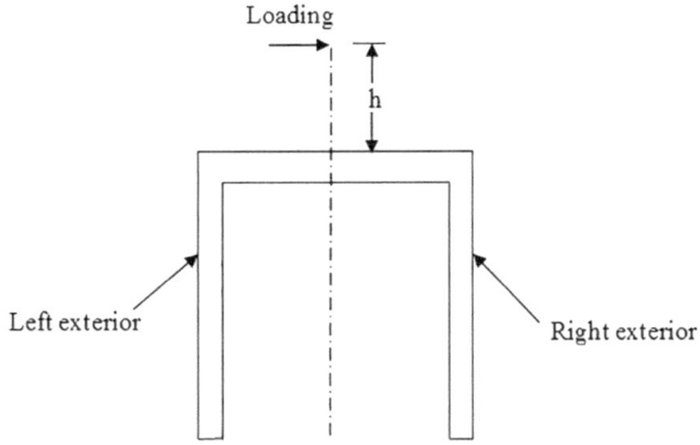

Fig. 5 Schematic diagram indicating bucket sides

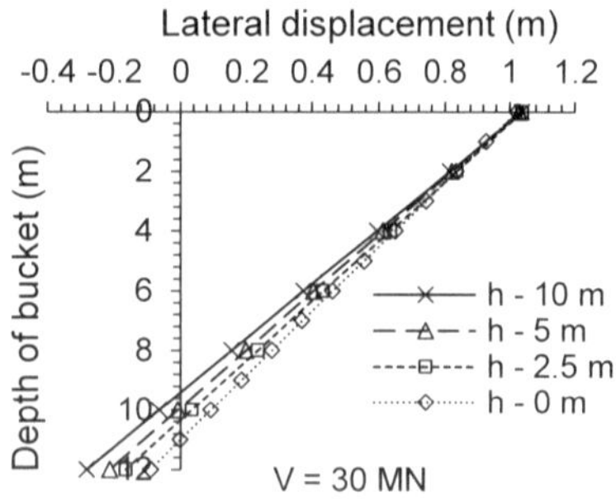

Fig. 6 Depth of point of rotation of bucket foundation (D = 10 m, L = 12 m)

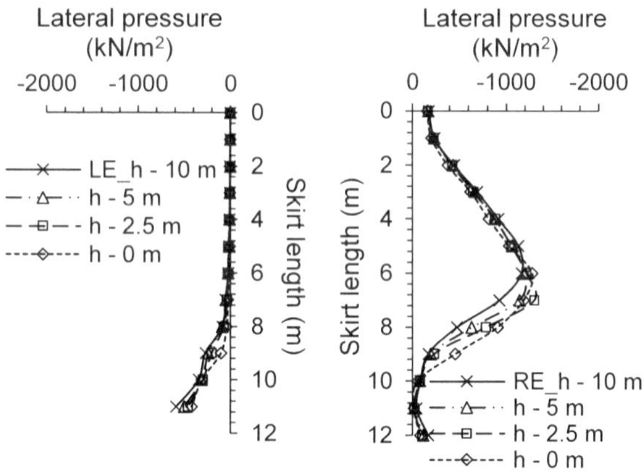

Fig. 7 Lateral pressure along bucket depth

As the point of rotation moves upwards, a greater soil reaction force is exerted below this point on the left exterior face. In contrast, on the right exterior face, the resultant horizontal soil reaction force decreases, which is the reason that the ultimate lateral load capacity reduces with an increase of the eccentricity.

3.5 *Lateral Load-Overturning Moment Interaction Diagram*

In case of bucket foundations for wind turbines, deflection control at sea bed level is vital from stability point of view. In the present study, an angular rotation of 0.5° has been considered to arrive at the permissible lateral load capacity.

The capacity of the bucket foundation can be described graphically by using lateral load-overturning moment interaction diagram. To plot this diagram, the values of lateral load and overturning moment corresponding to 0.5° rotation of bucket lid are obtained for each load eccentricity of the three geometries used.

These values have been used to obtain an interaction diagram as shown in Fig. 8, consisting of three envelopes for the three geometries and superstructure load of

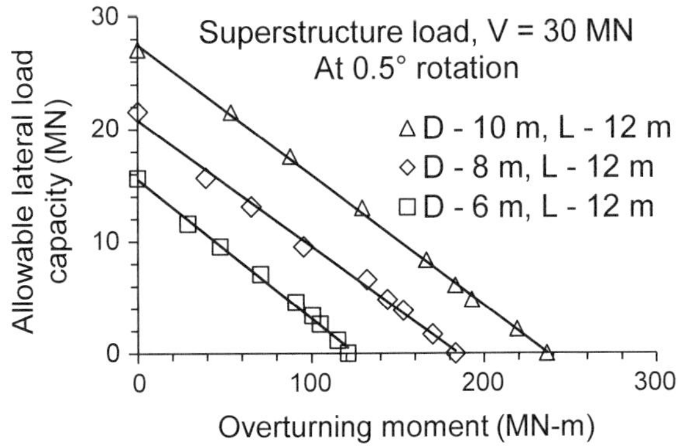

Fig. 8 Interaction diagram at 0.5° rotation

30 MN. Any combination of lateral load and overturning moment, located outside the envelope of respective bucket geometry, will cause instability of the bucket foundation. In that situation, the next larger geometry can be selected in a step-wise manner till the stability requirement is fulfilled.

4 Conclusions

Based on the three-dimensional numerical analysis of bucket foundation embedded in very dense sand subjected to lateral loads, the following conclusions have been made:

- The ultimate lateral load capacity of the bucket foundation decreases with load eccentricity. The increase in the capacity with a larger geometry is noticeable, but the influence of superstructure load is marginal.
- The initial stiffness of the bucket foundation is higher for a lower load eccentricity.
- The depth of point of rotation decreases with an increase of the load eccentricity.
- For a particular load combination of superstructure weight and resultant lateral load from water waves and wind currents, the lateral load-overturning moment interaction diagram enables a selection of the required geometry of the bucket foundation as a preliminary design.

References

ABAQUS. (2010). User's Manual 6.10.

Achmus, M., Akdag, C. T., & Thieken, K. (2013). Load-bearing behaviour of suction bucket foundations in sand. *Applied Ocean Research, 43,* 157–165.

Byrne, B. W., & Houlsby, G. T. (1999). Drained behaviour of suction caisson foundations on very dense sand. *Offshore Technology Conference,* OTC 10994, pp. 18.

Byrne, B. W., & Houlsby, G. T. (2004). Experimental investigations of response of suction caissons to transient vertical loading. *Journal of Geotechnical and Geoenvironmental Engineering, 128*(11), pp. 926–939 (*ASCE*).

Ibsen, L. B., Barari, A., & Larsen, K. A. (2015). Effect of embedment on the plastic behaviour of bucket foundations. *Journal of Waterway, Port, Coastal, and Ocean Engineering, 141*(6), pp. 06015005, 9 (*ACSE*).

Kelly, R. B., Houlsby, G. T., & Byrne, B. W. (2006). A comparison of field and laboratory tests of caisson foundations in sand and clay. *Géotechnique, 56*(9), 617–626.

Sun, X.Y., Tang, X.W., & Luan, M.T. (2009). Study on the horizontal bearing capacity of bucket foundations on saturated soft clay ground. *Electronic Journal of Geotechnical Engineering, 14*, p. 11 (Bund.N).

Effect of Soil Structure Interaction Analysis on the Response of Fixed Offshore Jacket Structure

Seeram Madhuri and M. G. Muni Reddy

Abstract Offshore jacket platforms are being extensively used for the exploration of oil and natural gas. Nowadays these structures are being suggested for supporting offshore wind turbines. The jacket structures are fixed to sea bed using piles. These piles penetrate into the soil and transfers structural loads to the soil. The soil structure interaction of the piles should be studied for the safety of the structure. In the present study, the soil structure interaction (SSI) of the jacket structure is studied by developing a MATLAB code and performed static analysis using stiffness method. Structural elements are modeled using three dimensional beam elements. The modulus of sub-grade reaction of the surrounding soil is estimated using Vesic equation. The spring stiffness of the surrounding soil along the length of the piles of the jacket is estimated using Newmark distribution. Wave force on the structural elements of the jacket structures is estimated using Morison's equation. The deflection of the topside and at the sea bed is presented in this paper. The effect of the marine growth on the static and dynamic response is also studied and concluded that the response of the structure and the natural period of the structure are increased when the marine growth is present.

Keywords Jacket structures · Soil springs · Morison equation
Finite element analysis · Three-dimensional beam element · Response

---

S. Madhuri (✉)
Department of Civil Engineering, University College of Engineering,
JNTU Kakinada, Kakinada, India
e-mail: madhuri.seeram@gmail.com

M. G. Muni Reddy
Department of Civil Engineering, University College of Engineering,
Andhra University, Visakhapatnam, India
e-mail: munireddy9@rediffmail.com

A. I. V. and V. B. Maji (eds.), *Geotechnical Applications*, Lecture Notes in Civil Engineering 13, https://doi.org/10.1007/978-981-13-0368-5_34

1 Introduction and Brief Literature Review

Mainly offshore jacket structures are used for gas and oil exploration, processing, and supporting structures for offshore wind turbines, etc. Many of the offshore jacket structures are constructed at water depths ranging from 30 to 300 m. The jacket structure has to resist different structural, equipment, environmental, and accidental loads, etc. Among all the loads wave, current and wind loads are continuous throughout the life of the structure. The stability of the jacket structure mainly depends on foundation system. The surrounding soil and soil structure interaction plays major role in the response of the jacket structure. Incremental dynamic analysis of jacket platform including the pile soil interaction was performed under earthquake loading by using Winkler beam foundation technique (Behrouz and Alireza 2008). Rehman et al. (2012) performed a nonlinear finite element analysis by using two-dimensional beam elements for modeling the structural members under wind, wave, and combination of wind and wave loads. The results indicated considerable effect of current velocity on the response of the jacket and nonlinear displacement increment was observed. Asgarian and Flouz (2008) studied the effect of soil–pile structure interaction on dynamic characteristics of a prototype jacket structure by performing experimental and numerical investigations for hinged and skirt pile boundary conditions using ABAQUS and SACS software's. Soil–pile interaction was modeled in SACS using simplified model. Based on the study it was observed that the soil–pile interaction had significant effect on the dynamic characteristics of the jacket structure. Shi et al. (2015) performed soil structure interaction study of a jacket type offshore wind turbine by developing flexible foundation, p-y model of the pile groups and fixed foundation. Based on the study it was observed that the flexible foundation system estimated well the loads on the offshore wind turbine and suggested to consider the pile group in the fatigue analysis. Ghassemi Zadeh et al. (2015) studied the nonlinear response of fixed offshore platform under the combined wave, wind and current loading for one year and one hundred year return periods using SAP2000 software. The direction of the wave also studied and concluded that the loading and direction has significant effect on the fixed offshore structure by investigating bending moments, axial forces, and shear forces on the members. The marine growth develops on the jacket structure after several months of the installation. Due to marine growth, the mass of the structure and diameter increases. Based on the literature review it is understood that the soil structure interaction analysis of jacket structure with marine growth on the structural elements under environmental loading is limited. Hence the main objective of the present study is to study the effect of soil structure interaction under wave and current loading when the jacket members are having marine growth on the structural members.

2 Analysis of Fixed Jacket Structure

The present-studied four-legged jacket structure is having a deck size of 46 m × 46 m is located at a water depth of 153 m. The foundation level of the jacket piles is up to a depth of 84 m from the sea bed. The stiffness of the soil springs of the surrounding soil is estimated using Vesic (1961) equation and Newmark (1942) distribution. The soil profile considered in the present study and estimated soil spring stiffness of the piles is given in Table 1. The outer diameter and the wall thickness of the jacket legs and inclined members are about 1.2 m 15 mm and 0.8 m × 12 mm respectively. The foundation of the jacket structure is assumed with steel piles of 1.524 m outer diameter and 60 mm wall thickness. The thickness of the marine growth is taken as 100 mm up to the water depth of 40 m and 50 mm below 40 m water depth to sea bed level. The density of marine growth is considered as 1400 kg/m^3. Topside weight of 8.15 t/m^2 is considered and applied as nodal loads. The density of steel is assumed as 7850 kg/m^3. A MATLAB code is generated to analyze the fixed offshore jacket structure. All the structural members and foundation piles are modeled using three dimensional beam elements with six degrees of freedom at each node (i.e., three translations and three rotations). Coordinates, element connectivity, member properties, and constraint details are given as input. A finite element model is developed. Wave (wave height of 8 m wave period of 12 s) and current (velocity of 1 m/s) loading on the structural members is estimated using Morison equation (Eq. 1). The water particle velocity and accelerations along the water depth are estimated using Airy's wave theory. The wave loading is estimated by descritizing the wave into time steps. The wave load at each time step is estimated. The varying wave loads per unit length of structural members below the water line are estimated and converted to nodal loads. The maximum wave loading in the wave cycle is considered for the prediction of the deflections (Eq. 2). The soil structure interaction analysis is performed for the considered jacket structure with and without marine growth and wave directions of 0°, 45°, and 90°. The drag coefficient is taken as 0.6 and 0.7 in the cases of without and with marine growth respectively, where as inertia coefficient is taken as 2.0. The stiffness of the surrounding soil is added to the elemental stiffness matrix of the structural members of the piles of the jacket from sea bed to the foundation level. Global stiffness matrix of the structure and load vector is assembled based on connectivity, orientation of the element with the global axis and constraints. The generated finite element model of the jacket is shown in Fig. 1. A static analysis is performed using stiffness method (Eq. 2). The static analysis of the jacket structure is also performed by considering fixity depth (as per IS 2911 (Part 1/Sec 2):1979). The deflections at sea bed and top of the deck for the considered cases are listed in Table 2. A modal analysis is also performed by considering soil structure interaction and marine growth. The natural period's up to sixth modes are listed in Table 3 and first mode shapes with and without marine growth are shown in Fig. 2.

Table 1 Soil profile of the jacket structure

Soil type	Depth from seabed (m)	Modulus of sub-grade reaction (kN/m^2)	Soil spring stiffness (kN/m)
Medium dense sand	0–2	15,000	7027.115
Firm to stiff clay	4–6	75,000	142765.3
Soft to firm clay	8–12	45,000	90235.53
Stiff clay	14–28	80,000	157930.1
Very stiff to hard clay	30–52	85,000	183436.5
Medium dense sand	52–58	68,000	132226.8
Hard clay	58–68	68,000	127415.8
Strong limestone	68–86	100,000	194558.5

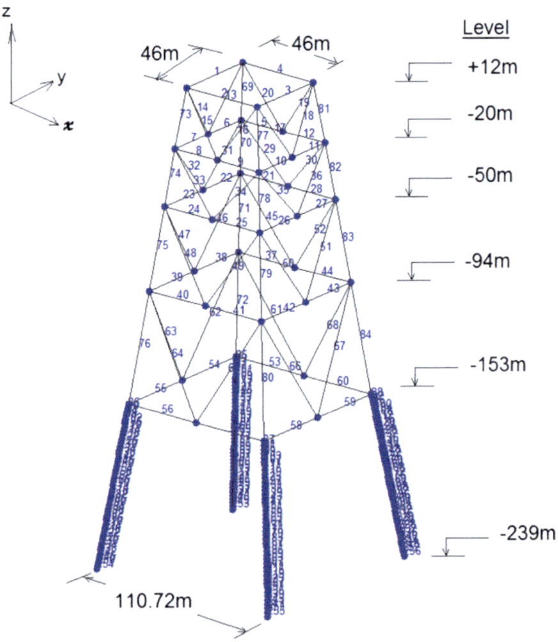

Fig. 1 Finite element model of jacket structure with soil structure interaction

$$df = \frac{1}{2}\rho_w D C_d (u_f + u_c)|u_f + u_c| + \rho_w A_c C_m \dot{u}_f \quad (1)$$

$$[K]\{X\} = \{f(t)\}, \quad (2)$$

Table 2 Deflections at topside and seabed of the jacket structure

	Topside (mm)			Seabed (mm)		
Deflection along	*X*	*Y*	*Z*	*X*	*Y*	*Z*
0° wave direction						
SSI	12.1	11.2	−1.8	0.4	0	−1.8
SSI and MG	16.8	12.8	−1.8	0.4	0	−1.7
Fixity and MG	−16.3	14.8	−0.1	−0.03	0.004	−0.0002
45° wave direction						
SSI	−12.2	11.4	−1.9	0.4	0	−1.8
SSI and MG	−17	13	−1.9	0.4	0	−1.7
Fixity and MG	−16.7	14.7	0.1	−0.03	0.004	0.0006
90° wave direction						
SSI	−12.1	12	−1.9	0.4	0	−1.8
SSI and MG	−16.8	13.6	−1.9	0.4	0	−1.8
Fixity and MG	−16.8	14.2	0.2	0.03	0.04	0.0007

Table 3 Natural periods of the jacket structure including soil structure interaction

Mode No.	Natural period (s) (SSI)		Natural period (s) (without SSI)	
	Without marine growth	With marine growth	Without marine growth	With marine growth
1	4.8224	5.6683	5.1067	6.0024
2	3.8003	4.6519	5.0345	5.9177
3	3.8003	4.6519	5.0345	5.9177
4	3.0387	3.5891	4.8225	5.6685
5	3.0387	3.3687	3.0541	3.5891
6	3.0541	2.8101	2.2601	2.6872

where,

df	force per unit length
C_d	Drag coefficient
D	Diameter of the element
u_f	Fluid velocity along the water depth
u_c	Current velocity along the water depth
$\dot{u}_f$	Fluid acceleration along the water depth
C_m	Inertia coefficient = $C_a + 1$
C_a	Added mass coefficient
A_c	Cross sectional area of the tube
ρ_w	Fluid density
$[K]$	Global Stiffness matrix of the structure
$\{X\}$	Displacement of the structure
$\{f(t)\}$	Wave Force exerted on structure

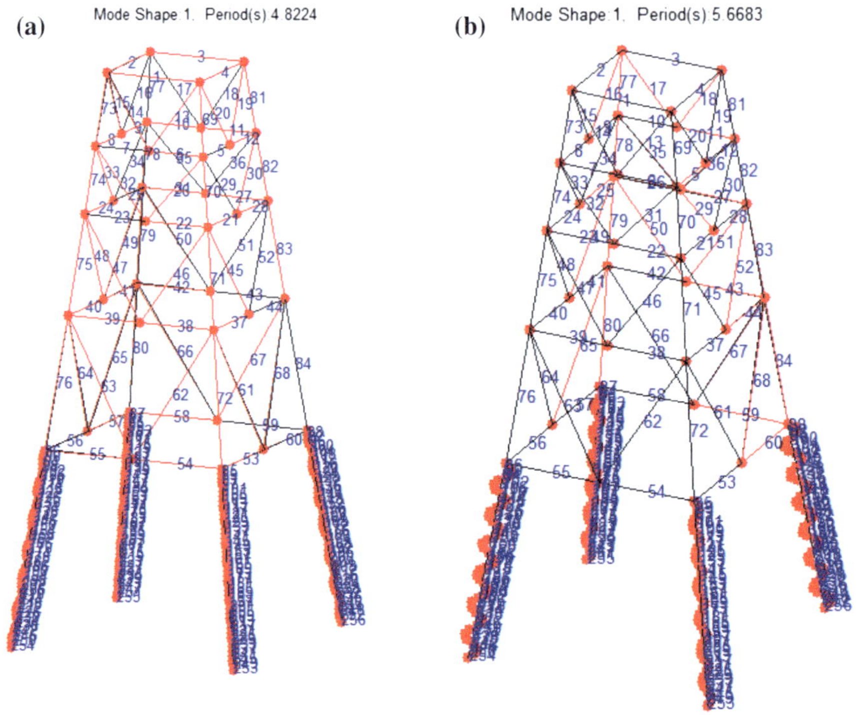

Fig. 2 **a** Mode shape of the jacket structure without marine growth, **b** mode shape of the jacket structure with marine growth

3 Results and Discussions

The observed deflections of the topside and at sea bed in all the wave directions considered indicate that the deflection at sea bed is lesser than the deflection at top side indicating the effect of soil. The topside deflection is observed 38% higher when the marine growth is considered than the deflection when marine growth is not considered. The top side deflection is observed lesser in fixity depth method than the SSI. The deflection at sea bed is observed lesser in all the considered cases, indicating the contact of structure with the soil. The dynamic analysis results indicate that the natural period of the jacket structure increases when the marine growth is considered due to the increase in structural mass.

4 Conclusions

Based on the present study it is observed that the response of the structure and natural period is increased when marine growth is considered due to the additional structural mass and no significant change in the structural stiffness. The modal analysis without SSI indicated higher natural period than the SSI analysis results. This is observed due to the reduction in the stiffness of the structure when SSI in not considered. The lateral deflections are observed higher when SSI and MG is compared with SSI and Fixity analyses. In order to predict the structural response accurately, the soil structure interaction analysis shall be performed. The difference in deflection is observed at 0° and 90° wave direction even though the structure is symmetric. This is due to the change in wave direction.

References

Asgarian, B., Flouz, A. (2008). Incremental dynamic analysis considering pile soil structure interaction for the jacket type offshore platforms. In *Proceedings of the ASME 27th International Conference on Offshore Mechanics and Arctic Engineering*, OMAE 2008, June 15–20, Estoril, Portugal.

Ghassemi Zadeh, S. M., Shojayee, B. R., & Vaziri, K. O. S. M. S. (2015). Finite element numerical method for nonlinear interaction response analysis of offshore jacket affected by environment marine forces. *Open Journal of Marine Science, 5,* 422–442.

IS 2911(Part1/ Sec2). (1979). *Indian standard code of practice for design and construction of pile foundation, part-1*, BIS, New Delhi.

Newmark, N. M. (1942). Numerical procedure for computing deflections, moments and buckling loads. In *Proceedings of ASCE* (Vol. 68, p. 679). New York, N.Y.: ASCE.

Raheem, S. E., Abdel Aal, E. M. A., Abdel Shafy, A. G. A., & Abdel Seed, F. K. (2012). Nonlinear analysis of offshore structures under wave loadings. In: *15th World Conference on Earthquake Engineering*, LISBOA.

Shi, W., Park, H. C., Chung, C. W., Shin, H. K., Kim, S. H., Lee, S. S., et al. (2015). Soil structure interaction on the response of jacket type offshore wind turbine. *International Journal of Precision Engineering and Manufacturing Technology, 2*(2), 139–148.

Vesic, A. B. (1961). Bending beam resting on isotropic elastic solid. *Journal of Engineering Mechanics Division, ASCE, 87*(EM2), 35–33.

Effect of Stiffness Degradation of Clay in the Dynamic Response of Monopile-Supported Offshore Wind Turbines

K. A. Abhinav and Nilanjan Saha

Abstract Stiffness degradation studies for monopile-supported offshore wind turbines (OWTs) are usually limited to the geotechnical domain, largely ignoring the dynamic loads from the wind and the waves. This paper makes use of a time-domain approach, coupling aerodynamic and hydrodynamic loads, to investigate the influence of stiffness degradation in clay on the response of a monopile-supported OWT in a water depth of 20 m. *p-y* curves are used to represent the soil–structure interaction (SSI) in the lateral direction and a suitable degradation method is applied to consider the effects of cyclic loading. It is observed that the influence of stiffness degradation wanes with increasing number of load cycles. OWT's being highly dynamic structures; the debilitating effects of stiffness degradation cannot be entirely discounted.

1 Introduction

Offshore wind is fast emerging as a source of clean energy, capable of addressing the issue of depleting fossil fuel reserves. Majority of the OWT structures have been installed in shallow waters (depth < 30 m) and are supported on monopiles. They are large diameter steel pipe piles of diameters varying from 4 to 6 m. With about 50% of the total cost of an offshore wind farm spent on the foundations, it becomes essential to study the SSI problem in detail.

The design philosophy for offshore structures like oil platforms cannot be directly applied to offshore wind turbines. They are vulnerable to resonance from wind, waves and rotor effects and have to be designed to fall within a narrow safe frequency zone. There exists no specific design procedure for the foundations of

K. A. Abhinav (✉) · N. Saha
Department of Ocean Engineering, IIT Madras, Chennai, India
e-mail: abhinavka@gmail.com

N. Saha
e-mail: nilanjan@iitm.ac.in

A. I. V. and V. B. Maji (eds.), *Geotechnical Applications*, Lecture Notes in Civil Engineering 13, https://doi.org/10.1007/978-981-13-0368-5_35

monopile-supported OWTs. The widely used-*p*-*y* curves suffer from the drawback that they have been experimentally validated only for smaller diameters. While small diameter piles show flexible behavior, monopiles are of large diameters and exhibit rigid body rotation. Latest design standards for OWTs (DNV OSJ101 2014) specify that *p*-*y* curves may be used with caution, for large diameter monopiles.

There have been a few studies on cyclic loading and stiffness degradation effects on monopiles. Achmus et al. (2009) proposed a degradation model for cyclic loading in sand. From experiments in sand, LeBlanc et al. (2010) concluded that cyclic loading increased the pile stiffness. Carswell et al. (2015) compared the effect of different *p*-*y* methods for monopile in clay. Most of these works were limited to the soil domain.

This paper attempts to investigate the influence of stiffness degradation of clayey soil under cyclic wind and wave loads, on the response of a monopile-supported OWT, within a time-domain framework. Aerodynamic, hydrodynamic and geotechnical domains were considered and the finite element (FE) method was used. A stiffness degradation scheme by Rajashree and Sundaravadivelu (1996), has been applied on *p*-*y* curves for clay by Matlock (1970). As the serviceability criteria are governing, for OWTs, the results are presented in terms of displacements and rotations at the head of the monopile.

2 Finite Element Model

2.1 NREL 5-MW Offshore Wind Turbine

The present study makes use of the NREL 5-MW OWT (Jonkman et al. 2009). It is a three-bladed horizontal axis wind turbine with an upwind rotor configuration. The hub and rotor diameters are 3 and 126 m respectively. The tower extends from 10 m above the mean sea level (MSL) to 87.6 m above MSL. The diameter and wall thickness of the tower varies from 6 m and 27 mm at the base, to 3.87 m and 19 mm at the top. The pile has a diameter of 6 m and wall thickness of 60 mm. The water depth is 20 m.

2.2 Soil Conditions

A uniform clay soil profile is used in the study. The effective unit weight (γ') and undrained shear strength (c_u) are 8 kN/m^3 and 100 kPa respectively. The experimental coefficient (*J*) has a value of 0.25 and the strain at 50% max. stress ε_{50} is 0.005.

Table 1 Load cases used in the study

LC No.	V (m/s)	H_s (m)	T_p (s)
1	11.4	3.1	10.1
2	39	9.5	12.8

2.3 Load Cases

Two load cases are considered—LC1, is an operating case at the rated wind speed of 11.4 m/s and LC2 is an extreme load case, with a wind speed of 39 m/s. They correspond to the design load cases (DLC) 1.2 and 6.4 of IEC (2009) respectively. Sea-states are correlated on wind speeds and are derived using a probabilistic formulation (Johannessen et al. 2002). The met-ocean conditions are given in Table 1. Wind and waves are considered to be co-directional.

Here, V is the hub-height 10-min mean wind speed, H_s is the significant wave height and T_p is the peak spectral period.

2.4 Loads on the OWT

The OWT is subjected to dynamic loading, from wind and waves. For fixed OWT's the coupling effect between wind and wave loading can be neglected (Gao et al. 2010). A superposition method is adopted to model their combined influence, using two computer codes. Initially, coupled aerodynamic–hydrodynamic analysis is performed using the aeroelastic programme FAST (Jonkman and Buhl 2005) and six-component time series of the loads acting at the hub are obtained. They are now applied as external forces on the monopile model in USFOS (Søreide et al. 1993), a FE programme for offshore structures, and coupled hydrodynamic-geotechnical analyses are carried out.

NREL's TurbSim (Jonkman 2009) is used to generate 3-D stochastic wind fields. The frequency content of the wind velocity is described using the Kaimal spectrum. Aerodynamic forces on the blades are computed by means of the blade element momentum theory. Here, the total load on the blade is obtained by summing up the component loads on the individual blade elements long its span. Ocean waves are irregular and their surface profile is generated from the JONSWAP spectrum, using an equal area method. The wave loads on the monopile are computed using the Morison equation (Morison et al. 1950), as the sum of inertia and drag forces.

$$f = \rho C_M \frac{\pi D^2}{4} \dot{u} + \frac{1}{2} \rho \, C_D |u| u \tag{1}$$

Here, f is the horizontal force per unit length, D represents the diameter of the pile and u stands for the relative water particle velocity in the horizontal direction.

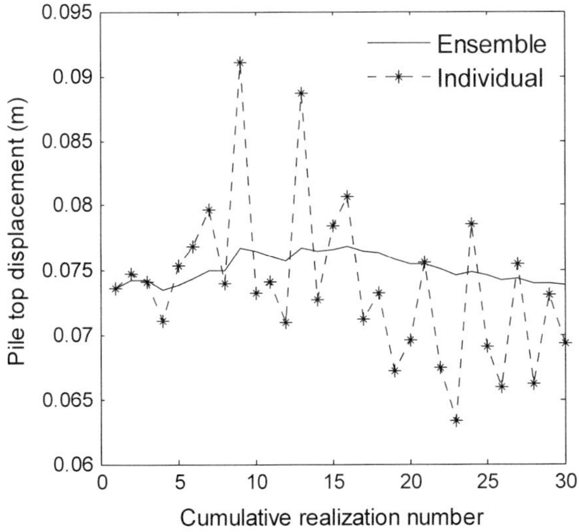

Fig. 1 Eliminating uncertainty effects

C_M and C_D are the empirical coefficients for inertia and drag, respectively and ρ is the density of water. The upper dot represents the time derivative.

Random phases are used to simulate the time series of wind and waves result in statistical uncertainty. Varying the random phases would result in the generation of different time series for the same met-ocean state. Such an uncertainty can be eliminated by increasing the number of simulations for the same load case. This is illustrated in Fig. 1. The pile top displacement at mudline under LC2 is studied for 30 individual realizations of the load process. It is observed that the use of individual estimates results in significant variations of the response. However, ensemble averaging is a reliable method for computing the response, with the values converging after about 25 realizations. In the present study, 30 different simulations are performed for each load case and the ensemble mean values are reported.

3 Modelling

3.1 Structural Model

USFOS makes use of the Idealized Structural Unit Method (Yukio and Rashed 1984), where the structure is discretized into actual physical units. The Hilber, Hughes and Taylor (HHT)-α method (Hilber et al. 1977) is used for numerical time integration. Each simulation is run for 600 s.

3.2 Soil Model

The soil spring properties in the lateral direction are modelled using the static *p-y* curves by Matlock (1970). Here, *p* is the soil resistance and *y* is the pile deflection at a particular depth (*x*) below the mudline. The static *p-y* curves are developed as follows:

$$p = 0.5p_{\text{ult}}\left(\frac{y}{y_c}\right)^{\frac{1}{3}}, \tag{2}$$

where $y_c = 2.5\,\epsilon_{50}b$. p remains constant beyond a value of $y = 8y_{50}$. 'b' is the diameter of the pile. p_{ult} is the ultimate soil resistance per unit length of the pile, taken as the smaller of the following:

$$p_{\text{ult}} = \left(3 + \frac{\gamma'}{c_{\text{u}}}x + \frac{J}{b}x\right)c_{\text{u}}b;\; p_{\text{ult}} = 9c_{\text{u}}b \tag{3}$$

Stiffness degradation is implemented using the model by Rajashree and Sundaravadivelu (1996). The ultimate soil resistance is degraded based on the number of cycles *N*, as follows:

$$p_{\text{ult}N} = (1 - \lambda_N)p_{\text{ult}} \tag{4}$$

The degradation factor λ_N is given by

$$\lambda_N = \frac{y_1}{0.2b}\log N \le 1 \tag{5}$$

y_1 is the static displacement of the soil spring and is assumed as 1% of the pile diameter, in the present study (Carswell et al. 2015). Spring elements are inserted along the depth of the pile, at a spacing of 3 m. The soil stiffness is represented by the initial slope of the *p-y* curve. Figure 2 shows the static and the degraded *p-y* curves for a depth of 1.5 m.

4 Results

Figures 3 and 4 show the variation in the pile head displacement and rotation, with respect to the increasing number of load cycles. '0' number of cycles refers to the initial case, where stiffness degradation is disregarded. At high wind speeds, the power production is suspended and the blades are parked. However, due to increased wave height and loads, the response of the monopile at LC2 is significantly higher than that at LC1. The influence of stiffness degradation reduces the resistance of the *p-y* curves and results in increased responses.

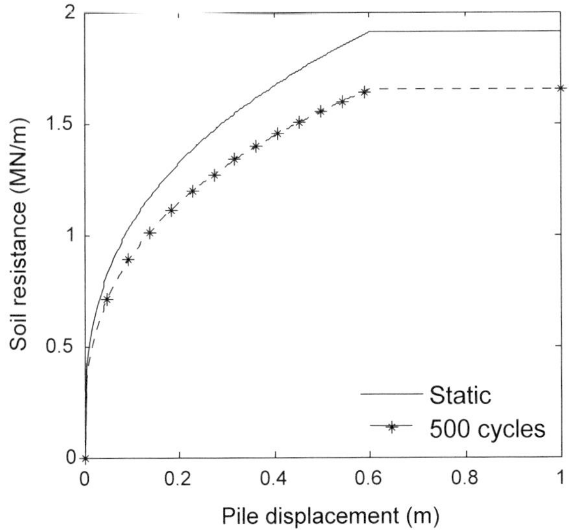

Fig. 2 *p-y* curves for static and cyclic cases

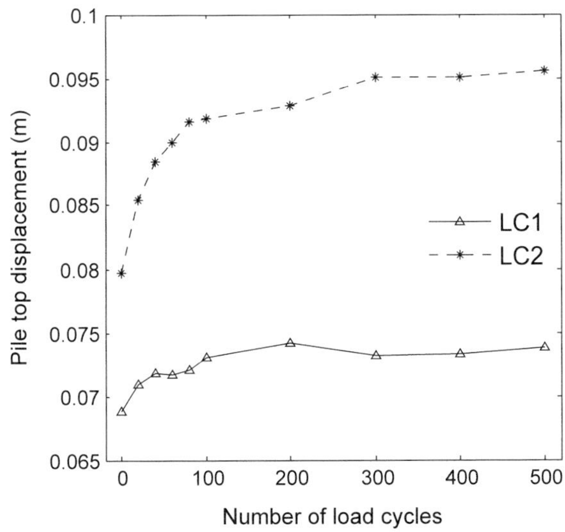

Fig. 3 Variation of pile head displacement

The effect of stiffness degradation is predominant only during the initial stages (for load cycles <100 in number). With further increase in the number of load cycles, the percentage reduction in the strength of the soil curves is reduced and beyond 300 cycles, the response levels off, as seen in Figs. 3 and 4. The pile top rotation is safe with respect to the serviceability criteria of 0.5° (Achmus et al. 2009).

Figure 5 shows the profile of maximum displacement, along the depth of the pile, for both the static case and after 500 load cycles. SSI for large diameter

Fig. 4 Variation of pile head rotation

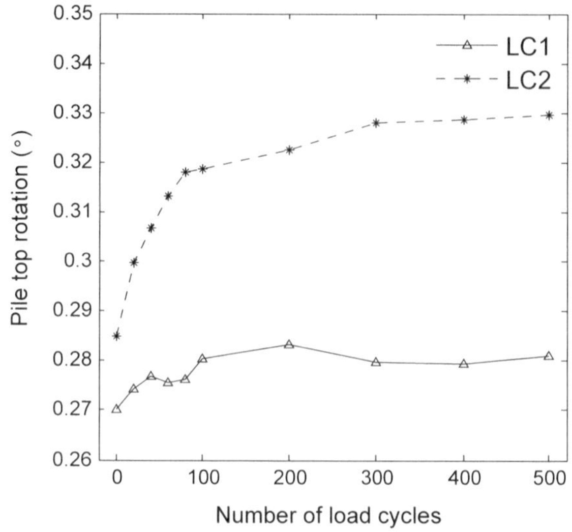

monopiles in clay shows flexible behavior with a 'toe-kick' phenomenon, whereas the toe of the pile rotates in the negative direction. The point of inflection is observed at a depth of 22 m below the mudline. The pile in the degraded soil exhibits displacements with a percentage increase of 7 and 13% at the top and toe respectively, over the static case.

OWT foundations are usually designed for 50-year load cases and are rarely subjected to even 100 load cycles. However, stiffness degradation studies are aimed at understanding the performance of these highly dynamic structures under cyclic

Fig. 5 Pile displacement profile

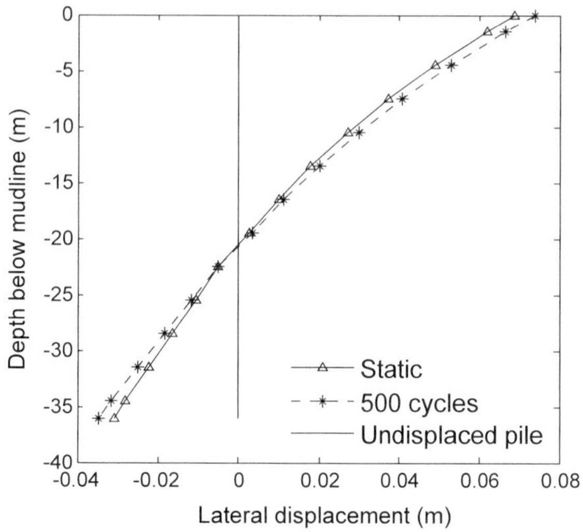

loads, in a bid to ensure that the safe natural frequency criteria or the serviceability criteria are not violated.

5 Conclusions

The influence of stiffness degradation of clay on the lateral response of a monopile OWT in a water depth of 20 m has been investigated by means of the finite element method. Aerodynamic, hydrodynamic and geotechnical domains are considered together. A method proposed by Rajashree and Sundaravadivelu (1996) have been used to account for the reduction in soil strength due to cyclic loading.

The influence of stiffness degradation in increasing the lateral response of the pile top is visible during the initial stages (<100 cycles), but tails off with further increase in the number of cycles. Large diameter monopiles in clay exhibit flexible behavior. While the limit state criteria with respect to pile top rotation is satisfied even at heavy sea-states, such studies may still serve as a safety check for monopiles designed using the *p*-*y* method.

Acknowledgements This work was supported by a grant from the Ministry of Human Resource Development, Govt. of India. The authors would like to acknowledge the help of Tore Holmas, with USFOS and Jason Jonkman, with FAST.

References

Achmus, M., Kuo, Y.-S., & Abdel-Rahman, K. (2009). Behavior of monopile foundations under cyclic lateral load. *Computers and Geotechnics, 36,* 725–735.

Carswell, W., Fontana, C., Arwade, S. R. & DeGroot, D. J. (2015). Comparison of cyclic P-Y methods for offshore wind turbine monopiles subjected to extreme storm loading. In *Proceedings of 34th International Conference on Ocean, Offshore and Arctic Engineering*, Newfoundland, Canada.

DNV-OS-J101. (2014). *Design of offshore wind turbine structures*. AS, Norway: Det Norske Veritas.

Gao, Z., Saha, N., Moan, T., & Amdahl, J. (2010). Dynamic analysis of offshore fixed wind turbines under wind and wave loads using alternative computer codes. In *Proceedings of 3rd EAWE Conference, TORQUE 2010: The Science of Making Torque from Wind*, Crete.

Hilber, H. M., Hughes, T. J., & Taylor, R. L. (1977). Improved numerical dissipation for time integration algorithms in structural dynamics. *Earthquake Engineering and Structural Dynamics, 5*(3), 283–292.

IEC. (2009). IEC 61400-3: Wind turbines part 3: Design requirements for offshore wind turbines. International Electrotechnical Commission.

Johannessen, K., Meling, T. S., & Haver, S. (2002). Joint distribution for wind and waves in the northern North Sea. *International Journal of Offshore and Polar Engineering, 12* (1), 1–8.

Jonkman, B. J. (2009). Turbsim User's guide: V. 1.50.

Jonkman, J. M., & Buhl Jr., M. L. (2005). *FAST—User's guide* (Technical Report NREL/EL-500-38230). Golden, CO: National Renewable Energy Laboratory.

Jonkman, J. M., Butterfield, S., Musial, W., & Scott, G. (2009). *Definition of a 5-MW reference wind turbine for offshore system development* (Technical Report NREL/TP-500-38060). Golden, CO: National Renewable Energy Laboratory.

Matlock, H. (1970). Correlations for design of laterally loaded piles in soft clay. In *Proceedings of Offshore Technology Conference*, Dallas, Texas.

Morison, J. R., O'Brien, M. P., Johnson, J. W., & Schaaf, S. A. (1950). The force exerted by surface waves on piles. *Petroleum Transactions, 189,* 149–154.

Rajashree, S. S., & Sundaravadivelu, R. (1996). Degradation model for one-way cyclic lateral load on piles in soft clay. *Computers and Geotechnics, 19*(4), 289–300.

Søreide, T. H., Amdahl, J., Eberg, E., Holmås, T., Hellan Ø. (1993). USFOS-A computer program for progressive collapse analysis of steel offshore structures. *Theory Manual*, SINTEF.

Yukio, U., & Rashed, S. M. (1984). The idealized structural unit method and its application to deep girder structures. *Computers and Structures, 18*(2), 277–293.

MIX
Papier aus verantwortungsvollen Quellen
Paper from responsible sources
FSC® C105338
FSC
www.fsc.org

If you have any concerns about our products,
you can contact us on
ProductSafety@springernature.com

In case Publisher is established outside the EU,
the EU authorized representative is:
Springer Nature Customer Service Center GmbH
Europaplatz 3, 69115 Heidelberg, Germany

Printed by Libri Plureos GmbH
in Hamburg, Germany